AF443194

HYBRID METHODS OF MOLECULAR MODELING

Progress in Theoretical Chemistry and Physics

VOLUME 17

Honorary Editors:

W.N. Lipscomb *(Harvard University, Cambridge, MA, U.S.A.)*
Yves Chauvin *(Institut Français du Pétrole, Tours, France)*

Editors-in-Chief:

J. Maruani *(formerly Laboratoire de Chimie Physique, Paris, France)*
S. Wilson *(formerly Rutherford Appleton Laboratory, Oxfordshire, U.K.)*

Editorial Board:

V. Aquilanti *(Università di Perugia, Italy)*
E. Brändas *(University of Uppsala, Sweden)*
L. Cederbaum *(Physikalisch-Chemisches Institut, Heidelberg, Germany)*
G. Delgado-Barrio *(Instituto de Matemáticas y Física Fundamental, Madrid, Spain)*
E.K.U. Gross *(Freie Universität, Berlin, Germany)*
K. Hirao *(University of Tokyo, Japan)*
R. Lefebvre *(Université Pierre-et-Marie-Curie, Paris, France)*
R. Levine *(Hebrew University of Jerusalem, Israel)*
K. Lindenberg *(University of California at San Diego, CA, U.S.A.)*
M. Mateev *(Bulgarian Academy of Sciences and University of Sofia, Bulgaria)*
R. McWeeny *(Università di Pisa, Italy)*
M.A.C. Nascimento *(Instituto de Química, Rio de Janeiro, Brazil)*
P. Piecuch *(Michigan State University, East Lansing, MI, U.S.A.)*
S.D. Schwartz *(Yeshiva University, Bronx, NY, U.S.A.)*
A. Wang *(University of British Columbia, Vancouver, BC, Canada)*
R.G. Woolley *(Nottingham Trent University, U.K.)*

Former Editors and Editorial Board Members:

I. Prigogine (†)	I. Hubač (*)
J. Rychlewski (†)	M.P. Levy (*)
Y.G. Smeyers (†)	G.L. Malli (*)
R. Daudel (†)	P.G. Mezey (*)
H. Ågren (*)	N. Rahman (*)
D. Avnir (*)	S. Suhai (*)
J. Cioslowski (*)	O. Tapia (*)
W.F. van Gunsteren (*)	P.R. Taylor (*)

† : deceased; * : end of term

For other titles published in this series go to www.springer.com/series/6464

Hybrid Methods of Molecular Modeling

ANDREI L. TCHOUGRÉEFF

Division of Electrochemistry
Department of Chemistry
Lomonosov Moscow State University
Moskva 119899, Russia

 Springer

Andrei L. Tchougréeff
Division of Electrochemistry
Department of Chemistry
Lomonosov Moscow State University
Moskva 119899, Russia

ISBN 978-1-4020-8188-0   e-ISBN 978-1-4020-8189-7

Library of Congress Control Number: 2008925051

PROGRESS IN THEORETICAL CHEMISTRY
AND PHYSICS

A series reporting advances in theoretical molecular and material sciences, including theoretical, mathematical and computational chemistry, physical chemistry and chemical physics

Aim and Scope

Science progresses by a symbiotic interaction between theory and experiment: theory is used to interpret experimental results and may suggest new experiments; experiment helps to test theoretical predictions and may lead to improved theories. Theoretical Chemistry (including Physical Chemistry and Chemical Physics) provides the conceptual and technical background and apparatus for the rationalisation of phenomena in the chemical sciences. It is, therefore, a wide ranging subject, reflecting the diversity of molecular and related species and processes arising in chemical systems. The book series *Progress in Theoretical Chemistry and Physics* aims to report advances in methods and applications in this extended domain. It will comprise monographs as well as collections of papers on particular themes, which may arise from proceedings of symposia or invited papers on specific topics as well as initiatives from authors or translations.

The basic theories of physics – classical mechanics and electromagnetism, relativity theory, quantum mechanics, statistical mechanics, quantum electrodynamics – support the theoretical apparatus which is used in molecular sciences. Quantum mechanics plays a particular role in theoretical chemistry, providing the basis for the spectroscopic models employed in the determination of structural information from spectral patterns. Indeed, Quantum Chemistry often appears synonymous with Theoretical Chemistry: it will, therefore, constitute a major part of this book series. However, the scope of the series will also include other areas of theoretical chemistry, such as mathematical chemistry (which involves the use of algebra and topology in the analysis of molecular structures and reactions); molecular mechanics, molecular dynamics and chemical thermodynamics, which play an important role in rationalizing the geometric and electronic structures of molecular assemblies and polymers, clusters and crystals; surface, interface, solvent and solid-state effects; excited-state dynamics, reactive collisions, and chemical reactions.

Recent decades have seen the emergence of a novel approach to scientific research, based on the exploitation of fast electronic digital computers. Computation provides a method of investigation which transcends the traditional division between theory and experiment. Computer-assisted simulation and design may afford a solution to complex problems which would otherwise be intractable to theoretical analysis, and may also provide a viable alternative to difficult or costly laboratory experiments. Though stemming from Theoretical Chemistry, Computational Chemistry is a field of research in its own right, which can help to test theoretical predictions and may also suggest improved theories.

The field of theoretical molecular sciences ranges from fundamental physical questions relevant to the molecular concept, through the statics and dynamics of isolated molecules, aggregates and materials, molecular properties and interactions, and the role of molecules in the biological sciences. Therefore, it involves the physical basis for geometric and electronic structure, states of aggregation, physical and chemical transformation, thermodynamic and kinetic properties, as well as unusual properties such as extreme flexibility or strong relativistic or quantum-field effects, extreme conditions such as intense radiation fields or interaction with the continuum, and the specificity of biochemical reactions.

Theoretical chemistry has an applied branch – a part of molecular engineering, which involves the investigation of structure–property relationships aiming at the design, synthesis and application of molecules and materials endowed with specific functions, now in demand in such areas as molecular electronics, drug design or genetic engineering. Relevant properties include conductivity (normal, semi- and supra-), magnetism (ferro- or ferri-), optoelectronic effects (involving nonlinear response), photochromism and photoreactivity, radiation and thermal resistance, molecular recognition and information processing, and biological and pharmaceutical activities; as well as properties favouring self-assembling mechanisms, and combination properties needed in multifunctional systems.

Progress in Theoretical Chemistry and Physics is made at different rates in these various research fields. The aim of this book series is to provide timely and in-depth coverage of selected topics and broad-ranging yet detailed analysis of contemporary theories and their applications. The series will be of primary interest to those whose research is directly concerned with the development and application of theoretical approaches in the chemical sciences. It will provide up-to-date reports on theoretical methods for the chemist, thermodynamician or spectroscopist, the atomic, molecular or cluster physicist, and the biochemist or molecular biologist who wish to employ techniques developed in theoretical, mathematical or computational chemistry in their research programmes. It is also intended to provide the graduate student with a readily accessible documentation on various branches of theoretical chemistry, physical chemistry and chemical physics.

CONTENTS

PREFACE

Computer aided modeling of polyatomic molecular systems is one of the leading consumers of processor time and computer memory nowadays. Despite tremendous progress in both computer hardware and molecular modeling software, the complete quantum mechanics-based numerical study of a realistic model of any, say biologically or technologically, relevant system is out of the reach of the workers in the field. The problem, however, is not only the enormity of computational resources required for conducting such a study, but the absence of any clear proof "by construction"of the validity of the employed calculation methods and a lack of real understanding of the result. These two problems are related to each other and the situation may be described as follows: even if we get an answer by a quantum mechanical (QM) or quantum chemical (QC) modeling package, we are almost never able to say what the physical reasons are for it to be that or something else. We cannot add anything to that last number printed in the output. Chemists, however, generally think differently. They need more trends than numbers. The reason is of course that in many cases exact experimentally derived numbers are missing. This situation is by no means a new one. Yet at the dawn of numerical quantum chemistry, C.A. Coulson [1] made a point about the importance of qualitative understanding and commented that accuracy of quantum chemical calculation is "purchased very dearly" since "ab initioists abandon all conventional chemical concepts and simple pictorial quality in their results".

This situation, well known to the workers in the field, has occurred due to a factor external to quantum chemistry itself, namely the intense development of computational hardware during the past few decades. The numerical point of view, which reduces the subject of Quantum Chemistry to obtaining certain numbers, has thus become predominant. It might be acceptable, but the situation changes completely when we find ourselves in the realm of complex systems (for which, as we shall see, hybrid modeling is basically necessary): obtaining numerical results for the complex systems or their subsequent interpretation in the frame of the standard procedures becomes too costly if at all possible and the answer obtained numerically becomes unobservable (if some one does not understand just *one* number to be the answer e.g. the energy). Therefore hybrid QM/MM (QM is Quantum and MM is Molecular Mechanics) modeling requires development of the relevant concepts which could help to achieve decision making while singling out the relevant quantum and classical parts and establishing the adequate construction of the interface between them on a rational basis. This can be done by finding an alternative to pure numerics – the qualitative and theoretical approach, paying attention to the development of adequate concepts related to hybrid modeling and learning to perform the calculations using theoretical concepts relevant to the system under study.

The very possibility of hybrid modeling is intimately related to the idea of dividing the problem to be solved or the object to be studied into parts formalized in various techniques of separating variables. Dividing into parts is the most general method of studying the reality. As a philosophical maxim it was first formulated by René Descartes in his "Discours de la Méthode" [2] ("to divide each of the difficulties under examination into as many parts as possible, and as might be necessary for its adequate solution"). Separation of a complex system into parts has two aspects: the technical aspect, aimed at simplifying calculations by separating the variables describing the system under study, and the conceptual aspect, having as a purpose the development of qualitative concepts i.e. identifying the ideas which would describe the system in adequate, comprehensible terms. Clearly, the description of a system comprising numerous strongly interacting components in terms of some almost independent parts and/or variables describing these parts will be inevitably approximate and the art here is to select these parts and variables in such a way that the description of their terms is acceptable. One may be pretty sure that in the case of the complex problem that requires hybrid modeling, there will be no chance to "invent" adequate parts into which the modeled system has to be divided "from one's head". Fortunately, the quantum mechanical paradigm itself provides sufficient requirements, which allow the reasonable identification of the parts the system can and has to be divided into. The adequate parts must be observable. This very general requirement allows one to establish a relation between hybrid modeling and the rest of theoretical chemistry. Yet at the early stage of the development of chemical theory the idea of "chromophores"– some specific parts of the molecule responsible for the *color* of the substance – was proposed. This approach was not that naive as it can seem nowadays since it helped to make the problem tractable by significantly reducing the number of variables (those related to the chromophore only) and to take its environment as a weak perturbation. Particularly remarkable in this context however is the observability of the chromophore.

Regarding the problems of the electronic structure of molecular systems, we notice that in the past, the importance of the qualitative concepts and explanations has been stressed many times. In this context, V.A. Fock [3, 4] discussed the (basically metaphysical) problem of interrelation between "exact solution" and "approximate explanation". His point was that any approximation (more precisely, the general form of the trial electron wave function i.e. an *Ansatz* used for it) sets the system of qualitative concepts (restricted number of variables), which can only be used for interpreting the calculation results and for describing the experiments. A characteristic example for QC is provided by the orbital energies and the MO expansion coefficients coming from the Hartree-Fock-Roothaan (HFR) approximation. Although in a great number of cases they can be related to the observed ionization potentials, they are nevertheless only mental constructs, having a definite sense only within the HFR realm, becoming invalid beyond its scope.

The chromophores are obviously observable entities. Are there others? This question has been addressed by Ruedenberg who suggested a kind of extension of the standard quantum mechanical definition of observability from the *quantities* to the *entities*. The example he used had a rather unhappy destiny in quantum chemistry although it relates to the fundamental chemical concept – that of chemical bond. At quite an early stage it was decided that chemical bonds are not observable as "there is no quantum mechanical operator for the bond". This argument is, however, not acceptable as the "bond" is not assumed to be a quantity, but an entity and to deduce nonobservability of bonds from the fact that there is no operator for the bond is equivalent to concluding that there is no atomic nuclei as there are no operators for them. Nevertheless, something had to be done about the observability of entities and K. Ruedenberg [5] proposed the following definition: "fragments in a molecular system can be singled out if these latter are observable, so that they manifest a reproducible and natural behavior; if for a series of molecules variations of fragments fit to that or another curve and its parameters can be found empirically by considering enough of the series members this proves that singling out the fragments makes sense". This definition allows us to single out numerous fragments which can be two-center two-electron bonds, or conjugate π-systems, open d-shells, atomic cores, etc. An adequate theory must be constructed in terms of such observable objects. At first glance the current situation in quantum chemistry is in sharp contradiction to this requirement. However, as we show in Chapter 2, the real constructs of quantum chemistry rely heavily upon the above-mentioned observable objects. This allows us to consider the whole of quantum chemistry from the hybrid perspective. As a result the hybrid methods, instead of being an isolated and specific area of how to program junctions between classically and quantally[1] treated parts of complex systems, shift to the center of the theory. This allows us to talk about the usual QM/MM methods, as of the hybrid methods, in a narrow sense.

It also allows us to reach multiple goals. First, it allows sensible and natural interpretation of the result in chemical terms, and with the use of chemical concepts. Second, estimates of the correction (error) to the energy (or any other quantity) coming from the use of the approximate form of the wave function in this calculation

---

[1] Trying to find an adverb to be a counterpart to 'classically' the author faced certain problem: no adequate antonym had been designed so far. Merriam-Webster suggests 'quantal' as an adjective derived from 'quantum'. So we decided to use an adverb 'quantally' already used in the required meaning in *Handbook of Solvents* (*Chemicals*) by George Wypytch, Noyes Publications 2001 (p. 21), in *Modern Electrochemistry 2A: Fundamentals of Electrodics* by John O'M. Bockris, Amulya K.N. Reddy, and Maria E. Gamboa-Aldeco, Springer 2001 (p. 724), and in *Introduction to Computational Chemistry* by Frank Jensen, John Wiley & Sons 1998 (p. 393).

can be obtained. And last, but not the least, this approach allows us to carry out the entire calculation with relatively low computational costs using *effective electron Hamiltonians* for the important (i.e. observable) parts of the system – the "chromophores" – and leaving the defects of the restricted form of the trial wave function to be taken into account by renormalized matrix elements of these effective Hamiltonians.

From the above the reader may conclude that this book is largely devoted to the theory of hybrid methods. This it true to a large extent. Nevertheless, the author could not (and did not want to) ignore the existing hybrid QM/MM methods described in the literature and widely used for describing various aspects of the behavior of the complex molecular systems. The key practical problem when applying the QM/MM methodology, namely, the substantiated construction/selection of the junction between the parts of the system described at the QM and MM levels, respectively, is thoroughly discussed here. The author's feeling is that the "Sturm und Drang"period of the hybrid QM/MM modeling has come to an end and that it is time now to give an evaluation of the state-of-the-art reached during this period and to present a theory of this family of methods, capable of giving a general view of the field, to identify the fundamental problems characteristic of it and to propose physically better based and mathematically more sound approaches to these problems. In this context the theory is useful also because it allows us to introduce some order into the diversity of the junction forms present in the literature, which otherwise resembles the famous classification of animals given by J.-L. Borges [6].

This book offers a step by step derivation of the consistent theoretical picture of hybrid modeling methods and the thorough analysis of the underlying concepts. This forms a basis for classification and analysis of current practical methods of hybrid molecular modeling, including the narrow meaning of this term. Historical remarks are important here since they put the current presentation in a general context and establish a relation with other areas of theoretical chemistry. It presents its material paying attention both to the physical soundness of the approximations used and to mathematical rigor, which are necessary for the practical development of the robust modeling code and for a conscious use of either existing or newly developed modeling tools. The reader should have a knowledge of the basic concepts of quantum and computational chemistry and/or molecular modeling. Familiarity with vector spaces, operators, wave functions, electron densities, second quantization and other tools is also necessary. Short discussions of these topics are given only to establish the relation between the standard presentation of these items in the literature and their specific form as required in the context of the theory of hybrid modeling described in this book.

This book is intended both for practicing experts and students in molecular modeling and to those in related areas, such as Materials Science, Nanoscience, and Biochemistry, who are interested in making an acquaintance with the conceptual basis of hybrid modeling and its limitations, which possibly enables them to make educated decisions while choosing a tool appropriate for solving their specific problem and for interpreting the results of the modeling. It also contains a self-sufficient example of developing a targeted hybrid method designed for molecular modeling of transition metal complexes with open d-shells. This presentation allows the reader to specify on the spot all the significant elements of the general theory and to see how they work.

The theory described here was originally developed by the author, as also the specific targeted application of the theory to molecular modeling of the transition metal complexes. This and other original methods of molecular modeling described here have been implemented in FORTRAN program suits. They are a kind of "research software" available for use to other researchers through the Net Laboratory access system which provides sample input files and minimal reference information to start with, at http://www.qcc.ru/~netlab .

Some of the results presented in this book have been published in original research papers and in two reviews in the Springer series of Progress of Theoretical Chemical Physics based on materials of the Congress on Theoretical Chemical Physics and of the European Conference on Physics and Chemistry of Quantum Systems both edited by Prof. J. Maruani and Prof. S. Wilson. When the material was presented at these conferences, Prof. J. Maruani and Prof. S. Wilson suggested that I extend and reorganize it into a book. Without their kind suggestion and constant encouragement and support, this book would never have appeared. Prof. I. Mayer kindly agreed to read the manuscript and give his valuable comments. I am very grateful to him for his help in improving the manuscript significantly. All the errors are of course the author's fault.

The process of rearrangement and of translation into English of some results available only in Russian took somewhat longer time than originally planned. I beg pardon and hope on understanding of all those whom I promised to do something during this period and failed to perform it on time. I am particularly thankful to Ms Laura Chandler of Springer Verlag for her kind patience.

References

1. C.A. Coulson. *Rev. Mod. Phys.* **32**, 170, 1963.
2. R. Descartes. *Discours de la Méthode*. J'ai Lu, 2004.
3. V.A. Fock. *Usp. Fiz. Nauk.* **16**, 1070 [in Russian], 1936.
4. V.A. Fock. in *Philosophical Problems of Physics* [in Russian], 1974.
5. K. Ruedenberg. Physical Nature of Chemical Bond. *Rev. Mod. Phys.* **34**, 326, 1962.
6. J.-L. Borges. The Analytical Language of John Wilkins. in J.-L. Borges. *Labyrinths: Selected Stories and Other Writings.* New Directions Publishing Corporation, 1964.

June 2008 *Andrei L. Tchougréeff*

MOLECULAR MODELING: PROBLEM FORMULATION AND WRAPPING CONTEXTS

Abstract   In this chapter we start with a brief recap of the general setting of the molecular modeling problem and the quantum mechanical and quantum chemical techniques. It may be of interest for students to follow the description of nonstandard tools of quantum mechanics and quantum chemistry presented after that. These tools are then used to develop a general scheme for separating electronic variables in complex molecular systems, which yields the explicit form of its potential energy surface in terms of the electronic structure variables of the subsystem treated at a quantum mechanical level, of the force fields for the subsystem treated classically, and explicitly expressing the central object of any hybrid scheme – the inter-subsystem junction – in terms of the generalized observables of the classically treated subsystem: its one-electron Green's function and polarization propagator.

1.1. MOTIVATION AND GENERAL SETTING

Molecular modeling includes a collection of computer-based tools of varying theoretical soundness, which make it possible to explain, and eventually predict, the properties of molecular systems on the basis of their composition, geometry, and electronic structure. The need for such modeling arises while studying and/or developing various chemical products and/or processes. The raison d'être of molecular modeling is provided by chemical thermodynamics and chemical kinetics, the basic facts of which are assumed to be known to the reader.[1]

According to chemical thermodynamics the relative stability of chemical species and thus their basic capacity to transform to each other (understood in a very wide sense, for example, as the possibility to form solutions i.e. homogeneous mixtures with each other or undergo phase transitions e.g. from gas to liquid state) is described by the equilibrium constant of the (at this point) hypothetical process:

$$(1.1) \qquad R_1 + R_2 + ... \rightleftarrows P_1 + P_2 + ...$$

where $R_1, R_2, ...$ stand for the reactant species and $P_1, P_2, ...$ stand for the product species governed by the equilibrium constant $K_{eq} = K_{eq}(T, P, ...)$ dependent on

---

[1]The sources in physical chemistry are numerous. Elementary volumes to be known by heart are [1, 2]; the more the better.

temperature, pressure, and other external conditions so that the concentrations (or other quantities taking their part in the specific situations e.g. – partial pressures) of the species denoted as $[R_1], [R_2], ..., [P_1], [P_2], ...$ obey the ratio:

$$(1.2) \qquad K_{\mathrm{eq}} = \frac{[P_1] \cdot [P_2] \cdot ...}{[R_1] \cdot [R_2] \cdot ...}$$

If the set of species $R_1, R_2, ...$ on the left of eq. (1.1) is more stable than $P_1, P_2, ...$ on the right, the equilibrium constant shows that by being less than one: $K_{\mathrm{eq}} < 1$. In this case the system prepared as a mixture of species $R_1, R_2, ...$ with $[P_1] = [P_2] = ... = 0$ tends to stay in the state where the concentrations of reactants are generally larger than those of the products, although of course in the equilibrium described by eq. (1.2) all species are present in concentrations necessary to satisfy this equation. If the opposite happens and $K_{\mathrm{eq}} > 1$ the system prepared as a mixture of the species $R_1, R_2, ...$ tends to transform to that formed predominantly by the species $P_1, P_2, ...$, to the extent prescribed by the previous inequality and eq. (1.2).

The magnitude of K_{eq} is governed by the relative value $\Delta\Delta G_f$ for the left and right sides of eq. (1.1) of a single quantity – the Gibbs free energies of formation of the system under study in its left hand and right hand states:

$$(1.3) \qquad K_{\mathrm{eq}} = \exp\left(-\frac{\Delta\Delta G_f}{k_B T}\right) \text{ or}$$

$$k_B T \ln K_{\mathrm{eq}} = -\Delta\Delta G_f$$

(k_B is the Boltzmann constant and T stands for the temperature). The latter is expressed through the Gibbs free energies of formation of each species:

$$(1.4) \qquad \Delta\Delta G_f = \left(\Delta G_f(R_1) + \Delta G_f(R_2) + ... + \Delta G_f^{\mathrm{mix}}(R_1, R_2, ...)\right)$$
$$- \left(\Delta G_f(P_1) + \Delta G_f(P_2) + ... + \Delta G_f^{\mathrm{mix}}(P_1, P_2, ...)\right)$$

which are themselves related to their enthalpies ΔH_f and entropies ΔS_f of formation:

$$(1.5) \qquad \Delta G_f = \Delta H_f - T\Delta S_f$$

counted from the enthalpies and entropies of formation of chemical elements taken as reference (zero) points and on the contributions to the free energy $\Delta G_f^{\mathrm{mix}}$ coming from the interactions between the system components which may be different for their forms present on the left and right sides of eq. (1.1). In this setting, for two systems having the same composition – brutto formula describing the fractions of *atoms* of each chemical element present in the system – that one which has the smaller (more negative) value of ΔG_f is more stable. In this case the transformation of the system with a larger value of ΔG_f to that with the smaller one is thermodynamically allowed no matter how long it may take to perform this process in reality. So, in order to predict on the basis of a computer experiment the thermodynamic feasibility of the process, one has to be able to calculate the Gibbs free energies of formation for the system in two states with such precision that the difference between the two is also accurate enough. This is obviously a challenge since the quantity more or less directly describing the pragmatic situation – the equilibrium constant eq. (1.2) – depends on the quantity to be calculated exponentially, so that knowing K_{eq} up to a factor of two which makes quite a difference in terms concentrations requires knowing the free energy difference $\Delta\Delta G_f$ with the precision less than $k_B T$ which for ambient temperature (one relevant to modeling of biological processes) corresponds to the precision of 0.03 eV or 0.001 Hartree.

Chemical kinetics, in contrast to chemical thermodynamics, centers on the *rates* of chemical transformations:

$$(1.6) \qquad R_1 + R_2 + \ldots \to P_1 + P_2 + \ldots$$

According to it the reaction rate i.e. the rate of the concentration change of any participating species is given by the relation:

$$(1.7) \qquad \frac{d[P_1]}{dt} = \frac{d[P_2]}{dt} = \cdots = -\frac{d[R_1]}{dt} = -\frac{d[R_2]}{dt} = \cdots = k[R_1] \cdot [R_2] \ldots$$

where the concentration independent quantity k – rate constant – depends on temperature, pressure, and other external conditions. The good old transition state theory [1,2] allows us to relate the rate constants $k = k_{\text{TST}}$:

$$(1.8) \qquad k_{\text{TST}} = \kappa \frac{k_B T}{h} \exp\left(-\frac{\Delta G^{\neq}}{k_B T}\right)$$

with some thermodynamically-looking quantities, although it is somewhat less fundamental than thermodynamics.[2] Above $\Delta G^{\neq}$ is the Gibbs free energy difference between the *transition state* of the system and the reactants. By *transition state* one somewhat vaguely understands a certain specific state of the system when an individual assembly of the reacting molecules turns out in such a situation that it inevitably transforms to the assembly of product molecules. The multiplier κ – transmission coefficient – takes care of all possible deviations from this simple picture (formula). Of course, such a peculiar state cannot be easily represented and for that reason, during the modeling process, additional assumptions are made concerning its nature (see below). One can, however, see that the precision requirements derived from chemical kinetics are similar to those based on chemical thermodynamics: knowing the reaction rate up to a practically significant factor of two at ambient temperature requires knowing the somewhat vaguely defined quantity $\Delta G^{\neq}$ with the same precision as the thermodynamically sound free energies of formation.

---

[2]The "classical" TST dating back to the 30-ies as presented in say [10] is the simplest way of relating obseravble *macroscopic* rates of chemical transformations with the *microscopic* view of energy of molecues. It is not surprizing that within 70 years of development it has been criticized and improved. For more recent views of this topic see [11, 12].

The next question to be answered is where the Gibbs free energies may be obtained theoretically. Statistical thermodynamics[3] gives an answer, allowing one to express the thermodynamic quantities entering eqs. (1.3) and (1.8) in terms of a single quantity Q – the partition function of the system and its derivatives with respect to temperature and volume according to the formulae:

$$A = -k_B T \ln Q; \quad U = k_B T^2 \left(\frac{\partial \ln Q}{\partial T}\right)_V$$

$$(1.9) \quad P = -\left(\frac{\partial A}{\partial V}\right)_T; \quad C_V = \left(\frac{\partial U}{\partial T}\right)_V$$

$$H = U + PV; \quad S = \frac{U - A}{T}$$

$$G = H - TS$$

Here, U is the internal energy, A is the Helmholtz free energy, P is the pressure, C_V is the heat capacity at constant volume, H is the enthalpy, S is the entropy, and G is the Gibbs free energy itself.

The partition function Q is thus a fundamental quantity which in a way contains all the information about the system under study. Its general expression is:

$$(1.10) \quad Q(T) = \int_\Omega \exp\left(-\frac{\mathcal{E}(\omega)}{k_B T}\right) d\omega$$

where $\mathcal{E}(\omega)$ stands for the energy of an individual state ω of the system under consideration and the integration may include summation over discrete states of the system. The set of variables ω characterizing the system deserves some discussion. As we know, materials and substances (the primary topic of our interest) are formed of molecules, molecules of atoms, atoms of nuclei and electrons and nuclei of protons and neutrons. It is very probable that these latter (together with electrons), known today as elementary particles, are in some sense formed of smaller particles (quarks). The question is to what extent we must be concerned about all that and on what level of detail. The answer is determined by the energy range in which we are interested, which in its turn is set by the temperature T. Through the exponential factor it restricts the variety of the states of the system accessible in the conditions of our experiment. For the temperatures relevant for chemistry i.e. not higher than a couple of thousands Kelvin, all nuclei reside in their ground states (not talking about hypothetical quark excitations) unless their spin degrees of freedom relevant for the NMR and ESR/EPR experiments are addressed. The latter however, require an external magnetic field to manifest themselves. It is a separate area of research and we do not

---

[3] As previously, the sources on statistical thermodynamics are hardly numerable. Conciseness in them struggles with comprehensibility and both lose. Elementary information is given in physical chemistry courses already mentioned [1, 2]. More fundamental courses are [3] – a rather physical one and [4] – a classical text on this subject. An interesting approach based on consistent usage of a single quantity – the entropy – is described in [5].

pursue here problems related to it. Most electrons move around while being tightly attached to nuclei. That means that atoms are almost perfectly "elementary" particles for molecular modeling problems. Closer to a lower border of the described temperature region, that which better suits the living conditions of biological systems, one may safely think that not only do atoms form separate entities, but that relatively stable assemblies of atoms known as molecules move around preserving their integrity. The main content of chemistry is precisely those relatively rare events when stable atomic associates (molecules) exchange by atoms or change the mutual arrangement of the atoms they are formed of. In this temperature range the manifestations of any independent motion of electrons are extremely rare events, so that one can think that the only variables describing the individual states ω of the system are the positions $\mathbf{R}_\alpha$ and the momenta $\mathbf{P}_\alpha$ of the nuclei of all atoms $\alpha = 1 \div N$ composing the system under study. The points $\{\mathbf{R}_\alpha, \mathbf{P}_\alpha | \alpha = 1 \div N\}$ form what is called the phase space of a molecular system. In this case the energies $\mathcal{E}(\omega)$ entering eq. (1.10) are:

$$(1.11) \quad \mathcal{E}(\omega) = \mathcal{E}(\{\mathbf{P}_\alpha\}, \{\mathbf{R}_\alpha\}) = \mathcal{T}(\{\mathbf{P}_\alpha\}) + \mathcal{U}(\{\mathbf{R}_\alpha\})$$

where $\mathcal{T}$ and $\mathcal{U}$ are, respectively, the kinetic and potential energy of the system of atoms and notation $\{\mathbf{R}_\alpha\}, \{\mathbf{P}_\alpha\}$ refers to the entire set of atomic radius-vectors and momenta involved. The quantities $\mathbf{R}_\alpha$ and $\mathbf{P}_\alpha$ are vectors in the three-dimensional space and the integration in eq. (1.10) must be understood as a usual integration over the $6N$-dimensional phase space. In the discussed temperature range one can also safely limit oneself by a classical description of nuclear motions (see, however, below).

If the system under study consists of only weakly interacting or noninteracting molecules (gas phase) the thermodynamical (or other observable) quantities can be obtained from single molecule calculations by relying upon the corresponding statistical theory and the assumptions inherent for it. In this case the number of atoms N in the molecule can be thought to be the number of atoms in the entire system. The procedure is as follows. First, a search for local minima of the potential energy

$$(1.12) \quad \min_q \mathcal{U}(\{\mathbf{R}_\alpha\}(q))$$

is performed. These minima are related to more or less stable arrangements of the atoms of the molecule. In the above expression it is assumed that q is the set of independent nuclear coordinates characterizing the system so that for all atoms α in the system, their Cartesian coordinates $\mathbf{R}_\alpha = \mathbf{R}_\alpha(q)$ are some 3-vector functions of $3N - 6$ independent components of q ($3N - 5$ in the case of linear arrangement of nuclei). This corresponds to the fact that the potential energy surface (PES) $\mathcal{U}$ is invariant under the translation and rotation of the system (molecule) as a whole, so that it does not depend on the corresponding 6 functions of $\{\mathbf{R}_\alpha\}$ describing these motions (of which three are simply the Cartesian coordinates of the center of mass of the molecule, and the other three are the angles necessary to define the orientation of e.g. principal axes of the tensor of inertia – see below – of the molecule with respect to the laboratory reference frame; energy, clearly, does not depend on these angles). The deep minima of $\mathcal{U}$ (much deeper than k_BT) are then identified with the stable states of the molecule under consideration, its structure isomers, which is important information from the chemical point of view.

Then the hypothesis is applied that in the vicinity of each of the minima of potential energy surfaces (PES) the potential energy can be adequately represented by its expansion up to second power of the variations of the nuclear coordinates with respect to their equilibrium values $q^{(0)}$. In the vicinity of this minimum the appropriate coordinates are

$$(1.13) \quad \delta q = q - q^{(0)}$$

the $3N - 6$ independent shifts of the atoms relative to their equilibrium positions. In terms of these coordinates the PES acquires the form:

$$(1.14) \quad \mathcal{U}(\delta q) = \mathcal{U}_0 + \frac{1}{2} \sum_{i,j}^{3N-6} \left. \frac{\partial^2 \mathcal{U}}{\partial q_i \partial q_j} \right|_{q=q^{(0)}} \delta q_i \delta q_j$$

From the above assumption that $q = q^{(0)}$ is a true minimum it follows that the symmetric matrix $\left. \frac{\partial^2 \mathcal{U}}{\partial q_i \partial q_j} \right|_{q=q^{(0)}}$ can be transformed into diagonal form by going to the normal coordinates (modes) which are linear combinations of the basic nuclear shifts δq_i. In the classical treatment the $3N - 6$ normal modes describe independent oscillations of the molecular system with the frequencies ω_ν, $\nu = 1 \div 3N - 6$, which are square roots of the eigenvalues of the dynamic matrix in the atomic mass weighted coordinates.[4] Sometimes the situation can be more involved so that nontrivial relative motions of atoms or atomic groups in the molecules on which $\mathcal{U}$ depends cannot be adequately described as (small) oscillations, but by contrast represent motions of large amplitudes. They are then described by two (inversions) or by multiple well potentials (internal rotations).

The harmonic approximation reduces to assuming the PES to be a hyperparaboloid in the vicinity of each of the local minima of the molecular potential energy. Under this assumption the thermodynamical quantities (and some other properties) can be obtained in the close form. Indeed, for the ideal gas of polyatomic molecules the partition function Q is a product of the partition functions corresponding to the translational, rotational, and vibrational motions of the nuclei and to that describing electronic degrees of freedom of an individual molecule:

$$Q = Q_{\text{trans}} Q_{\text{rot}} Q_{\text{vib}} Q_{\text{el}}$$

$$Q_{\text{trans}} = \left(\frac{M}{2\pi \hbar^2 k_B T} \right)^{\frac{3}{2}} V$$

---

[4]The topic discussed here pertains to classical mechanics, namely to the theory of small oscillations. The fundamental sources are [6] and [7]. More chemistry (vibration spectroscopy) oriented is [8]. A concise and clear (not the same thing!) description of classical mechanics is presented in [9].

$$Q_{\text{rot}} = \left(\frac{2k_B T}{\hbar^2}\right)^{\frac{3}{2}} \frac{(8\pi I_x I_y I_z)^{\frac{1}{2}}}{\sigma}$$

$$(1.15) \quad Q_{\text{vib}} = \prod_{\nu=1}^{3N-6} \frac{\exp\left(-\frac{\hbar\omega_\nu}{2k_B T}\right)}{1 - \exp\left(-\frac{\hbar\omega_\nu}{k_B T}\right)}$$

$$Q_{\text{el}} = g_e \exp\left(-\frac{\mathcal{U}_0}{k_B T}\right)$$

where M is the molecular mass:

$$(1.16) \quad M = \sum_\alpha^N M_\alpha,$$

(here M_α is the atomic mass of atoms in the system); I_x, I_y, and I_z are the principal inertia momenta obtained by diagonalizing the 3×3 inertia tensor calculated at the equilibrium geometry:

$$(1.17) \quad \sum_\alpha^N M_\alpha \begin{pmatrix} y_\alpha^2 + z_\alpha^2 & -x_\alpha y_\alpha & -x_\alpha z_\alpha \\ -x_\alpha y_\alpha & x_\alpha^2 + z_\alpha^2 & -y_\alpha z_\alpha \\ -x_\alpha z_\alpha & -y_\alpha z_\alpha & x_\alpha^2 + y_\alpha^2 \end{pmatrix}$$

where x_α, y_α, and z_α are the equilibrium Cartesian coordinates of atoms relative to the center of masses of the molecule; σ is the so-called symmetry factor and ω_ν are the harmonic frequencies obtained by diagonalizing the second derivatives matrix of the potential energy eq. (1.14). The given forms of the rotational and vibrational partition functions take care of the quantum effects in molecular vibrations (having a nonvanishing zero vibration energy of $\dfrac{\hbar\omega_\nu}{2}$ associated with each vibration) as well as of the quantum behavior of molecular systems in general. The electronic contribution at this point can be taken as the number of degeneration of the electronic ground state g_e multiplied by the Boltzmann factor for the value of the potential energy in the corresponding minimum. The deepest minimum can be taken as a natural zero energy. Since the electronic excitations are supposed to be rare in the temperature range considered, different electronic states can be accessed only if they are degenerate with the ground one which is reflected by the multiplier g_e. For the ideal gas of some number of identical molecules (typically for the mole i.e. for the Avogadro number N_A of particles) the above multipliers must be taken in power of N_A and the whole must be divided by $N_A!$ to take care of the indistinguishability.

The described procedure applies to all available potential minima on the potential energy of the polyatomic system under consideration. Then, taking the necessary logarithms and performing the necessary differentiations, one arrives at the estimates of the thermodynamic quantities: enthalpies, entropies, and the Gibbs free energies, associated with a given minimum of the molecular potential energy. The differences between the thermodynamic quantities associated with different minima are considered as estimates of $\Delta\Delta H_f, \Delta\Delta S_f$, and $\Delta\Delta G_f$ entering eqs. (1.4) and (1.5). One can easily realize that in a majority of cases the $\Delta\Delta G_f$ is dominated by the values of

$\Delta \mathcal{U}_0$ – the differences between the depths of different minima: so going down from a shallower to a deeper minimum of the molecular PES means decreasing the Gibbs free energy and such a process generally has an equilibrium constant larger than unity. However, for precise calculations, the terms coming from translations, rotations and vibrations also must be taken into account.

Description of chemical reactivity (kinetics) also can be achieved in this setting.[5] When kinetics is addressed in its simplest TST form the key quantity of interest is the Gibbs free activation energy $\Delta G^{\neq}$. As we noticed previously, to calculate it one has to make some assumptions concerning the nature of transition state to which it is attributed. First of all we notice that molecular potential energy generally has multiple minima, which are usually interpreted as different stable states of the system. Then it is natural to think that molecular potential energy has also other critical points, like maxima and saddle points of different nature. It is assumed in the TST, that the transition state for the transformation of a reactant to a product (in, say, simplest monomolecular isomerization reaction) is such a saddle point by which one can pass from one local minimum of molecular PES to another by acquiring a minimal additional energy along the path. It means that if there exists more than one path from one minimum to another, only the one with the minimal height of the saddle point counts and all others should be discarded. After the transition state is identified, all the characteristics needed to calculate the different contributions to its free energy (like vibrational frequencies and inertia momenta) are sought and used according to general formulae eqs. (1.9) and (1.15). The most important consideration is that, at the defined saddle point, one of the vibrational normal modes (that crossing the ridge between the reactant and the product basins) corresponds to the negative eigenvalue of the dynamic matrix and thus to an imaginary frequency. It must be excluded from the count of vibrational modes while calculating the vibrational partition function and the vibrational contributions to the thermodynamic quantities. The electronic contribution to the partition function must be calculated using the molecular potential energy at the saddle point $\mathcal{U}^{\neq}$. With this after necessary transformations we get to a natural qualitative conclusion, that the higher barriers correspond to slower reactions, although precise calculations require much more subtle work, which is not a topic for the present introductory review.

The above results apply to the ideal gas of molecules. The objects addressed in the context of molecular modeling of complex systems are known in the form of macroscopic samples, mostly in the condensed phase. Thus the intermolecular degrees of freedom significantly contribute to the thermodynamical and other properties due to intermolecular interactions. For taking these latter into account the Monte-Carlo (MC) or molecular dynamics (MD) techniques are applied to model systems containing from hundreds to thousands of molecules and correspondingly tens and hundreds of thousands of atoms. These two approaches represent two more modern contexts where a demand for efficient methods of calculation of molecular potential

---

[5]Elementary introduction to chemical kinetics can be found in [1] and a more detailed one in [10]. An old, but eternally fresh description of the TST is given in [13].

energy appears. Incidentally this allows us to treat nonideal systems with interactions between molecules.[6]

Molecular dynamics is frequently portrayed as a method based on the ergodicity hypothesis which states that the trajectory of a system propagating in time through the phase space following the Newtonian laws of motion given by the equations:

$$(1.18) \quad M_\alpha \ddot{\mathbf{R}}_\alpha = -\nabla_\alpha \mathcal{U}(\{\mathbf{R}_\alpha\})$$

where ∇_α stands for the gradient with respect to spatial coordinates of the α-th atom; or in the Hamilton form:

$$(1.19) \quad \dot{\mathbf{P}}_\alpha = -\nabla_\alpha \mathcal{U}(\{\mathbf{R}_\alpha\})$$
$$\dot{\mathbf{R}}_\alpha = \frac{1}{M_\alpha}\mathbf{P}_\alpha,$$

comes infinitesimally close to each point in the phase space. From this hypothesis by very subtle considerations one can derive a conclusion that the thermodynamical averages $\langle A \rangle$ of an observable $A = A(\{\mathbf{P}_\alpha\}, \{\mathbf{R}_\alpha\})$ which is actually the only true observable quantity can be calculated as a time average of the same observable along some trajectory (solution of eq. (1.18))[7]:

$$(1.20) \quad \langle A \rangle = \lim_{\tau \to \infty} \frac{1}{\tau} \int_0^\tau A(\{\mathbf{P}_\alpha(t)\}, \{\mathbf{R}_\alpha(t)\})dt$$

Similarly the quantities which are functions describing the response of the system to external time dependent fields can be modeled by the MD approach.

---

[6]Detailed description of Monte Carlo and molecular dynamics techniques are given in [14]. Brief descriptions are given in many sources on computational chemistry e.g. [15] and [16].

[7]The sad truth is that the actually existing MD procedures implemented in respective computer codes do not really rely upon the ergodic theorem. In fact the using of various "thermostates" or "barostates" makes the sampling in the MD procedures to be performed not over the constant enenrgy hypersurface in the phase space as it must be if the true Newtonian trajectory had been used, but over a wider area of the phase space, which is in a way an advantage – using a "fat" trajectory instead of "thin" ones allows one to acquire the thermodynamical averages earlier than the infinite time assumed by the ergodic theorem. Incidentally using true Newtonian trajectories is in a way senseless as shows the simple estimate due to É. Borel [É. Borel, Introduction géométrique à quelques théories physiques, Paris: Gauthier-Villars, p. 97, 1914]. There Borel shows that under conditions taking place in a gas the uncertainty in the direction of the molecular motion coming from uncertainty of initial data amplifies such that a shift of one gram of matter by one centimeter in a star located at a separation of several light years from the Earth results in an error in the direction of 4π within a time of 10^{-6} seconds. Clearly much larger masses coming significantly closer to our planet are not either explicitly included in any MD experiment and thus their effect must be modeled by some randomness which is tacitly introduced by the above mentioned "barostates" and by for purpose using very crude algorithms for integrating Newton equations.

An alternative to MD, not relying upon the ergodicity hypothesis, is the Monte-Carlo procedure which yields the required thermodynamical average of observable A by performing the numerical estimate of the following integral (note that A does not depend on momenta):

$$(1.21) \quad \langle A \rangle = Z^{-1} \int_\Omega A(\{\mathbf{R}_\alpha\}) \exp(-\frac{\mathcal{U}(\{\mathbf{R}_\alpha(q)\})}{k_B T}) dq$$

$$Z = \int_\Omega \exp(-\frac{\mathcal{U}(\{\mathbf{R}_\alpha(q)\})}{k_B T}) dq$$

In the above expression Z stands for the configuration integral which differs from the partition function Q by being calculated only over the coordinates of the molecules and not over their momenta. This is possible because the coordinate and momentum parts of the whole phase space are run over independently throughout the integration and to a simple quadratic form of the kinetic energy, which thus can be integrated immediately.

All the examples of the wrapping contexts for molecular modeling can be characterized as major consumers of numerical methods of calculating molecular potential energy. We have mentioned several times that in the physical conditions assumed throughout molecular modeling, the independent motions of electrons are usually rare. However, it must be understood that in fact the required molecular potential energy function is defined by that state which electrons in the molecule acquire for any given positions $\{\mathbf{R}_\alpha\}$ of all atomic nuclei. This state can be found only from sufficiently quantum mechanical calculation of molecular electronic structures. On the other hand, the situation when only one electronic state is addressed in the chemical experiment and thus in the supporting molecular modeling is, although predominant, not the only possible one. In complex cases (which are in fact the true targets of the hybrid methods described here) several electronic states of molecular system can be accessed by experimentalists and thus must be covered by the modeling tools. The first step is rather transparent: in addition to the variables of the phase space $\{\mathbf{P}_\alpha\}$ and $\{\mathbf{R}_\alpha\}$ one has to add a discrete variable m distinguishing various electronic states of the polyatomic system which has to be used throughout in the thermodynamic calculations. This generalizes the definition of the partition function:

$$(1.22) \quad Q = \sum_m g_m \int_\Omega \exp(-\frac{\mathcal{T}(\{\mathbf{P}_\alpha\})+\mathcal{U}_m(\{\mathbf{R}_\alpha\})}{k_B T}) d\{\mathbf{P}_\alpha\} d\{\mathbf{R}_\alpha\}$$

The quantities $\mathcal{U}_m(\{\mathbf{R}_\alpha\})$ are inevitably obtainable only from a quantum mechanical calculation of molecular electronic structure for numerous points in the configuration space. In the following section we review this sophisticated problem.

1.2. MOLECULAR POTENTIAL ENERGY: QUANTUM MECHANICAL PROBLEM

In all the contexts of molecular modeling reviewed briefly above, it was taken for granted that the quantities on the right hand side of the above equations – the potential energies of molecular systems in their corresponding electronic states considered as functions of the system variables $\mathcal{U}_m(q) = \mathcal{U}_m(\mathbf{R}_{\{\alpha\}}(q))$ i.e. the PES – exist and are known. The concept of PES, however, is not an elementary one since in order to be defined it requires a certain construct known as the Born-Oppenheimer approximation.[8] Although a deeper analysis of this approximation is not a task of the present book, we briefly touch upon this problem here as its treatment shares an important common feature with our main topic, namely, variable separation in the quantum mechanical context.

The procedure begins with writing down the quantum mechanical Hamiltonian for a molecular system (electrons + nuclei) in the coordinate space:

$$\hat{H}(\{\mathbf{R}_\alpha\}, \{\mathbf{r}_i\}) = \hat{T}_n + \hat{T}_e + \hat{V}(\{\mathbf{R}_\alpha\}, \{\mathbf{r}_i\})$$

with

$$\hat{T}_n = -\tfrac{1}{2} \sum_{\alpha=1}^{N} \tfrac{1}{M_\alpha} \nabla_\alpha^2; \hat{T}_e = -\tfrac{1}{2} \sum_i \nabla_i^2$$

$$(1.23) \qquad \hat{V}(\{\mathbf{R}_\alpha\}, \{\mathbf{r}_i\}) = \hat{V}_{ne}(\{\mathbf{R}_\alpha\}, \{\mathbf{r}_i\}) + \hat{V}_{ee}(\{\mathbf{r}_i\}) + \hat{V}_{nn}(\{\mathbf{R}_\alpha\})$$

$$\hat{V}_{ne}(\{\mathbf{R}_\alpha\}, \{\mathbf{r}_i\}) = -\sum_{\alpha j} \frac{Z_\alpha}{|\mathbf{R}_\alpha - \mathbf{r}_j|}$$

$$\hat{V}_{ee}(\{\mathbf{r}_i\}) = \frac{1}{2} \sum_{i \neq j} \frac{1}{|\mathbf{r}_i - \mathbf{r}_j|}; \hat{V}_{nn}(\{\mathbf{R}_\alpha\}) = \frac{1}{2} \sum_{\alpha \neq \beta} \frac{Z_\alpha Z_\beta}{|\mathbf{R}_\alpha - \mathbf{R}_\beta|}$$

In these expressions written with use of so-called "atomic units" (elementary charge, electron mass and Planck constant are all equal to unity) $\mathbf{R}_\alpha$s stand as previously for the spatial coordinates of the nuclei of atoms composing the system; $\mathbf{r}_i$s for the spatial coordinates of electrons; M_αs are the nuclear masses; Z_αs are the nuclear charges in the units of elementary charge. The meaning of the different contributions is as follows: $\hat{T}_e$ and $\hat{T}_n$ are respectively the electronic and nuclear kinetic energy operators, $\hat{V}_{ne}$ is the operator of the Coulomb potential energy of attraction of electrons to nuclei, $\hat{V}_{ee}$ is that of repulsion between electrons, and $\hat{V}_{nn}$ that of repulsion between the nuclei. Summations over α and β extend to all nuclei in the (model) system and those over i and j to all electrons in it.

The variables describing electrons and nuclei are termed electronic and nuclear. For the majority of problems which arise in chemistry, the nuclear variables can be thought to be the Cartesian coordinates of the nuclei in the physical three-dimensional space. Of course the nuclei are in fact inherently quantum objects which manifest in such characteristics as nuclear spins – additional variables describing internal states of nuclei, which do not have any classical analog. However these latter variables enter into play relatively rarely. For example, when the NMR, ESR or Mössbauer experiments are discussed or in exotic problems like that of the ortho-para dihydrogen conversion. In a more common setting, such as the one represented by the

---

[8]The Born-Oppenheimer approximation is described in numerous sources on *quantum chemistry*; [17] being a standard text. Detailed derivation recommended to everyone who wants to understand it is presented in [18].

non-relativistic Hamiltonian eq. (1.23) for the molecular system, the nuclear spins do not enter.

The electronic variables are, however, unavoidably quantum and consist of Cartesian coordinates $\mathbf{r}$ of each electron in the system and of their respective projections of the spin s on some predefined axis $x = (\mathbf{r}, s)$. The electronic spin projections do not enter the above (nonrelativistic) Hamiltonian as well, but significantly affect the electronic structure of the system through the Pauli principle (see below).

Further reasoning leading to the PES concept, i.e. the Born-Oppenheimer construct, is based on the fact that the molecular systems are formed by the further indivisible (under the conditions of the chemical experiment) units of two sorts: nuclei and electrons. The mass of electron is smaller by at least three orders of magnitude than that of the lightest nucleus (proton), which suggests that at any current configuration of nuclei the electrons would feel only the current value of the electrostatic field induced by the nuclei, so that the electrons have enough time to adjust their state to any current position of the nuclei. When studying nuclear motion one may think that nuclei are followed by electrons with no delay and thus feel only the potential of the distribution of electrons defined by the current configuration of nuclei. Thus the variables describing the system or more precisely the spatial coordinates of the particles: those of nuclei and electrons enter in a nonequivalent manner. These notions are formalized in choosing a special *Ansatz* for the wave function of the molecular system described by the Hamiltonian eq. (1.23). As we know from quantum mechanics, the state of any quantum system is described by its wave function of coordinates of *all* particles composing the system. The wave functions corresponding to stationary states of the system satisfy the time independent Schrödinger equation of the form:

$$(1.24) \quad \hat{H}(\{\mathbf{R}_\alpha\}, \{\mathbf{r}_i\})\Psi(\{\mathbf{R}_\alpha\}, \{\mathbf{r}_i\}) = \varepsilon\Psi(\{\mathbf{R}_\alpha\}, \{\mathbf{r}_i\})$$

The *Ansatz* corresponding to the Born-Oppenheimer approximation for the wave function of all particles consists in separating the nuclear and electronic variables, i.e. in representing the total wave function in the form of a simple product:

$$(1.25) \quad \Psi(\{\mathbf{R}_\alpha\}, \{\mathbf{r}_i\}) \approx \Phi(\{\mathbf{r}_i\}; \{\mathbf{R}_\alpha\})\chi(\{\mathbf{R}_\alpha\})$$

of the electronic wave function $\Phi(\{\mathbf{r}_i\}; \{\mathbf{R}_\alpha\})$ calculated in the field of the *fixed* nuclei ($\{\mathbf{R}_\alpha\} = const.$), and $\chi(\{\mathbf{R}_\alpha\})$ is the nuclear wave function. Substituting this form into eq. (1.24) yields the following. Each nuclear configuration $\{\mathbf{R}_\alpha\}$ defines an electrostatic field acting upon the electronic subsystem of the entire molecular system. At each configuration of the nuclei this subsystem obeys the electronic Schrödinger equation:

$$(1.26) \quad \hat{H}_e\Phi(\{\mathbf{r}_i\}; \{\mathbf{R}_\alpha\}) = \mathcal{U}(\{\mathbf{R}_\alpha\})\Phi(\{\mathbf{r}_i\}; \{\mathbf{R}_\alpha\})$$

with the electronic Hamiltonian

$$(1.27) \quad \hat{H}_e = \hat{T}_e + \hat{V}(\{\mathbf{R}_\alpha\}, \{\mathbf{r}_i\}) = -\frac{1}{2}\sum_{i=1}^{N_e} \nabla_i^2 + \hat{V}(\{\mathbf{R}_\alpha\}, \{\mathbf{r}_i\})$$

formally including the energy of the nuclear–nuclear repulsion $\hat{V}_{nn}$, which in the present context is not any more an operator acting on the coordinates of the particles under study (electrons), but a so-called c-number – a constant (with respect to $\{\mathbf{r}_i\}$s) term added to the electronic Hamiltonian not affecting the functional form of the solution $\Phi(\{\mathbf{r}_i\}; \{\mathbf{R}_\alpha\})$ of the electronic Schrödinger equation eq. (1.26). In fact eq. (1.26) is not a unique equation, but the family of equations parametrized by the nuclear configurations $\{\mathbf{R}_\alpha\}$. At each configuration $\{\mathbf{R}_\alpha\}$ eq. (1.26) has multiple solutions numbered by integers $m = 0, \ldots$: the eigenvalues $\mathcal{U}_m(\{\mathbf{R}_\alpha\})$ and the eigenfunctions $\Phi_m(\{\mathbf{r}_i\}; \{\mathbf{R}_\alpha\})$ termed together as eigenstates. Of course they are different for the different values of coordinates $\{\mathbf{R}_\alpha\}$. This dependence is generally termed parametric dependence of the electronic eigenvalues and eigenfunctions of the molecular system on the nuclear coordinates. Propagating the m-th solution of eq. (1.26) through the nuclear configuration space determines a PES $\mathcal{U}_m(\{\mathbf{R}_\alpha\})$ for the given (ground if $m = 0$ or excited if $m > 0$) electronic state. It is then used as the potential energy for the nuclear motion described by the nuclear Schrödinger equation:

$$(1.28) \quad \hat{H}_n \chi(\{\mathbf{R}_\alpha\}) = \varepsilon \chi(\{\mathbf{R}_\alpha\})$$

where

$$(1.29) \quad \hat{H}_n = - \sum_{\alpha=1}^{N} \frac{1}{2M_\alpha} \nabla_\alpha^2 + \mathcal{U}_m(\{\mathbf{R}_\alpha\})$$

is the (effective) Hamiltonian for the nuclear motions occurring if the system resides in its m-th electronic state. As in the case of the electronic Schrödinger equation, the nuclear Schrödinger equation is not a single equation in the general sense. The potential energy and thus the Hamiltonian itself is specific for each of the electronic eigenstates numbered by the subscript m. A physical prerequisite for employing the Born-Oppenheimer approximation is that it is possible to consider only one (ground or excited) isolated electronic state of the system and to treat the nuclear motion as if it evolved on the single PES $\mathcal{U}_m(\{\mathbf{R}_\alpha\})$. In this case χ is a function of the nuclear coordinates *only*. Within this picture the effect of the electronic subsystem is condensed to a single function PES $\mathcal{U}_m(\{\mathbf{R}_\alpha\})$. In the vicinity of the PES minima the harmonic approximation allows one to get a simple quantum description of the nuclear motions of the molecular system. The nuclear wave function of the molecular system appears as a product of harmonic oscillator wave functions for all normal modes – eigenvectors of the dynamic matrix eq. (1.14).

The molecular modeling usually assumes the nuclear motions to be classical i.e. described by eq. (1.18) or eq. (1.19) rather than quantum described by eq. (1.28). Generally, molecular modeling consists in calculating the PES $\mathcal{U}_m(\{\mathbf{R}_\alpha\})$ of the molecular systems in their m-th electronic states, which are some functions of the entire set of independent nuclear coordinates $\{\mathbf{R}_\alpha\} = \{\mathbf{R}_\alpha(q)\}$. Sometimes other PES's elements are necessary, like its gradients, Hessians etc., although it is quite a rare case when the derivatives of PESs of the orders higher than two are required. The obtained PESs can then be used for calculations of thermodynamic or kinetic quantities, as described in the previous section. The theory of molecular electronic structure, on the other hand, gives basic procedures for obtaining $\mathcal{U}_m(q)$s by approximating solutions of the electronic Schrödinger equation eq. (1.26) which is done by using various quantum mechanical and quantum chemical techniques, which we will review briefly in subsequent sections.

1.3. BASICS OF THE QUANTUM MECHANICAL TECHNIQUE

Solving the families of electronic Schrödinger equations eqs. (1.26) and (1.27), which are the second order equations in partial derivatives of $3N_e$ variables, is an enormously complex problem for any system with a minimally realistic number of electrons. For that reason the practical means of solving (approximately) Schrödinger equations are based on alternative formulations which we review here briefly. These techniques can be further subdivided into quantum mechanical (QM) ones, common for all microscopic problems, and the quantum chemical ones which take into account specific aspects of the many-electron problem in a strong nonuniform field induced by nuclei. The relevant QM tools are described in numerous books at any required level of simplicity/complexity.[9] Here we recall these techniques only briefly, trying to pay attention to qualitative aspects and interrelations between their different forms which might be suitable in various situations, directing the reader to other sources for detailed formal descriptions.

1.3.1. The variation principle

The variational principle is the basis for a majority of practical methods of quantum mechanics. It is an implementation of a very general mathematical method: some very sophisticated problem – in our case solving the partial differential equation of an enormous number of variables – is equivalently reformulated to search for just one "controlling" number in some sense measuring how far we are from the solution of the original complex problem. The variational principle in quantum mechanics basically states several things: there is such a number, this number is the energy, the smaller is the energy the closer we are to the ground state of the system. It exists in several versions of which its formulation for the ground state is the most important.[10] It reads as follows: Among the wave functions Ψ normalized to unity and satisfying the boundary conditions of the problem under consideration, the expectation value of the energy E is the upper bound for the exact energy of the ground state E_0 i.e. for any trial function Ψ the inequality holds

$$(1.30) \quad E = \langle \Psi | \hat{H} | \Psi \rangle \geq E_0$$

---

[9]Books on quantum mechanics are numerous. Understandably, the author inclines towards the Russian school of presenting quantum mechanics, which may be subdivided into three major subschools (one of Leningrad – Fock [19] and two of Moscow: Landau [20] and Blokhintsev [21]). Classical works are also very useful [22–24].

[10]Literature on the variational principle is voluminous. In what relates to quantum mechanics the texts also contain necessary information on the variational principle. Here we add only [25] – a brilliant mathematical text on variations and [26] – a somewhat too casuistic, but useful, description of different aspects of the variational technique. I. Mayer [18] gives a good survey of what is necessary.

which turns to the equality only for the exact ground-state eigenfunction and eigenvalue. We have used in eq. (1.30) the so-called bracket notation dating back to P.A.M. Dirac to denote the expectation values – diagonal matrix elements or general matrix elements of the operators calculated over the wave functions. This is a handy notation allowing one to manipulate such quantities without addressing in detail what is actually needed to find them. For any pair of functions $\Psi = \Psi(x_1, x_2, ...)$, $\Phi = \Phi(x_1, x_2, ...)$ and any operator $\hat{A} = \hat{A}(x_1, x_2, ...)$ of an observable all taken in the coordinate representation (the only one we have been using so far) the expressions $\langle \Psi | \hat{A} | \Phi \rangle$ are understood as integrals:

$$(1.31) \quad \langle \Psi | \hat{A} | \Phi \rangle = \int \Psi^*(x_1, x_2, ...) \hat{A}(x_1, x_2, ...) \Phi(x_1, x_2, ...) dx_1 dx_2 ...$$

An important special case in such an expectation value is the scalar product of two functions, where the operator $\hat{A}$ is the identity operator $\hat{I} = \hat{I}(x_1, x_2, ...) \equiv 1$. Then one can write:

$$(1.32) \quad \langle \Psi | \Phi \rangle = \langle \Psi | \hat{I} | \Phi \rangle = \int \Psi^*(x_1, x_2, ...) \Phi(x_1, x_2, ...) dx_1 dx_2 ...$$

In the many-electron system, the system variables x are the pairs of spatial coordinates of an electron $\mathbf{r}$ and its spin projection s. In this case, the symbols $\int dx_i$ must be understood as $\sum_s \int d\mathbf{r}_i$ i.e. as integration over spatial coordinates of each electron and a summation over its two possible spin-projections. An important gain of the above notation is that it allows one to "postpone for tomorrow" the unpleasant integrations involving a knowledge of calculus and reduces everything to simple algebraic manipulations with formal quantities, some of which may be known e.g. by assumptions. A simple example is given here:

$$(1.33) \quad \langle \Psi | \lambda_1 \Phi_1 + \lambda_2 \Phi_2 \rangle = \lambda_1 \langle \Psi | \Phi_1 \rangle + \lambda_2 \langle \Psi | \Phi_2 \rangle$$

Similar relations hold for operators which will be explained below and widely used throughout the book.

With these notation the *proof* evolves as follows: The energy is the quantum mechanical expectation value $E = \langle \Psi | \hat{H} | \Psi \rangle$ of the Hamiltonian calculated for any wave function Ψ normalized to unity ($\langle \Psi | \Psi \rangle = 1$). The Schrödinger equation is a linear equation of the form:

$$(1.34) \quad \hat{H}\Psi = \varepsilon\Psi$$

with an Hermitean Hamiltonian $\hat{H}$. Thus its solutions, eigenvectors Ψ_i and corresponding eigenvalues E_i, form a complete orthonormalized set of functions satisfying the boundary conditions of the problem. For the eigenvectors and eigenvalues of an Hermitean operator $\hat{H}$, the following holds by definition:

$$(1.35) \quad \hat{H}\Psi_k = E_k\Psi_k,$$

$$\langle \Psi_k | \Psi_l \rangle = \delta_{kl}$$

Since the operator is limited from below it has a minimal eigenvalue E_0 which is by definition that of the ground state Ψ_0. Then any wave function Ψ satisfying the same boundary conditions can (completeness of the set $\{\Psi_k\}$) be expanded in a series

$$(1.36) \qquad \Psi = \sum_k u_k \Psi_k$$

As one can see normalization to unity gives:

$$(1.37) \qquad \langle \Psi | \Psi \rangle = \sum_k |u_k|^2 = 1$$

Then the quantum mechanical expectation value of the Hamiltonian over the function Ψ reads:

$$(1.38) \qquad E = \langle \Psi | \hat{H} | \Psi \rangle = \sum_k |u_k|^2 E_k$$

Since $E_k \geq E_0$, we get

$$(1.39) \qquad E \geq \sum_k |u_k|^2 E_0 = E_0$$

Subtracting E_0 from E and assuming that the ground state is nondegenerate we get the following inequality where all excitation energies ΔE_i are strictly positive:

$$(1.40) \qquad E - E_0 = \sum_k |u_k|^2 (E_k - E_0) = \sum_{k>0} |u_k|^2 \Delta E_k \geq 0$$

so that equality can be reached only when all $|u_k|^2 = 0$ for $k > 0$. Then the normalization condition yields the result $u_0 = 1$ and thus $\Psi = \Psi_0$ (the equality in fact holds up to immaterial phase factor of absolute value of unity $e^{i\alpha}$). This provides the required proof (possible degeneracy of the ground state is not a great problem here).

1.3.2. The linear variational method (Ritz method)

The variation principle as formulated above does not seem to be practical as it relies upon solutions of the Schrödinger equation – the eigenvectors and eigenvalues of the Hamiltonian, which in fact have to be found. It can be used, however, as a starting point for constructing more practical methods. They are based on the concept of the trial wave function of the system. Let us consider a family of functions dependent on the system variables (in the case of electrons, on their spatial coordinates and spin projections $\{x_i\} = \{(\mathbf{r}_i, s_i)\}$), satisfying the boundary conditions of the problem, and in whatever sophisticated manner depending on some other set of variables ξ – variational variables – i.e. having the form $\Psi(\xi|\{x_i\})$. Function $\Psi(\xi|\{x_i\})$, which may be subject to additional conditions of the form $S(\xi) = 0$, of which the most frequently occurring in practice is again the normalization condition $S(\xi) = \langle \Psi(\xi) | \Psi(\xi) \rangle - 1 = 0$), (in the bracket notation the integration is assumed over the *system* variables x_i and not on the *variation* variables ξ) is then called the trial wave function for the system. For any value of ξ it can be expanded over the eigenfunctions of the Hamiltonian, such that the expansion amplitudes (coefficients) u_k themselves become some functions of ξ i.e. $u_k = u_k(\xi)$. If the functions of the family $\Psi(\xi|\{x_i\})$ are constructed so that for each value of ξ they are normalized to unity then the sets of all expansion coefficients $u_k(\xi)$ also satisfy the normalization condition eq. (1.37) automatically for all values of ξ. Substituting $u_k(\xi)$ into eq. (1.38) yields an estimate for the energy $E = E(\xi)$ for which the inequality eq. (1.39): $E(\xi) \geq E_0$ also holds. This is a variational estimate for the ground state energy obtained with the trial wave functions of the selected class. Minimizing $E(\xi)$ with respect to ξ brings the best estimate for the ground state energy within the selected class of trial functions.

Among the classes of the trial wave functions, those employing the form of the linear combination of the functions taken from some predefined *basis* set lead to the most powerful technique known as the linear variational method. It is constructed as follows. First a set of M normalized functions Φ_k, each satisfying the boundary conditions of the problem, is selected. The functions Φ_k are called the "basis functions" of the problem. They must be chosen to be linearly independent. However we do not assume that the set of $\{\Phi_k\}$ is complete so that any Ψ can be exactly represented as an expansion over it (in contrast with exact expansion eq. (1.36)); neither is it assumed that the functions of the basis set are orthogonal. *A priori* they do not have any relation to the Hamiltonian under study – only boundary conditions must be fulfilled. Then the trial wave function Ψ is taken as a linear combination of the basis functions Φ_k:

$$(1.41) \quad \Psi(\xi) = \sum_{k=1}^{M} u_k \Phi_k$$

so that the expansion amplitudes themselves take the part of the variational variables ξ and must be subject to the normalization condition $\langle \Psi(\xi)|\Psi(\xi)\rangle = 1$. The best estimate for the energy with the trial wave function of the above form in the chosen basis $\{\Phi_k\}$ is then obtained by selecting the coefficients u_k such that the expectation value for the energy is minimal:

$$(1.42) \quad \min_{\{u_k\}} E = \min_{\{u_k\}} \langle \Psi(\xi)|\hat{H}|\Psi(\xi)\rangle$$

with the additional condition

$$(1.43) \quad \langle \Psi(\xi)|\Psi(\xi)\rangle = 1$$

Since the functions Φ_i are fixed, the expectation value of the energy E considered as a function of the variational parameters u_i is a quadratic form:

$$(1.44) \quad E(\{u_l\}) = \langle \Psi|\hat{H}|\Psi\rangle = \sum_{k,l=1}^{M} u_k^* H_{kl} u_l$$

(expansion amplitudes are in general case complex and u_k^* is a complex conjugate of u_k.) Here the notation is introduced for the matrix elements of the Hamiltonian with respect to the selected functional basis:

$$(1.45) \quad H_{kl} = \langle \Phi_k | \hat{H} | \Phi_l \rangle$$

The extremum for this form must be found taking into account the normalization condition

$$(1.46) \quad \langle \Psi | \Psi \rangle = \sum_{k,l=1}^{M} u_k^* M_{kl} u_l$$

where the notation for the matrix elements of the metric characterizing the basis set

$$(1.47) \quad M_{kl} = \langle \Phi_k | \Phi_l \rangle$$

is introduced.

In order to find extrema of $E(\{u_l\})$, subject to the normalization condition, standard moves known as the Lagrange multipliers' method are applied, which readily lead us to the well-known form of the generalized matrix eigenvalue/eigenvector problem:

$$(1.48) \quad \sum_{l=1}^{M} (H_{kl} - E M_{kl}) u_l = 0$$

The above equality must hold for each k, so that finding extrema of the auxiliary quadratic function is equivalent to a system of m linear equations. Assembling the quantities H_{kl} and M_{kl}, into $M \times M$ matrices $\mathbf{H}$ and $\mathbf{M}$ representing the Hamiltonian and the metric, respectively, and the amplitudes u_i into a column-vector $\mathbf{u}$, we rewrite the system of linear equations (eq. (1.48)) in the form

$$(\mathbf{H} - E\mathbf{M})\mathbf{u} = 0$$

or

$$\mathbf{H}\mathbf{u} = E\mathbf{M}\mathbf{u}$$

which is known as a "generalized matrix eigenvalue problem". If the functional basis $\{\Phi_k\}$ is taken to be orthonormalized, $M_{kl} = \delta_{kl}$; the metric matrix in this basis becomes the unity matrix, $\mathbf{M} = \mathbf{1}$. This reduces to a "standard" matrix eigenvalue problem:

$$(1.49) \quad \mathbf{H}\mathbf{u} = E\mathbf{u}$$

Solving the eigenvalues problem for a Hermitian matrix is equivalent to diagonalizing it by performing the similarity transformation of the matrix $\mathbf{H}$ by the unitary matrix $\mathbf{U}$ composed of its eigenvectors. The reverse also holds: the columns of a matrix $\mathbf{U}$ diagonalizing the Hermitian matrix $\mathbf{H}$ are the eigenvectors of the latter. In fact

$$(1.50) \quad \mathbf{U}^\dagger \mathbf{H} \mathbf{U} = \mathbf{D}, \, D_{kl} = E_k \delta_{kl}$$

If all the eigenvalues are different, the transformation matrix $\mathbf{U}$ is unique up to the order of the eigenvalues/eigenvectors and to the arbitrary phase factors $e^{i\alpha_k}$ for k-th column (eigenvector). Otherwise, i.e. when some eigenvalues coincide – are degenerate – the eigenvectors matrix is defined up to arbitrary unitary transformations of the eigenvectors spanning the subspaces which belong to each of the degenerate eigenvalues.

The minimal eigenvalue of the matrix $\mathbf{H}$ should be taken as an upper bound for the exact ground state eigenvalue of the original Schrödinger equation written in terms of differential operator and coordinate wave functions (in the complete and thus infinite basis). They obviously *never* coincide unless the exact ground state wave function Ψ_0 is by chance one of the basis functions $\{\Phi_k\}$. This obviously can hardly happen unless the problem is very simple and does not require any serious numerical treatment. Otherwise the amount of noncoincidence depends of course on the choice of the basis set $\{\Phi_k\}$. One can easily imagine the situation when the exact ground state wave function is orthogonal to the subspace spanned by the basis set $\{\Phi_k\}$. Then both the minimal eigenvalue of the matrix problem and the eigenvector corresponding to it have nothing to do with the exact ground state of the original problem. By this, one can see that the choice of the basis set is very important. In practice of course much effort is spent in finding a better basis or getting some equivalent of this. It can be formalized by introducing an additional (multidimensional) parameter ω, which in a concise form represents the subset of the entire functional space spanned by the trial functions of the chosen form. In the case of the linear Ritz method for the finite basis described in this section the parameter ω defines this basis $\{\Phi_k\}$. In a more general case it may include information concerning the *form* of the trial wave function, which on some physical grounds can be chosen to be much more sophisticated than a simplistic expansion eq. (1.41) due to allowing a better coverage of the vicinity of the exact ground state. This will be exemplified later.

1.3.3. Perturbation methods

The variation methods, particularly the linear variational method of solving the Schrödinger equation reducing to diagonalization of its matrix representation described briefly above, serve largely as a basis for developing numerical procedures. Qualitative theories are by contrast based on a different type of reasoning. Since in the present book we are concerned with the general theoretical constructs necessary for developing methods of hybrid modeling, we briefly review the relevant perturbative techniques.

Diagonalizing a general Hermitian matrix (of an acceptable dimensionality), being feasible numerically, does not provide any insight on the qualitative nature of the result. The general strategy of perturbational methods is based on the idea of obtaining estimates of necessary quantities, including eigenvalues and eigenvectors of Hamiltonians of interest, on the basis of an a priori knowledge of the eigenvalues and eigenvectors of a simpler, but in a sense close (unperturbed), Hamiltonian. Of

20 *Andrei L. Tchougréeff*

course from a certain angle it can be viewed also as a formula for constructing an adequate basis set for the linear variation method (see above). We consider here only the simplest Rayleigh–Schrödinger perturbation theory, as it will be the only version used in this book.

Let us consider the situation when the solutions (i.e., the orthonormalized eigenvectors $\Psi_i^{(0)}$ and the corresponding eigenvalues $E_i^{(0)}$) of the Schrödinger equation

$$(1.51) \quad \hat{H}^{(0)}\Psi_i^{(0)} = E_i^{(0)}\Psi_i^{(0)}$$

with an unperturbed Hamiltonian $\hat{H}^{(0)}$ are known. This information can be used for obtaining estimates of the eigenvalues and eigenvectors of a "perturbed" Schrödinger equation, containing a Hamiltonian $\hat{H}(\lambda)$ depending on a smallness parameter λ. It is "close" to $\hat{H}^{(0)}$ provided λ is small enough. The perturbed equation reads:

$$(1.52) \quad \hat{H}\Psi = (\hat{H}^{(0)} + \lambda\hat{W})\Psi = E\Psi$$

Perturbation theory (PT) tries to represent the eigenvectors and the eigenvalues of the perturbed Schrödinger equation eq. (1.52) as power series:

$$(1.53) \quad \Psi_k = \sum_{m=0}^{\infty} \lambda^m \psi_k^{(m)}, E_k = \sum_{n=0}^{\infty} \lambda^n \varepsilon_k^{(n)}$$

with respect to the "perturbation strength parameter" λ. Alternatively one can write

$$(1.54) \quad \Psi_k = \Psi_k^{(0)} + \sum_{m=1}^{\infty} \Psi_k^{(m)}, E_k = E_k^{(0)} + \sum_{n=1}^{\infty} E_k^{(n)}$$

where

$$\Psi_k^{(m)} = \lambda^m \psi_k^{(m)} \;; \quad E_k^{(n)} = \lambda^n \varepsilon_k^{(n)}$$

The problem of convergence of the series eq. (1.53) and eq. (1.54) will not be addressed here. In general terms it can be said that only the so-called asymptotic convergence of the latter can possibly be assumed. Moreover, in most cases of quantum chemical interest, λ is not a true variable of the problem as it cannot be changed in any physical experiment, but is only an accounting tool. In this case the result is obtained by setting $\lambda = 1$ at the end of the calculation. Then of course

$$\Psi_k^{(m)} = \psi_k^{(m)} \;; \quad E_k^{(n)} = \varepsilon_k^{(n)}$$

holds in all orders of the PT. In any case we are going to use only the first nontrivial terms of these expansions.

Formal derivation evolves as follows: substituting the expansions eq. (1.53) for the k-th eigenstate into the Schrödinger equation eq. (1.52) yields:

$$(1.55) \quad (\hat{H}^{(0)} + \lambda\hat{W}) \sum_{m=0}^{\infty} \lambda^m \psi_k^{(m)} = \sum_{n=0}^{\infty} \lambda^n \varepsilon_i^{(n)} \sum_{m=0}^{\infty} \lambda^m \psi_k^{(m)}$$

Then equating the coefficients at the equal powers of λ (accounting!) we get:

$$\left(\hat{H}^{(0)} - E_k^{(0)}\right) \Psi_k^{(0)} = 0 \, ,$$

$$\left(\hat{H}^{(0)} - E_k^{(0)}\right) \Psi_k^{(1)} = \lambda \left(\varepsilon_k^{(1)} - \hat{W}\right) \Psi_k^{(0)} \, ,$$

$$(1.56) \quad \left(\hat{H}^{(0)} - E_k^{(0)}\right) \Psi_k^{(2)} = \lambda \left(\varepsilon_k^{(1)} - \hat{W}\right) \Psi_k^{(1)} + \lambda^2 \varepsilon_i^{(2)} \Psi_k^{(0)}$$

$$\cdots\cdots\cdots\cdots\cdots$$

$$\left(\hat{H}^{(0)} - E_k^{(0)}\right) \Psi_k^{(m)} = \lambda \left(\varepsilon_k^{(1)} - \hat{W}\right) \Psi_k^{(m-1)} + \dots + \lambda^m \varepsilon_k^{(m)} \Psi_k^{(0)}$$

This is a system of inhomogeneous linear equations for the functions (vectors) $\Psi_i^{(m)}$ (the mixed notation for the perturbation corrections to eigenvalues and eigenvectors is used above). The 0-th order in λ yields the unperturbed problem and thus is satisfied automatically. The others can be solved one by one. For this end we multiply the equation for the first order function by the zeroth-order wave function and integrate which yields:

$$(1.57) \quad \varepsilon_k^{(1)} = \langle \Psi_k^{(0)} | \hat{W} | \Psi_k^{(0)} \rangle = \langle \psi_k^{(0)} | \hat{W} | \psi_k^{(0)} \rangle$$

Applying the same move to the further equations we get:

$$\varepsilon_k^{(n)} = \langle \psi_k^{(0)} | \hat{W} | \psi_k^{(n-1)} \rangle$$

As we can see, to get the correction of the n-th order to the eigenvalue (energy) one has to know the correction on the $(n-1)$-th order to the eigenvector (wave function). In fact a much stronger statement is valid, namely, that knowing the correction of the n-th order to the wave function allows us to know the correction of the $(2n+1)$-th order to the energy. Since we are interested here only in lower order corrections, we do not elaborate on this further. One can find proofs and detailed discussions in books by I. Mayer [18] and by L. Zülicke [27].

Further development i.e. obtaining the corrections to the wave functions (eigenvectors) depends on the character of the spectrum of the eigenvalues of the unperturbed Hamiltonian $\hat{H}^{(0)}$. Two major cases are distinguished: when all eigenvalues of the zero order problem eq. (1.51) are different it is referred to as the nondegenerate case. When some of the eigenvalues of the unperturbed problem coincide, it is referred to as a degenerate case. These cases are generally considered separately.

1.3.3.1. Nondegenerate case

The problem of finding a vector is usually solved by representing the required vector as an expansion with respect to some natural set of basis vectors. Following this method one can expand the vector of the n-th order correction to the k-th unperturbed vector $- \psi_k^{(n)}$ in terms of the solutions $\Psi_k^{(0)}$ (eigenvectors) of the unperturbed problem eq. (1.51):

$$(1.58) \quad \Psi_k^{(n)} = \lambda^n \sum_{l \neq k} u_{il}^{(n)} \Psi_l^{(0)} \qquad (n \geq 1)$$

By this, the expansion coefficients $u_{il}^{(n)}$ are themselves of the 0-eth order in λ. The restriction $l \neq k$ indicates that the correction is orthogonal to the unperturbed vector. In order to get the corrections to the k-th vector, we find the scalar product of the perturbed Schrödinger equation for it written with explicit powers of λ with one of the eigenvectors of the unperturbed problem $\Psi_j^{(0)*}$ $(j \neq k)$. For the first order in λ we get:

$$(1.59) \quad \lambda W_{jk} + E_j^{(0)} \lambda u_{kj}^{(1)} = E_k^{(0)} \lambda u_{kj}^{(1)}$$

where we used the fact that:

$$\langle \Psi_k^{(0)} | \hat{H}^{(0)} | \Psi_j^{(0)} \rangle = E_j^{(0)} \delta_{kj}$$

and the notation

$$(1.60) \quad W_{kj} = \langle \Psi_k^{(0)} | \hat{W} | \Psi_j^{(0)} \rangle$$

Collecting the correction to the wave function on the left side and all the perturbation terms on the right side we get

$$(E_j^{(0)} - E_k^{(0)}) \lambda u_{kj}^{(1)} = -\lambda W_{kj}$$

which can be resolved for the coefficients at λ on the two sides:

$$(1.61) \quad u_{kj}^{(1)} = -\frac{W_{jk}}{E_j^{(0)} - E_k^{(0)}}$$

The first order correction to the k-th eigenvector (wave function) then reads as follows:

$$(1.62) \quad \Psi_k^{(1)} = -\lambda \sum_{j \neq k} \left(\frac{W_{jk}}{E_j^{(0)} - E_k^{(0)}} \right) \Psi_j^{(0)}$$

$$\psi_k^{(1)} = -\sum_{j \neq k} \left(\frac{W_{jk}}{E_j^{(0)} - E_k^{(0)}} \right) \Psi_j^{(0)}$$

This result indicates that the Rayleigh–Schrödinger PT is expected to be well applicable in those cases in which these fractions are small. Inserting eq. (1.62) into expressions for the energy corrections we get the well-known explicit expression for the second-order ones:

$$\varepsilon_k^{(2)} = \sum_{j \neq k} \left(-\frac{W_{jk}}{E_j^{(0)} - E_k^{(0)}} \right) W_{kj} = -\sum_{j \neq k} \frac{|W_{kj}|^2}{E_j^{(0)} - E_k^{(0)}}$$

$$E_k^{(2)} = -\lambda^2 \sum_{j \neq k} \frac{|W_{kj}|^2}{E_j^{(0)} - E_k^{(0)}}$$

The second-order correction to the energy of a ground state is always negative as all $E_j^{(0)} > E_0^{(0)}$ for $j \neq 0$ (nondegenerate case).

We do not elaborate further on this as the results concerning higher orders of the PT can be found in many sources (see e.g. [18]). One more remark can be given: restricting to the second order correction to the energy and the first order correction to the wave function allows us to treat the perturbation operator as strictly off-diagonal with $W_{kk} \equiv 0$ for all k, as the diagonal matrix elements of the perturbation do not affect the wave functions. This allows us to simplify some further general formulae.

1.3.3.2. Expectation values: linear response

In the previous section we described the result of "turning on" a perturbation on the wave functions (eigenvectors) of the unperturbed Hamilton operator with non-degenerate spectrum in the lowest order when this effect takes place. In quantum mechanics the wave function is an intermediate tool, not an observable quantity. The general requirement of the theory is, however, to represent the interrelations between the observables. For this we give here the formulae describing the effect of a perturbation upon an observable. Let us assume that in one of its unperturbed states $\Psi_k^{(0)}$ the system is characterized by the expectation value of an observable $\hat{A}$:

$$\left\langle \hat{A} \right\rangle_k^{(0)} = \left\langle \Psi_k^{(0)} \left| \hat{A} \right| \Psi_k^{(0)} \right\rangle$$

Turning on the perturbation $\lambda \hat{W}$ produces the correction to the wave functions (eigenvectors) of the system described by eq. (1.62). Inserting it into the definition of the expectation value of A yields:

$$\left\langle \hat{A} \right\rangle_k^{(\lambda)} = \left\langle \Psi_k^{(0)} + \Psi_k^{(1)} \left| \hat{A} \right| \Psi_k^{(0)} + \Psi_k^{(1)} \right\rangle$$

Assembling the terms linear in λ and introducing the notation

$$A_{kj} = \left\langle \Psi_k^{(0)} \left| \hat{A} \right| \Psi_j^{(0)} \right\rangle$$

we get the augment in the observable A linear in the strength of the above perturbation

$$(1.63) \quad \delta \langle A \rangle_k^{(\lambda)} = \langle A \rangle_k^{(\lambda)} - \langle A \rangle_k^{(0)} =$$

$$= -\lambda \sum_{j \neq k} \left(\left(\frac{W_{jk}}{E_j^{(0)} - E_k^{(0)}} \right) A_{kj} + \left(\frac{W_{jk}^*}{E_j^{(0)} - E_k^{(0)}} \right) A_{jk} \right)$$

The coefficient at λ describes the linear response of the quantity A to the perturbation $\hat{W}$. It can be given a rather more symmetric form. Indeed the amplitude of the j-th unperturbed state in the correction to the k-th state is proportional to some skew Hermitian operator (the perturbation matrix $\mathbf{W}$ is Hermitian, but the denominator changes its sign when the order of the subscripts changes). With this notion and assuming that $W_{kk} \equiv 0$ (see above) we can remove the restriction in the summation and write:

$$(1.64) \quad \delta \langle A \rangle_k^{(\lambda)} = -\lambda \sum_j \left(A_{kj} \left(\frac{W_{jk}}{E_j^{(0)} - E_k^{(0)}} \right) - \left(\frac{W_{kj}}{E_j^{(0)} - E_k^{(0)}} \right) A_{jk} \right)$$

Introducing the notation

$$(1.65) \qquad \frac{W_{kj}}{E_j^{(0)} - E_k^{(0)}} = (1 - \delta_{kj}) K_{kj}$$

we immediately get:

$$(1.66) \qquad \delta \langle A \rangle_k^{(\lambda)} = -\lambda \left\langle \Psi_k^{(0)} \left| \left[\hat{K}, \hat{A} \right] \right| \Psi_j^{(0)} \right\rangle$$

where

$$(1.67) \qquad [\hat{A}, \hat{B}] = \hat{A}\hat{B} - \hat{B}\hat{A}$$

is the commutator of the operators $\hat{A}$ and $\hat{B}$.

It is of interest to learn more about the operator $\hat{K}$ which plays that remarkable rôle. It is defined by the perturbation operator $\hat{W}$ and the unperturbed Hamiltonian $\hat{H}^{(0)}$. As a function of $\hat{W}$ it is obviously linear in the sense that if $\hat{W} = \omega_1 \hat{W}_1 + \omega_2 \hat{W}_2$ then $\hat{K} = \omega_1 \hat{K}_1 + \omega_2 \hat{K}_2$ where $\hat{K}_1$ and $\hat{K}_2$ are generated by $\hat{W}_1$ and $\hat{W}_2$ according to the rule eq. (1.65). Thus the operation giving $\hat{K}$ by $\hat{W}$ can be considered a linear operator in the space of the operators themselves, which can be called *superoperator*. To study it and the relation of $\hat{K}$ to $\hat{H}^{(0)}$ let us consider the commutators of any linear operator $\hat{B}$ with $\hat{H}^{(0)}$. For all eigenvectors of $\hat{H}^{(0)}$ in the original space where $\hat{H}^{(0)}$ and $\hat{B}$ themselves act we can write:

$$(1.68) \qquad \left\langle \Psi_k^{(0)} \left| \left[\hat{H}^{(0)}, \hat{B} \right] \right| \Psi_j^{(0)} \right\rangle = \left\langle \Psi_k^{(0)} \left| \hat{H}^{(0)} \hat{B} - \hat{B} \hat{H}^{(0)} \right| \Psi_j^{(0)} \right\rangle =$$

$$= \left(E_k^{(0)} - E_j^{(0)} \right) \left\langle \Psi_k^{(0)} \left| \hat{B} \right| \Psi_j^{(0)} \right\rangle =$$

$$= \left(E_k^{(0)} - E_j^{(0)} \right) B_{kj}$$

This operation is linear with respect to $\hat{B}$ so that it defines some superoperator in the space of operators. This superoperator has a kernel – the subspace of operators that become zero upon its action: these are the operators diagonal in the basis of the eigenvectors of the selected operator $\hat{H}^{(0)}$. These are the only elements of the kernel, provided all the eigenvalues of $\hat{H}^{(0)}$ are different (nondegenerate case). The operators from the complement to this kernel are one-to-one transformed by the above multiplication of their matrix elements by eq. (1.68) by the differences of the eigenvalues of $\hat{H}^{(0)}$. The Hermitian operators (with zero diagonal) are transformed to the skew-Hermitian ones and *vice versa*. The superoperator defined by the operator $\hat{H}^{(0)}$ and performing the described transformation is called an adjoint superoperator of $\hat{H}^{(0)}$ and is denoted in mathematics as $\mathrm{Ad}_{\hat{H}^{(0)}}$. It can be inverted on the subspace complementary to its kernel. The action of the inverse superoperator $\mathrm{Ad}_{\hat{H}^{(0)}}^{-1}$ on off-diagonal operators is easily seen from the above relations, namely:

$$(1.69) \qquad \left\langle \Psi_k^{(0)} \left| \mathrm{Ad}_{\hat{H}^{(0)}}^{-1} \hat{B} \right| \Psi_j^{(0)} \right\rangle = \left(E_k^{(0)} - E_j^{(0)} \right)^{-1} B_{kj}$$

With its use the response of the quantity A in the k-th state of the operator $\hat{H}^{(0)}$ to the perturbation given by the operator $\hat{W}$ is conveniently written as:

$$(1.70) \quad \delta \langle A \rangle_k^{(\lambda)} = -\lambda \left\langle \Psi_k^{(0)} \left| \left[\mathrm{Ad}_{\hat{H}^{(0)}}^{-1} \, \hat{W}, \hat{A} \right] \right| \Psi_k^{(0)} \right\rangle$$

In the literature the response theory is largely described in the time-dependent form which requires a somewhat complicated technique of time ordering of the operators and Fourier transformations between time and frequency domains. The static responses which are largely needed in the present book appear as a result of subtle limit procedures for the frequencies flowing to zero. Here we have developed the necessary static results within their own realm.

1.3.3.3. Degenerate case[11]

The formulae presented in the previous sections do not apply if the unperturbed eigenstate to which the corrections are to be found belongs to a degenerate eigenvalue of the unperturbed Hamiltonian, because some of the denominators become zero. For that reason the degenerate case of the theory strongly differs from the nondegenerate case. The evanescence of the energy denominators is, however a more formal manifestation of a deeper reason for the difference between the two cases. From the substantial point of view the degenerate and nondegenerate cases are qualitatively distinct by the results obtained: in the nondegenerate case each of the eigenvalues acquires some corrections, but the overall form of the eigenvalues' spectrum does not change. In the degenerate case the perturbation changes (as we shall see) the character of the spectrum: the degeneracy is lifted and the relation between the perturbation and the result it produces is explained below.

Let us consider the situation when the eigenvalues $E_k^{(0)}$ of the unperturbed Hamiltonian $\hat{H}^{(0)}$ are respectively g_k-fold degenerate. In this case the unperturbed Schrödinger equation reads:

$$\hat{H}^{(0)} \Psi_{k,i}^{(0)} = E_k^{(0)} \Psi_{k,i}^{(0)} ; \quad i = 1, 2, \ldots, g_k$$

where we denote by $\Psi_{k,i}^{(0)}$ the g_k orthonormalized eigenvectors of the unperturbed problem, belonging to the eigenvalue E_k^0. The perturbed problem is:

$$(\hat{H}^{(0)} + \lambda \hat{W}) \Psi = E \Psi$$

Expanding Ψ and E as power series of λ, we have

$$(\hat{H}^{(0)} + \lambda \hat{W}) \sum_{n=0}^{\infty} \lambda^n \psi_k^{(n)} = \sum_{m=0}^{\infty} \lambda^m \varepsilon_k^{(m)} \sum_{n=0}^{\infty} \lambda^n \psi_k^{(n)}$$

---

[11]The author is greatly indebted to Prof. I. Mayer for his suggestions, which significantly improved the presentation in this section.

Equating as previously coefficients at the different powers of λ on the two sides we get for 0-eth power of λ

$$\hat{H}^{(0)}\Psi_k^{(0)} = E_k^{(0)}\Psi_k^{(0)}$$

which brings us back to the unperturbed Schrödinger equation which is satisfied by any arbitrary linear combination of the orthonormal basis vectors belonging to the eigenvalue $E_k^{(0)}$:

$$\Psi_k^{(0)} = \sum_{i=1}^{g_k} u_i \Psi_{k,i}^{(0)}$$

For the first power of λ we have just the same equation as for the nondegenerate case:

$$\hat{W}\Psi_k^{(0)} + \hat{H}^{(0)}\psi_k^{(1)} = \varepsilon_k^{(1)}\Psi_k^{(0)} + \varepsilon_k^{(0)}\psi_k^{(1)}$$

but the treatment must be quite different. Collecting on the left the terms containing the unperturbed vector $\Psi_k^{(0)}$ we get

$$(\hat{W} - \varepsilon_k^{(1)})\Psi_k^{(0)} = -(\hat{H}^{(0)} - E_k^{(0)})\psi_k^{(1)}$$

In the non-degenerate case it was an inhomogeneous equation with the nonvanishing right hand part, which could be used to determine the first order energy $\varepsilon^{(1)}$ and the expansion coefficients of the first order wave function $\Psi_k^{(1)}$. It is not like this in the degenerate case. It is easy to see by substituting expansion of $\Psi_k^{(0)}$ over the basis in the degenerate manifold:

$$(\hat{W} - \varepsilon_k^{(1)}) \sum_{i=1}^{g_k} u_i \Psi_{k,i}^{(0)} = -(\hat{H}^{(0)} - E_k^{(0)})\psi_k^{(1)}$$

The point is that the vectors $\Psi_{k,i}^{(0)}$ satisfying the unperturbed Schrödinger equation, if used to expand $\psi_k^{(1)}$, make the right hand side disappear and the equation becomes a uniform one. The only thing we can do is to use it to determine the proper expansion coefficients of the *zeroth* order wave function $\Psi_k^{(0)}$ in terms of the degenerate subspace as well as the first order energy. (The first order wave function is usually not calculated/considered in the degenerate case.)

This can be done by going to the eigenvalue equation which can be obtained simply by forming the scalar products from the left with the different unperturbed vectors $\Psi_{k,j}^{(0)*}$, and then the terms in the right hand side except the diagonal ones become zero:

$$\langle \Psi_{k,j}^{(0)} | \hat{W} - \varepsilon_k^{(1)} | \sum_{i=1}^{g_k} u_i \Psi_{k,i}^{(0)} \rangle = 0; \quad j = 1, 2, \ldots, g_k$$

and introducing the notation

$$W_{ji}^{(k)} = \langle \Psi_{k,j}^{(0)} | \hat{W} | \Psi_{k,i}^{(0)} \rangle$$

we get

$$\sum_{i=1}^{g_k} W_{mi}^{(k)} u_i = \varepsilon_k^{(1)} u_m; \qquad m = 1, 2, \ldots, g_k$$

which is precisely the eigenvalue/eigenvector problem for the operator $\hat{W}$ taken in the subspace spanned by the vectors $\Psi_{k,i}^{(0)}$ belonging to the degenerate eigenvalue of the original unperturbed Hamiltonian. In this case $\hat{W}$ takes the part of the Hamiltonian itself (in the degenerate subspace). This corresponds to the shift of the energy reference point by $E_k^{(0)}$. The energy correction in the degenerate first order, as we see, requires diagonalization of the matrix representation $\mathbf{W}^{(k)}$ in the subspace spanned by the eigenvectors of the degenerate eigenvalue of the unperturbed Hamiltonian. This correction is not unique, and generally g_k different first order corrections $\varepsilon_{k,m}^{(1)}$ appear after diagonalization and the degeneracy is lifted. This result clarifies the sense in which degenerate perturbation theory is somewhat discontinuous. Of course, as one can easily see, the amount of splitting of originally degenerate eigenvalues is proportional to λ: $E_{k,m}^{(1)} = \lambda \varepsilon_{k,m}^{(1)}$, and at zero perturbation the splitting vanishes. However the wave functions of each of the split states are the same irrespective of the strength of the perturbation. They are defined by the matrix block $\mathbf{W}^{(k)}$ and not by λ.

One more interesting example of the degenerate perturbation theory appears if one considers a possibility of having two degenerate eigenvalues of the unperturbed Hamiltonian $E_k^{(0)}$ and $E_l^{(0)}$ with the degeneracy numbers g_k and g_l respectively, subject to such a perturbation which has vanishing matrix elements at least in one of the degenerate manifolds:

$$(1.71) \qquad W_{mi}^{(k)} = 0$$

but nevertheless has nonvanishing ones between the states coming from *different* degenerate manifolds:

$$(1.72) \qquad W_{mi}^{(kl)} = \langle \Psi_{k,m}^{(0)} | \hat{W} | \Psi_{l,i}^{(0)} \rangle \neq 0$$

Then as previously inserting the expansion for eigenvalues and eigenvectors in powers of λ for the manifold related to the k-th degenerate eigenvalue of $\hat{H}^{(0)}$ and equating separately the terms up to the second order in λ on the left and on the right sides of eq. (1.56) we get:

$$\hat{H}^{(0)} \Psi_{k,i}^{(0)} = E_k^{(0)} \Psi_{k,i}^{(0)},$$

$$(1.73) \qquad \hat{W} \psi_k^{(0)} + \hat{H}^{(0)} \psi_k^{(1)} = \varepsilon_k^{(1)} \psi_k^{(0)} + \varepsilon_k^{(0)} \psi_k^{(1)},$$

$$\hat{H}^{(0)} \psi_k^{(2)} + \hat{W} \psi_k^{(1)} = \varepsilon_k^{(0)} \psi_k^{(2)} + \varepsilon_k^{(1)} \psi_k^{(1)} + \varepsilon_k^{(2)} \psi_k^{(0)}$$

The first equation satisfies trivially, and in the two following, we regroup the terms and get:

$$(1.74) \qquad \begin{aligned} \left(\hat{W} - \varepsilon_k^{(1)} \right) \psi_k^{(0)} &= - \left(\hat{H}^{(0)} - \varepsilon_k^{(0)} \right) \psi_k^{(1)}, \\ \left(\hat{H}^{(0)} - \varepsilon_k^{(0)} \right) \psi_k^{(2)} + \left(\hat{W} - \varepsilon_k^{(1)} \right) \psi_k^{(1)} &= \varepsilon_k^{(2)} \psi_k^{(0)} \end{aligned}$$

The first equation as in the nondegenerate case, can be satisfied by taking the correction $\psi_k^{(1)}$ to be orthogonal to the unperturbed ground state vector which is satisfied by any linear combination of vectors from the manifold related to the l-th degenerate eigenvalue. Inserting the required expansion we get:

$$(1.75) \quad \left(\hat{W} - \varepsilon_k^{(1)} \right) \psi_k^{(0)} = - \left(\hat{H}^{(0)} - \varepsilon_k^{(0)} \right) \sum_m u_{km}^l \Psi_{lm}^{(0)}$$

which must be satisfied for any vector $\psi_k^{(0)}$ in the k-th degenerate manifold. Let us take one of the basis functions $\Psi_{k,i}^{(0)}$ for it and multiply from the left by some other basis vector $\Psi_{k,j}^{(0)}$. Both the left and right parts of the equality vanish trivially for $i \neq j$ whereas for $i = j$ the right side vanishes and the left side equals to $\varepsilon_k^{(1)}$, which is only possible if

$$(1.76) \quad \varepsilon_k^{(1)} = 0$$

By contrast, forming from the scalar product from the left with the basis vectors $\Psi_{l,j}^{(0)}$ belonging to the l-th degenerate eigenvalue yields:

$$(1.77) \quad W_{ij}^{(kl)} = - \left(\varepsilon_l^{(0)} - \varepsilon_k^{(0)} \right) u_{kj}^l$$

which means that the i-th basis vector in the k-th degenerate manifold gets a first order correction:

$$(1.78) \quad \psi_{ki}^{(1)} = - \sum_j \frac{W_{ij}^{(kl)}}{\left(\varepsilon_l^{(0)} - \varepsilon_k^{(0)} \right)} \Psi_{lj}^{(0)}$$

for the eigenvector.

The energy is so far uncorrected. For obtaining this we consider the equation for the second order corrections for the energy. Inserting in it the equation for the vectors in the k-th manifold obtained in the previous step and noticing that the term containing the unperturbed Hamiltonian and its k-th eigenvalue vanishes as it should in the degenerate situation we arrive at an eigenvector/eigenvalue problem for the matrix in the k-th manifold:

$$(1.79) \quad \sum_{i=1}^{g_k} \sum_{n=1}^{g_l} \frac{W_{mn}^{(lk)} W_{ni}^{(kl)}}{\left(\varepsilon_k^{(0)} - \varepsilon_l^{(0)} \right)} u_i = \varepsilon_k^{(2)} u_m; \quad m = 1, 2, \ldots, g_k$$

which is analogous to the simple degenerate case described above, but with a difference that a new (second order) Hamiltonian serving to define the eigenvectors and eigenvalues $\varepsilon_k^{(2)}$ in the k-th manifold is now given as a sum over the states in the l-th manifold:

$$(1.80) \quad h_{mi}^{(k)} = \sum_{n=1}^{g_l} \frac{W_{mn}^{(lk)} W_{ni}^{(kl)}}{\left(\varepsilon_k^{(0)} - \varepsilon_l^{(0)} \right)}$$

We shall have a chance to see more elegant representations of this situation in the future.

Something more can be obtained if the perturbation $\hat{W}$ vanishes also within the l-th manifold so that the only nonvanishing matrix elements occur between the functions belonging to *different* manifolds. In this case applying the singular value decomposition allows us to state the following: There exist two unitary matrices $\mathbf{U}^{(k)}$ and $\mathbf{U}^{(l)}$ of the sizes $g_k \times g_k$ and $g_l \times g_l$, respectively, which when respectively applied on the left and on the right to the $g_k \times g_l$ matrix $\mathbf{W}^{(kl)}$ formed by the matrix elements $\langle \Psi_{k,m}^{(0)} | \hat{W} | \Psi_{l,i}^{(0)} \rangle$ of the operator $\hat{W}$ produce the matrix $\mathbf{U}^{(k)} \mathbf{W}^{(kl)} \mathbf{U}^{(l)}$ which has only $\min(g_k, g_l)$ nonvanishing elements $w_i = (\mathbf{U}^{(k)} \mathbf{W}^{(kl)} \mathbf{U}^{(l)})_{ii}$ $i = 1, 2, \ldots,$ $\min(g_k, g_l)$. Obviously the matrices $\mathbf{U}^{(k)}$ and $\mathbf{U}^{(l)}$ perform some unitary transformations in the respective manifolds. If the results of these transformations are taken for the new basis functions $\tilde{\Psi}_{k,i}^{(0)}, \tilde{\Psi}_{l,i}^{(0)}$ the original eigenvector/eigenvalue problem reduces to $\min(g_k, g_l)$ 2×2 eigenvector/eigenvalue problems of the form:

$$(1.81) \qquad \begin{pmatrix} \varepsilon_k^{(0)} & w_i \\ w_i^* & \varepsilon_l^{(0)} \end{pmatrix}$$

which can be easily solved even analytically.

1.4. ALTERNATIVE REPRESENTATIONS OF QUANTUM MECHANICS

The derivations made so far were done in terms of the wave functions (vectors). In fact all the basic tools of the quantum theory used throughout this book are covered by this brief account. However, it may be practical to have a variety of representations for the same set of basic techniques in different incarnations. Not adding too much either to pragmatic numerical tools or to a deeper understanding of what is going on, these tools are useful for getting general relations which are an important part of the present book and for a more economical representation of the variables of the problems considered here. For that reason we review them below.

1.4.1. Projection operators[12]

Taking a vector (function) $|\varphi\rangle$ in the space of the allowable wave functions (e.g. integrable with its square and normalized to unity) allows us to construct an operator $\hat{P}_\varphi$ which acts on an arbitrary vector $|\Psi\rangle$ (function) as:

$$(1.82) \qquad \hat{P}_\varphi |\Psi\rangle = |\varphi\rangle \langle \varphi | \Psi \rangle$$

i.e. yields the projection of $|\Psi\rangle$ on $|\varphi\rangle$. One can easily check that the operator $\hat{P}_\varphi$ for any $|\varphi\rangle$ possesses the properties

---

[12]Presentation in this section follows the route presented in the brilliant lectures delivered by one of the author's teachers Dr. V.I. Pupyshev of the Chemistry Department of the Moscow State University during last 30 years, but which appeared in a printed form [V.I. Pupyshev, Additional chapters of molecular quantum mechanics, Parts 1–3. Moscow University Publishers [in Russian], 2008] during the time of proofreading of the present book.

$$(1.83) \quad \hat{P}_\varphi^\dagger = \hat{P}_\varphi$$
$$\hat{P}_\varphi^2 = \hat{P}_\varphi \hat{P}_\varphi = \hat{P}_\varphi$$

– hermiticity and idempotency. This resolves one of the concerns related to the wave function picture of quantum mechanics. The wave function is not an observable quantity whereas the projection operator uniquely related to it in principle corresponds to an observable due to its hermiticity. The eigenvalues of $\hat{P}_\varphi$ can take two values 0 and 1 where the function $|\varphi\rangle$ is the eigenvector of $\hat{P}_\varphi$ with the eigenvalue 1 and any vector orthogonal to $|\varphi\rangle$ is an eigenvector with the eigenvalue 0. For a pair of projection operators generated by two orthogonal vectors $|\varphi\rangle$ and $|\psi\rangle$ the following holds:

$$(1.84) \quad \hat{P}_\varphi \hat{P}_\psi = \hat{P}_\psi \hat{P}_\varphi = 0$$

Any vector $|\Psi\rangle$ can be presented as a sum of its projection on $|\varphi\rangle$ and of its orthogonal complement:

$$|\Psi\rangle = \hat{P}_\varphi |\Psi\rangle + (1 - \hat{P}_\varphi)|\Psi\rangle$$

If an orthonormalized basis $\{\Phi_i\}$ is given, the vector $|\varphi\rangle$ is defined by its expansion over it:

$$(1.85) \quad |\varphi\rangle = \sum_i |\Phi_i\rangle\langle\Phi_i|\varphi\rangle$$

Then the projection operator $\hat{P}_\varphi$ acquires the form:

$$(1.86) \quad \hat{P}_\varphi = |\varphi\rangle\langle\varphi| = \sum_{ij} |\Phi_i\rangle\langle\Phi_i|\varphi\rangle\langle\varphi|\Phi_j\rangle\langle\Phi_j|$$

which yields its matrix representation $\mathbf{P}_\varphi$:

$$(1.87) \quad \langle\Phi_i|\hat{P}_\varphi|\Phi_j\rangle = (\mathbf{P}_\varphi)_{ij} = \langle\Phi_i|\varphi\rangle\langle\varphi|\Phi_j\rangle$$

For a pair of orthogonal vectors $|\varphi\rangle$ and $|\psi\rangle$ the operator

$$(1.88) \quad \hat{P} = \hat{P}_\varphi + \hat{P}_\psi$$

is also a projection operator in the sense that

$$(1.89) \quad \hat{P}^2 = (\hat{P}_\varphi + \hat{P}_\psi)^2 = \hat{P}_\varphi^2 + \hat{P}_\psi^2 = \hat{P}_\varphi + \hat{P}_\psi = \hat{P}_{\varphi \oplus \psi}$$

It projects to the subspace spanned by $|\varphi\rangle$ and $|\psi\rangle$. This construct is extended to any number of orthonormal vectors.

Now, let us return to an eigenvalue/eigenvector problem:

$$\hat{H}\Psi = E\Psi$$

Let $\{\Psi_i | i = 0, ...\}$ be the set of its eigenvectors corresponding to the eigenvalues E_i, respectively. The operators projecting to its eigenvectors are:

$$(1.90) \quad \hat{P}_i = |\Psi_i\rangle\langle\Psi_i|$$

Now we consider

$$(1.91) \quad \hat{H}\hat{P}_i = \hat{H}|\Psi_i\rangle\langle\Psi_i| = E_i|\Psi_i\rangle\langle\Psi_i|$$
$$\hat{P}_i\hat{H} = |\Psi_i\rangle\langle\Psi_i|\hat{H} = E_i|\Psi_i\rangle\langle\Psi_i|$$

Thus we get:

$$(1.92) \quad \hat{H}\hat{P}_i = \hat{P}_i\hat{H}$$

On the other hand the first equation can be rewritten as:

$$(1.93) \quad \hat{H}\hat{P}_i = E_i\hat{P}_i$$

Now, taking a sum of the projection operators $\hat{P} = \sum_i \hat{P}_i$ to any subset of the eigenvectors of the operator $\hat{H}$ and multiplying by the projection operators in different orders we get:

$$(1.94) \quad \hat{H}\hat{P} = \sum_i \hat{H}|\Psi_i\rangle\langle\Psi_i| = \sum_i E_i|\Psi_i\rangle\langle\Psi_i|$$
$$\hat{P}\hat{H} = \sum_i |\Psi_i\rangle\langle\Psi_i|\hat{H} = \sum_i E_i|\Psi_i\rangle\langle\Psi_i|$$

so that the rightmost parts of the above equations coincide and the left ones must do the same, which yields the Schrödinger equation in terms of the projection operators:

$$(1.95) \quad \hat{H}\hat{P} = \hat{P}\hat{H}$$

whose solutions are all operators $\hat{P}$ projecting to different possible subspaces spanned by eigenvectors of the operator $\hat{H}$.

1.4.2. Resolvent

Let us return to the Schrödinger equation for the projection operators $\hat{P}_i$ eq. (1.93):

$$\hat{H}\hat{P}_i = E_i\hat{P}_i$$

We notice that summing up projection operators corresponding to all individual eigenvectors yields the identity operator:

$$(1.96) \quad \sum_i \hat{P}_i = \hat{I}$$

This is simply the completeness relation for the eigenvectors of an Hermitian operator $\hat{H}$. We introduce formally the resolvent of the operator $\hat{H}$ as a function of a complex variable z:

$$(1.97) \quad \hat{R}(z) = (z\hat{I} - \hat{H})^{-1}$$

This definition is, however, legal since all the eigenvalues of a Hermitian $\hat{H}$ are real and thus for any nonreal z the operator in the brackets is not degenerate and thus can be inverted. On the other hand the following holds:

$$(1.98) \quad (z\hat{I} - \hat{H})\hat{P}_i = (z - E_i)\hat{P}_i$$

where on the left side there is an operator in the brackets although on the right side it is a (complex) number. Then we get a chain of equalities:

$$(1.99) \qquad \hat{R}(z)\left(z\hat{I} - \hat{H}\right)\hat{P}_i = \hat{P}_i = \hat{R}(z)\left(z - E_i\right)\hat{P}_i = \left(z - E_i\right)\hat{R}(z)\hat{P}_i$$

so that

$$(1.100) \qquad \hat{R}(z)\hat{P}_i = \frac{\hat{P}_i}{z - E_i}$$

Summing this over all i's yields, because of eq. (1.96):

$$(1.101) \qquad \hat{R}(z) = \sum_i \frac{\hat{P}_i}{z - E_i}$$

which reveals the structure of the resolvent of an Hermitian operator $\hat{H}$: it is a matrix (operator) function of the complex variable z; it has simple poles E_i on the real axis which are the eigenvalues of $\hat{H}$.

Summing over all i's the first equality of the above chain yields:

$$(1.102) \qquad \hat{R}(z)\left(z\hat{I} - \hat{H}\right) = \hat{I}$$

which is the Schrödinger equation in terms of the resolvent.

The residues theorem allows treating the resolvent as a formal solution of the eigenvector/eigenvalue problem. Indeed, taking a contour integral over any path $\mathcal{C}_i$ enclosing each of the poles one gets:

$$(1.103) \qquad \hat{P}_i = \frac{1}{2\pi i} \oint_{\mathcal{C}_i} \hat{R}(z)dz$$

Taking an integral over a path enclosing several poles yields the operator projecting to the subspace spanned by the corresponding eigenvectors.[13]

1.4.3. Approximate techniques for alternative representations of quantum mechanics

As mentioned earlier, alternative representations of quantum mechanics may be useful for deriving general relations and representing approximate treatments of quantum mechanical problems in a concise and convenient form. Here we briefly review the tools – largely the versions of the perturbation theory – presented in terms of projection operators, resolvents, and wave operators. As is usually done in the perturbation context, it is assumed that the eigenvectors and eigenvalues of some unperturbed Hamiltonian $\hat{H}^{(0)}$ are known. It means also that the projection operators, resolvents and wave operators (see below) representing the unperturbed eigenvalues and eigenvectors are known as well.[14]

---

[13]The best description of projection operators and resolvents is given by Kato [28]. A brief account is given in [27] also.

[14]The best description of perturbation techniques in terms of projection operators and resolvents is given by Kato [28].

1.4.3.1. Projection operators

Let the operator $\hat{P}^{(0)}$ project to some subspace spanned by several eigenvectors of the unperturbed Hamiltonian $\hat{H}^{(0)}$. It is known that a set of operators projecting to a subspace of the same dimensionality and including $\hat{P}^{(0)}$ can be parametrized in the following form [29–31]:

$$\hat{P} = (\hat{P}^{(0)} + \hat{V})(\hat{I} + \hat{V}^\dagger \hat{V})^{-1}(\hat{P}^{(0)} + \hat{V}^\dagger),$$

$$\dim \operatorname{Im} P = \dim \operatorname{Im} \hat{P}^{(0)}$$

(1.104)

by the matrices $\hat{V}$ and their Hermitian conjugate $\hat{V}^\dagger$ satisfying the conditions:

$$\hat{P}^{(0)}\hat{V} = 0; \hat{V}\hat{P}^{(0)} = \hat{V}; (\hat{I} - \hat{P}^{(0)})\hat{V}\hat{P}^{(0)} = \hat{V};$$

$$\hat{P}^{(0)}\hat{V}^\dagger = \hat{V}^\dagger; \hat{P}^{(0)}\hat{V}^\dagger(\hat{I} - \hat{P}^{(0)}) = \hat{V}^\dagger; \hat{V}^\dagger \hat{P}^{(0)} = 0$$

(1.105)

The above relations formally present the block-off-diagonal structure of the matrices $\hat{V}$ and $\hat{V}^\dagger$. Matrix $\hat{V}$ has $\dim \operatorname{Im} \hat{P}^{(0)} \times \dim \operatorname{Im}(\hat{I} - \hat{P}^{(0)})$ elements. The projection operator $\hat{P}^{(0)}$ itself corresponds to $\hat{V} = \hat{V}^\dagger = 0$. The dimensionality of the square matrix $\hat{V}^\dagger \hat{V}$ equals $\dim \operatorname{Im} \hat{P}^{(0)}$ so that for modest dimensions of the subspace under consideration (or eventually for the ground state) the matrix $\hat{I} + \hat{V}^\dagger \hat{V}$ can be easily inverted. As an alternative, one can use a recurrent relation:

$$(1.106) \quad (\hat{I} + \hat{V}^\dagger \hat{V})^{-1} = \hat{I} - \frac{\hat{V}^\dagger \hat{V}}{\hat{I} + \hat{V}^\dagger \hat{V}}$$

which yields the following form of the projection operator eq. (1.104):

$$(1.107) \quad \hat{P} = \hat{P}^{(0)} + (\hat{V} + \hat{V}^\dagger) + \hat{V}\hat{V}^\dagger -$$

$$-\frac{\hat{V}^\dagger \hat{V}}{\hat{I} + \hat{V}^\dagger \hat{V}} - \frac{\hat{V}\hat{V}^\dagger \hat{V}}{\hat{I} + \hat{V}^\dagger \hat{V}} - \frac{\hat{V}^\dagger \hat{V}\hat{V}^\dagger}{\hat{I} + \hat{V}^\dagger \hat{V}} - \frac{\hat{V}\hat{V}^\dagger \hat{V}\hat{V}^\dagger}{\hat{I} + \hat{V}^\dagger \hat{V}}$$

Iterating this move one can easily get an expansion of $\hat{P}$ into a power series in $\hat{V}$ and $\hat{V}^\dagger$.

Using either the exact form of the projection operator as a function of $\hat{V}$ and $\hat{V}^\dagger$, or cutting the series expansion, allows us to construct different approximate variation schemes with matrix elements of $\hat{V}$ as variables. On the other hand, inserting the expansion for the projection operator eq. (1.107) in the Schrödinger equation for the projection operator eq. (1.95) with the perturbed Hamiltonian gives in the first order:

$$(1.108) \quad (\hat{H}^{(0)} + \lambda\hat{W})\left(\hat{P}^{(0)} + (\hat{V} + \hat{V}^\dagger)\right) = \left(\hat{P}^{(0)} + (\hat{V} + \hat{V}^\dagger)\right)(\hat{H}^{(0)} + \lambda\hat{W})$$

Since the unperturbed Hamiltonian commutes with the unperturbed projection operator we obtain for terms linear in λ:

$$(1.109) \quad \lambda\hat{W}\hat{P}^{(0)} + \hat{H}^{(0)}(\hat{V} + \hat{V}^\dagger) = (\hat{V} + \hat{V}^\dagger)\hat{H}^{(0)} + \lambda\hat{P}^{(0)}\hat{W}$$

which can be rewritten in a form of a commutator equation:

$$(1.110) \quad \lambda \left[\hat{W}, \hat{P}^{(0)} \right] = - \left[\hat{H}^{(0)}, (\hat{V} + \hat{V}^{\dagger}) \right]$$

On the right we already have a familiar superoperator adjoint to the unperturbed Hamiltonian $\hat{H}^{(0)}$. However, the conditions of its inversibility are considerably eased. It is now enough only if the images of complementary projection operators Im $\hat{P}^{(0)}$ and $\mathrm{Im}(\hat{I} - \hat{P}^{(0)})$ contain no eigenvectors with common eigenvalues of $\hat{H}^{(0)}$. In this case the $\mathrm{Ad}_{\hat{H}^{(0)}}$ acting restricted to the space of matrices $\hat{V} + \hat{V}^{\dagger}$ is nondegenerate and thus can be inverted and the first order correction to the projection operator acquires the form:

$$(1.111) \quad \hat{V} + \hat{V}^{\dagger} = -\lambda \, \mathrm{Ad}_{\hat{H}^{(0)}}^{-1} \left[\hat{W}, \hat{P}^{(0)} \right]$$

If the perturbation operator $\hat{W}$ has itself the block structure given by projection operators $\hat{P}^{(0)}$ and $\hat{I} - \hat{P}^{(0)}$ so that

$$(1.112) \quad \hat{W} = \hat{w} + \hat{w}^{\dagger}$$
$$\hat{w} = \hat{P}^{(0)} \hat{w} (\hat{I} - \hat{P}^{(0)})$$

the result is further simplified as:

$$(1.113) \quad \hat{V} = -\lambda \, \mathrm{Ad}_{\hat{H}^{(0)}}^{-1} \left[\hat{w}, \hat{P}^{(0)} \right]$$

1.4.3.2. Resolvent

As mentioned earlier, the resolvent is a tool allowing one to formally write down the solution of an eigenvalue/eigenvector problem. It is also useful for developing perturbation expansions, which, as we saw previously, require somewhat tedious work when done in terms of vectors (wave functions).

According to eq. (1.102) the perturbed Schrödinger equation for the resolvent reads:

$$(1.114) \quad \hat{R}(z)(z\hat{I} - \hat{H}^{(0)} - \lambda \hat{W}) = \hat{I}$$

where the resolvent for the unperturbed Schrödinger equation reads:

$$(1.115) \quad \hat{R}^{(0)}(z) = (z\hat{I} - \hat{H}^{(0)})^{-1}$$

The simplest relation which can be written concerns the *inverses* of the resolvents:

$$(1.116) \quad \hat{R}^{-1}(z) = \hat{R}^{(0)-1}(z) - \lambda \hat{W}$$

This is not of great use by itself, as, according to eq. (1.103), the answer can be expressed through the resolvent itself, not the inverse of it. However, multiplying the above equation by $\hat{R}^{(0)}(z)$ from the left and by $\hat{R}^{-1}(z)$ from the right and regrouping terms yields:

$$(1.117) \quad \hat{R}(z) = \hat{R}^{(0)}(z) + \lambda \hat{R}^{(0)}(z) \hat{W} \hat{R}(z)$$

which is known as the Dyson equation. The simplest nontrivial approximation to its solution is to replace $\hat{R}(z)$ by $\hat{R}^{(0)}(z)$ in the right hand side, thus obtaining the linear correction as:

$$(1.118) \quad \hat{R}(z) = \hat{R}^{(0)}(z) + \lambda \hat{R}^{(0)}(z)\hat{W}\hat{R}^{(0)}(z)$$

in a close form. By contrast, inserting the Dyson equation itself in its right hand side gives:

$$(1.119) \quad \hat{R}(z) = \hat{R}^{(0)}(z) + \lambda \hat{R}^{(0)}(z)\hat{W}\hat{R}^{(0)}(z) + \lambda^2 \hat{R}^{(0)}(z)\hat{W}\hat{R}(z)$$

which can be iterated yielding a series:

$$\hat{R}(z) = \hat{R}^{(0)}(z) + \lambda \hat{R}^{(0)}(z)\hat{W}\hat{R}^{(0)}(z)$$
$$(1.120) \qquad\qquad + \lambda^2 \hat{R}^{(0)}(z)\hat{W}\hat{R}^{(0)}(z)\hat{W}\hat{R}^{(0)}(z) + ...$$

with a fairly simple form of the terms. It can be rewritten in a twofold manner:

$$(1.121) \quad \hat{R}(z) = \hat{R}^{(0)}(z)\left(\hat{I} + \sum_{n=1}^{\infty} \lambda^n \left(\hat{W}\hat{R}^{(0)}(z) \right)^n \right)$$

$$\hat{R}(z) = \left(\hat{I} + \sum_{n=1}^{\infty} \lambda^n \left(\hat{R}^{(0)}(z)\hat{W} \right)^n \right) \hat{R}^{(0)}(z)$$

which can be obtained also by direct expansion of the definition of the perturbed resolvent.

The sums in the brackets are those of the geometric series of operators (matrices). If λ is small enough they can be summed up:

$$(1.122) \quad \hat{R}(z) = \left(\hat{I} - \lambda \hat{R}^{(0)}\hat{W} \right)^{-1} \hat{R}^{(0)}$$

which can be also derived from eq. (1.117).

1.4.3.3. Wave operator and Van-Vleck transformation

One more representation for the perturbation technique is based on the so-called Van-Vleck transformation as applied to the exact Hamiltonian to exclude the interaction matrix elements and by this to approximately diagonalize it. As we remember the finding eigenvectors and eigenvalues of a Hamiltonian reduces to searching a unitary (orthogonal) matrix $\mathbf{U}$ transforming the original Hamiltonian matrix $\mathbf{H}$ into diagonal form $\mathbf{D}$ according to:

$$\mathbf{D} = \mathbf{U}^{-1}\mathbf{H}\mathbf{U} = \mathbf{U}^{\dagger}\mathbf{H}\mathbf{U}$$

(due to orthogonality or unitarity of $\mathbf{U}$ the relation $\mathbf{U}^{-1} = \mathbf{U}^{\dagger}$ holds). If, as it is done in the context of perturbation theory, the Hamiltonian matrix acquires the form:

$$\mathbf{H} = \mathbf{H}^{(0)} + \lambda \mathbf{W}$$

where the eigenvectors and the eigenvalues of the matrix $\mathbf{H}^{(0)}$ are assumed to be known, one can conclude that events evolve on the basis of these eigenvectors, so the matrix $\mathbf{H}^{(0)}$ is diagonal. The perturbation $\mathbf{W}$ is nondiagonal and the unperturbed eigenvectors and eigenvalues can be considered "zero approximations" to the exact ones of the Hamiltonian matrix $\mathbf{H}$. In this context, the sought matrix $\mathbf{U}$ is called the "wave operator" and is verbally described as a matrix transforming the approximate eigenstates (those of the matrix $\mathbf{H}^{(0)}$) into the exact ones (those of the matrix $\mathbf{H}$).

Obviously the zero order wave operator in this case equals the unity matrix:

$$(1.123) \quad \mathbf{U}^{(0)} = \mathbf{I}$$

It is easy to see that an arbitrary unitary matrix can be represented in the form:

$$(1.124) \quad \mathbf{U} = \exp(i\mathbf{\Lambda})$$

where $\mathbf{\Lambda}$ is Hermitian and the exponent has to be understood as the corresponding series expansion everywhere convergent. So for the lower orders we can write:

$$(1.125) \quad \mathbf{U} \approx \mathbf{I} + i\mathbf{\Lambda} - \frac{1}{2}\mathbf{\Lambda}^2 + \dots$$

Applying the approximate expansion for $\mathbf{U}$ from the right and the Hermitian conjugate from the left to the perturbed Hamiltonian matrix yields:

$$(1.126) \quad \mathbf{D} = \mathbf{U}^{\dagger}\mathbf{H}\mathbf{U} = (\mathbf{I} - i\mathbf{\Lambda} + \dots)\left(\mathbf{H}^{(0)} + \lambda\mathbf{W}\right)(\mathbf{I} + i\mathbf{\Lambda} + \dots) =$$
$$= \mathbf{H}^{(0)} + i[\mathbf{H}^{(0)}, \mathbf{\Lambda}] + \lambda\mathbf{W} + \dots$$

$\mathbf{H}^{(0)}$ is already diagonal, $\lambda\mathbf{W}$ by contrast is not; the Hermitian matrix $\mathbf{\Lambda}$ is to be determined from the relation

$$(1.127) \quad i[\mathbf{H}^{(0)}, \mathbf{\Lambda}] + \lambda\mathbf{W} = 0$$

which assures that the off-diagonal matrix elements have the order at least higher than the first in λ. Employing in a new context the concept of the adjoint superoperator (supermatrix) we can write:

$$(1.128) \quad \mathbf{\Lambda} = -i\lambda\,\mathrm{Ad}_{\mathbf{H}^{(0)}}^{-1}\,\mathbf{W}$$

which gives the solution of the problem. If the matrix $\mathbf{U}$ is presented up to the linear term in $\mathbf{\Lambda}$, this coincides (as one can see) with standard formulae for the nondegenerate perturbation theory for wave functions. The main problem with them is that these functions are not normalized to unity and for that reason the energy estimates obtained in the perturbation theory are not the variational estimates "from above". However, one can take the exponential expansion seriously and by this arrive at a unitary matrix; then the approximate eigenvectors are normalized. Although they diagonalize the Hamiltonian approximately (as mentioned earlier the remnant off-diagonal elements are of the order of λ^2) but the expectation value of the Hamiltonian matrix over the ground state produced by $\mathbf{U}$ provides a true upper bound for the ground state energy.

1.4.3.4. Löwdin partition technique

In this book we employ a range of quantum mechanical techniques. Most of them are reflected or employed in numerical methods implemented in QC packages. One of the general ways significantly simplifying the problem of diagonalizing matrices of large dimensionality is the Löwdin partition. From a slightly different point of view it can be considered a tool generating the whole variety of perturbative treatments in quantum mechanics and quantum chemistry. Let us consider this formalism.

Following Löwdin [32] we assume that we know a subspace of vectors which contains a good approximation of the exact ground state vector. In addition we can think that the low-energy excited states of interest also belong to this same subspace. Then let $\hat{P}$ be the projection operator onto this subspace and $\hat{Q} = \hat{I} - \hat{P}$ be the complementary projection operator satisfying the following conditions:

$$(1.129) \quad \hat{P}^2 = \hat{P}, \hat{Q}^2 = \hat{Q}, \hat{P}\hat{Q} = 0, \hat{P} + \hat{Q} = \hat{I}$$

These are nothing but the conditions of orthogonality of the subspace of interest $\mathrm{Im}\hat{P}$ ($\mathrm{Im}\hat{P}$ – image $\hat{P}$ – stands here for the set of vectors of a linear space which are obtained by action of the linear operator $\hat{P}$ upon all vectors of the linear vector space) and its complementary subspace $\mathrm{Im}\hat{Q}$.

To apply the partition of the whole vector space to the solution of the Schrödinger equation with the exact Hamiltonian $\hat{H}\Psi = E\Psi$ we multiply this equation from the left in turn by $\hat{P}$ and Q and making use of the fact that

$$(1.130) \quad \Psi = \hat{P}\Psi + Q\Psi$$

we arrive at a pair of equations:

$$(1.131) \quad \begin{aligned} \hat{P}\hat{H}\hat{P}\hat{P}\Psi + \hat{P}\hat{H}QQ\Psi &= E\hat{P}\Psi \\ \hat{Q}\hat{H}\hat{P}\hat{P}\Psi + \hat{Q}\hat{H}\hat{Q}\hat{Q}\Psi &= E\hat{Q}\Psi \end{aligned}$$

The second equation in this pair can be solved for $Q\Psi$:

$$(1.132) \quad \hat{Q}\Psi = (E\hat{Q} - \hat{Q}\hat{H}\hat{Q})^{-1}\hat{Q}\hat{H}\hat{P}\hat{P}\Psi$$

which is only a formal solution as existence of the inverse operator (matrix) in the right hand side is not guaranteed. Inserting the solution for $\hat{Q}\Psi$ in the first equation of the pair we get for the function $P\Psi$:

$$(1.133) \quad \left[\hat{P}\hat{H}\hat{P} + \hat{P}\hat{H}\hat{Q}(E\hat{Q} - \hat{Q}\hat{H}\hat{Q})^{-1}\hat{Q}\hat{H}\hat{P}\right]\hat{P}\Psi = E\hat{P}\Psi$$

The expression in the square brackets is an $\hat{H}^{\text{eff}}(E)$ and itself depends on energy. Its most important characteristic is that it acts in the subspace defined by the projection operator $\hat{P}$ ($\hat{P}$-block) $\mathrm{Im}\,\hat{P}$, but its eigenvalues by construction coincide with the eigenvalues of the exact Hamiltonian. The eq. (1.133) represents a pseudoeigenvalue problem as the operator in the right hand part is $\hat{H}^{\text{eff}}(E)$, where "pseudo" indicates its own dependence on the sought energy eigenvalue. In principle such a problem has to be solved iteratively until self-consistency is reached for all eigenvalues of interest. One usually does that if the partitioning of the entire space into subspaces is used for constructing approximate diagonalization schemes directed to the lower eigenvalues [33] of the matrices of higher dimensionality. On the other hand, expanding the inverse operator in the definition of the effective Hamiltonian in the series yields different perturbation series. For more details on this one can take a look at [58].

1.5. BASICS OF QUANTUM CHEMISTRY

The approximation techniques described in the earlier sections apply to any (non-relativistic) quantum system and can be universally used. On the other hand, the specific methods necessary for modeling molecular PES that refer explicitly to electronic wave function (or other possible tools mentioned above adjusted to describe electronic structure) are united under the name of quantum chemistry (QC).[15] Quantum chemistry is different from other branches of theoretical physics in that it deals with systems of intermediate numbers of fermions – electrons, which preclude on the one hand the use of the "infinite number" limit – the number of electrons in a system is a sensitive parameter. This brings one to the position where it is necessary to consider wave functions dependent on spatial $\mathbf{r}$ and spin s variables of all N electrons entering the system. In other words, the wave functions sought by either version of the variational method or meant in the frame of either perturbational technique – the eigenfunctions of the electronic Hamiltonian in eq. (1.27) are the functions $\Psi(x_1, \ldots, x_N)$ where x_i stands for the pair of the spatial radius vector of i-th electron $\mathbf{r}_i$ and its spin projection s_i to a fixed axis. These latter, along with the boundary conditions (in this case reducing to the square integrability requirement), must satisfy also symmetry conditions known as the Pauli principle.[16] Namely, the wave function $\Psi(x_1, \ldots, x_N)$, to be correctly formed, must change its sign when its arguments referring to whatever pair of the electrons interchange their place in the argument list. This is the formulation of the Pauli principle in terms of the electronic wave function in the coordinate representation. Quantum chemistry uses different representations for the electronic structure, each requiring a slightly different appearance of the Pauli principle, which is built in the structure of the specific theoretical tools used in each specific representation. Due to sophistications brought about by the necessity to simultaneously bear in mind different representations of quantum chemistry in the context of hybrid modeling, we present here the most important ones used in this book.

---

[15] Modern quantum chemistry is described in numerous books of which we mention [17, 18, 27, 29, 30]. They differ in detail and depth.

[16] The nature of these conditions is still under dispute. The fact that too small particles cannot be ordered derives from quantum mechanics [21]. But the question whether the Pauli principle is an *independent* axiom of quantum mechanics or not is still unclear [34].

1.5.1. Many-electron wave functions

The most direct way to represent the electronic structure is to refer to the electronic wave function dependent on the coordinates and spin projections of N electrons. To apply the linear variational method in this context one has to introduce the complete set of basis functions Φ_K for this problem. The complication is to guarantee the necessary symmetry properties (antisymmetry under transpositions of the sets of coordinates referring to any two electrons). This is done as follows.

1.5.1.1. One-electron basis

Let us assume that a complete set of the orthonormalized functions $\varphi_\mathbf{n}(\mathbf{r})$ of the spatial coordinates $\mathbf{r}$ is known. They form a basis in the space $L = L^2(\mathbb{R}^3)$ of the square integrable functions of $\mathbf{r}$, known in this context as *orbitals*. The completeness condition means that the following holds:

$$(1.134) \quad \sum_\mathbf{n} \varphi_\mathbf{n}^*(\mathbf{r})\varphi_\mathbf{n}(\mathbf{r}') = \delta(\mathbf{r} - \mathbf{r}')$$

where the summation extends to the whole set of subscripts $\mathbf{n}$. (We use boldsface for the orbital index since the quantum numbers necessary to label basis functions in $L^2(\mathbb{R}^3)$ naturally organize in a three-component entity).

It is much easier to introduce the complete basis in the space of functions depending on the spin variable of one electron. Allowable values of the spin projection s (in the units of $\hbar$) are $\pm 1/2$. Corresponding functions have the form:

$$(1.135) \quad \begin{aligned} &\alpha(\tfrac{1}{2}) = 1; \alpha(-\tfrac{1}{2}) = 0 \\ &\beta(\tfrac{1}{2}) = 0; \beta(-\tfrac{1}{2}) = 1 \end{aligned}$$

This explains the commonly used terms according to which α electrons are those with "spin-up" whereas β-electrons have "spin-down". The orthonormality and completeness of the set of functions $\alpha(s)$ and $\beta(s)$ can be checked easily.

If one now takes a set of orthonormalized functions $\varphi_\mathbf{n}(\mathbf{r})$ and forms the products of the spin functions $\alpha(s)$ and $\beta(s)$ by $\varphi_\mathbf{n}(\mathbf{r})$'s, one obtains the complete set of functions:

$$(1.136) \quad \phi_k(x) = \phi_{\mathbf{n}\sigma}(x) = \phi_{\mathbf{n}\sigma}(\mathbf{r}, s) = \varphi_\mathbf{n}(\mathbf{r})\sigma(s)$$

with $\sigma = \alpha, \beta$. These functions are habitually termed *spin-orbitals*. They are functions of the electronic variables $x = (\mathbf{r}, s)$.

1.5.1.2. Slater determinants

Now let us select an ordered "tuple" of N subscripts referring to spin-orbitals: $K = \{k_1 < k_2 < \ldots < k_N\}$. The N-tuple of spin-orbitals defines uniquely the Slater determinant of N electrons as a functional determinant

$$
(1.137) \quad \Phi_K = \frac{1}{\sqrt{N!}}
\begin{vmatrix}
\phi_{k_1}(x_1) & \cdots & \phi_{k_1}(x_N) \\
\vdots & \ddots & \vdots \\
\phi_{k_N}(x_1) & \cdots & \phi_{k_N}(x_N)
\end{vmatrix}
$$

which is obviously a function of coordinates of N electrons. It is formed like any other determinant of a matrix with the only specificity that the rows of the latter are numbered by functions (in one row, all the elements are the values of the same function) and the columns are numbered by the electrons so that in each column the values of all entering spin-orbitals are calculated for the coordinates of that same electron). Apparently, if two spin-orbitals coincide, then for arbitrary values of coordinates two rows of such a determinant coincide and the determinant itself vanishes. On the other hand, if coordinates of two electrons coincide, then two columns of such a determinant also coincide and the determinant vanishes as well.

It can be proven [31] that all possible Slater determinants of N particles constructed from a complete system of orthonormalized spin-orbitals ϕ_k form a complete basis in the space of normalized antisymmetric (satisfying the Pauli principle) functions, of N electrons i.e. for any antisymmetric and normalizable Ψ one can find expansion amplitudes so that:

$$
(1.138) \quad
\begin{aligned}
\Psi(x_1,\ldots,x_N) &= \sum_K C_K \Phi_K \\
\sum_K C_K^2 &= 1
\end{aligned}
$$

Thus the basis of Slater determinants can be used as a basis in a linear variational method eq. (1.42) when the Hamiltonian dependent or acting on coordinates of N electrons is to be studied. The problem with this theorem is that for most known choices of the basis of spin-orbitals used for constructing the Slater determinants of eq. (1.137) the series in eq. (1.138) is very slow convergent. We shall address this problem later.

The general setting of the electronic structure description given above refers to a complete (and thus infinite) basis set of one-electron functions (spin-orbitals) $\phi_{n\sigma}(x)$. In order to acquire the practically feasible expansions of the wave functions, an additional assumption is made, which is that the orbitals entering eq. (1.136) are taken from a finite set of functions somehow related to the molecular problem under consideration. The most widespread approximation of that sort is to use the atomic orbitals (AO).[17] This approximation states that with every problem of molecular electronic structure one can naturally relate a set of functions $\chi_\mu(\mathbf{r})$, $|\{\mu\}| = M > N -$ atomic orbitals (AOs) centered at the nuclei forming the system. The orthogonality in general does not take place for these functions and the set $\{\chi_\mu\}$ is characterized

---

[17]It may not seem mandatory now, with the advent of plane wave basis sets. However, to give a better description, these latter are variously "augmented" to reproduce the behavior of electrons in the vicinity of nuclei. For more detailed description see [35] and reference therein.

by the metric matrix (known in this context as the overlap integrals matrix) S with the elements

$$(1.139) \quad S_{\mu\nu} = \int d^3\mathbf{r}\chi_\mu^*(\mathbf{r})\chi_\nu(\mathbf{r})$$

The finite set $\{\varphi_\mu\}$ of the orthonormalized functions required in the above construct eq. (1.137) of the Slater determinants can be obtained in numerous ways, say by applying the transformation $\mathbf{S}^{-\frac{1}{2}}$ to the set of initial AOs. This is called the Löwdin orthogonalization. It is not a unique way of constructing an orthonormal basis of orbitals as, obviously, any further orthogonal transform of the Löwdin orthogonalized basis set gives another basis set that is orthonormal as well. The above construct applies only when the metric matrix is not degenerate and thus the inverse square root can be calculated. It may become degenerate if the entire set of the AOs $\{\chi_\mu\}$ is linearly dependent. Of course, in practice, it never happens exactly, but a situation close to degeneration of the basis set occurs rather frequently when the eigenvalues of the $\mathbf{S}$ matrix become close to zero. Constructing the Slater determinants of a set of spin-orbitals containing a linearly dependent one results in a vanishing determinant (by this the set of the Slater determinants itself becomes linearly dependent). For that reason the linear combinations of χ_μs corresponding to too small eigenvalues of the metric matrix must be excluded. Thus the number of orthonormal orbitals φ_μ may be smaller than the original number of AOs. From now on we assume that M is the number of the linearly independent orthogonal combinations of AOs.

1.5.1.3. Implementations of AO basis sets

The functional form of the AOs is best described by an exponential function of the separation between the nucleus of the atom to which the AO is assigned (or centered upon) and the electron multiplied by a polynomial function of the same separation and by the angular part – the spherical function of the polar and azimuthal angles together with the separation describing the position of an electron in the spherical coordinate system centered at the nucleus. For example the hydrogen-like AOs – the solutions of the Schrödinger equation for a one-electron atom with the nuclear charge Z i.e. one with

$$\hat{V}_{ne} = -\frac{Z}{r}$$

have the form:

$$\chi_{Znlm}^H (r,\theta,\varphi) = R_{nl}^Z(r)Y_{lm}(\theta,\varphi)$$

$$R_{nl}^Z(r) = \left\{ \left(\frac{2Z}{n}\right)^3 \frac{(n-l-1)!}{2n\left[(n+l)!\right]^3} \right\}^{\frac{1}{2}} \times$$

$$\times \left(\frac{2Zr}{n}\right)^l \exp\left(-\frac{Zr}{n}\right) L_{n+l}^{2l+1}\left(\frac{2Zr}{n}\right)$$

$$L_n^k(x) = \frac{d^k}{dx^k}\left[\exp(x)\frac{d^n}{dx^n}(x^n\exp(-x))\right]$$

$$Y_{lm}(\theta,\varphi) = (-1)^{\frac{m+|m|}{2}}\left[\frac{1}{2\pi}\frac{2l+1}{2}\frac{(l-|m|)!}{(l+|m|)!}\right]^{\frac{1}{2}}P_l^{|m|}(\cos\theta)\exp(im\varphi)$$

$$P_l^{|m|}(x) = \frac{1}{2^l l!}(1-x^2)^{\frac{|m|}{2}}\frac{d^{l+|m|}}{dx^{l+|m|}}(x^2-1)^l$$

Here the triple nlm corresponds to the "vector" subscript $\mathbf{n}$ in the definition and the quantities $L_n^k(x)$ and $P_l^{|m|}(x)$ called respectively adjoint Laguerre and adjoint Legendre polynomials and can be checked immediately for their polynomial form. The main features correctly reproduced here by the hydrogen-like functions are the exponential decay of all wave function at large electron-nuclear separations and the fulfillment of the nuclear cusp condition for the s-states (ones with $l = 0$) in the coordinate origin ($r = 0$). One has to realize, however, two interrelated aspects: the hydrogen like AOs for the bound states (with the negative energy) do not form the complete set of one-electron functions in $L^2(\mathbb{R}^3)$ since the complete set is formed by the entire set of the solutions of the Schrödinger equation, which in the case of the hydrogen-like atom contains also the states with positive energies forming the continuous spectrum with energies above the dissociation limit of the atom. On the other hand, there is no reason to think that the exponents describing the decay of the electronic states with increase of the electron-nuclear separation are equal to $\frac{Z}{n}$ in a general case of a many-electron atom. By contrast, a general argumentation leads to the conclusion that in a many-electron atom the decay of the AO must follow the rule

$$\exp(-\zeta r) \text{ with } \zeta \sim \sqrt{IP}$$

where IP stands for the ionization potential – the energy necessary for an electron to be extracted from this AO (see below). The experimental values of ionization potential do not have too much to do with the squared nuclear charge, unless it goes about the hydrogen-like atoms. Thus many hydrogen-like AOs may be necessary to decently approximate a single exponent function with more or less arbitrary value of the *orbital exponent* ζ. For this reason the hydrogen-like functions are never used in practice for constructing AOs basis sets.

Slater functions   The general arguments concerning the physically sound form of the states to be included in the AOs basis sets given above have been implemented in the Slater type AOs:

$$\chi_{\zeta nlm}^{STO}(r,\theta,\varphi) = \frac{(2\zeta)^{n+\frac{1}{2}}}{\sqrt{(2n)!}}r^{n-1}\exp(-\zeta r)$$

Here again the triple nlm corresponds to the "vector" subscript $\mathbf{n}$ in the definition. Formally they can be obtained as solutions of the electronic Schrödinger equation with the potential of the form:

$$\hat{V}_{ne} = -\frac{n\zeta}{r} + \frac{n(n-1) - l(l+1)}{2r^2}$$

The original Slater rules for selecting the values of ζ conform to the idea of being related to the atomic ionization potentials (see above). However other schemes are also in use.

Gaussians The Gaussian type AOs are the most widespread basis set in the area of *ab initio* quantum chemistry:

$$\chi^{GTO}_{\alpha nlm}(r, \theta, \varphi) = \left[\frac{2^{2n}(n-1)!}{(2n-1)!}\sqrt{\frac{(2\alpha)^{2n+1}}{n}}\right]^{\frac{1}{2}} r^{n-1} \exp\left(-\alpha r^2\right) Y_{lm}(\theta, \varphi)$$

These functions can be understood as solutions of the Schrödinger equation with the potential

$$\hat{V}_{ne} = 2\alpha^2 r^2 + \frac{n(n-1) - l(l+1)}{2r^2}$$

In variance with the hydrogen-like and Slater functions the potential employed to formally construct the gaussian basis states has nothing to do with the real potential acting upon an electron in an atom. On the other hand the solutions of this (actually three-dimensional harmonic oscillator problem) form a complete discrete basis in the space of orbitals in contrast to the hydrogen-like orbitals.

An alternative (and in fact dominating) representation uses the possibility to represent the Gaussian function of the squared distance in a form of a product of three Cartesian Gaussian orbitals. Indeed as one can easily see

$$r^2 = x^2 + y^2 + z^2$$

so that

$$\exp\left(-\alpha r^2\right) = \exp\left(-\alpha x^2\right)\exp\left(-\alpha y^2\right)\exp\left(-\alpha z^2\right)$$

and taking into account the definitions of the spherical coordinates

$$x = r\sin\theta\cos\varphi$$

$$y = r\sin\theta\sin\varphi$$

$$z = r\cos\theta$$

one can easily figure out that the spherical function multiplied by r^{n-1} with $l \leq n-1$ becomes a polynomial in x, y, and z (in fact a uniform one – such that all monoms in x, y, and z entering it have the same overall power $n - 1$ in x, y, and z together). This allows one to equivalently represent a Gaussian supplied by the angular part in the form of a spherical function as a combination of Cartesian Gaussians:

$$\chi_{\alpha pqr}^{GTO}(x, y, z) = N_{\alpha p} N_{\alpha q} N_{\alpha r} x^p y^q z^r \exp\left(-\alpha x^2\right)$$

$$\times \exp\left(-\alpha y^2\right) \exp\left(-\alpha z^2\right)$$

$$N_{\alpha p} = \left\{ \sqrt{\frac{\pi}{2\alpha}} \frac{(2p-1)!!}{2^{2p}\alpha^p} \right\}^{-\frac{1}{2}} \text{etc}$$

Here the triple pqr corresponds to the "vector" subscript n in the definition.

The above Gaussian functions are termed primitive ones. They are used largely not by themselves but as a basis over which the Slater AOs are expanded. In this case, the expansion coefficients are called contraction coefficients and are fixed. However, with the passage of time, this restriction is eased step by step to assure flexibility of the AOs in response to variations of the atomic environment.

1.5.2. Full configuration interaction: exact solution of approximate problem

At this point we are sufficiently equipped to consider briefly the methods used to approximate the wave functions constructed in the restricted subspace of orbitals. So far the only approximation was to restrict the orbital basis set. It is convenient to establish something that might be considered to be the exact solution of the electronic structure problem in this setting. This is the full configuration interaction (FCI) solution. In order to find one it is necessary to construct all possible Slater determinants for N electrons allowed in the basis of $2M$ spin-orbitals. In this context each Slater determinant bears the name of a basis configuration and constructing them all means that we have their *full* set. Then the matrix representation of the Hamiltonian in the basis of the configurations Φ_K is constructed:

$$(1.140) \quad H_{KK'} = \left\langle \Phi_K \left| \hat{H} \right| \Phi_{K'} \right\rangle$$

and the FCI problem itself reduces to finding the lower eigenvalues and corresponding eigenvectors of the Hamiltonian matrix in the chosen basis by the linear variation method.

The FCI approach, if one had decided to actually use it to study any realistic problem, would require enormous computing power since the dimensionality of the FCI problem increases factorially with the increase of the size of the system: for N electrons in $2M$ spin-orbitals the number of basis configurations amounts to:

$$(1.141) \quad C_{2M}^N = \frac{(2M)!}{N!(2M - N)!}$$

Various symmetry constraints (mainly due to the requirement that the total spin of the considered system of N electrons must acquire some definite value – see below) reduce the number of basis vectors to be treated in a given eigenvector problem significantly, but even after that their dimensionalities remain too high. Of course nowadays even millions of configurations do not represent an unsolvable problem for numerical treatment, but our concern here is to develop a theory which helps to reduce at least that unnecessary computing that could in principle be avoided. In any case when it goes about a "complex system" of thousands of atoms, there is no hope of actually performing the described procedure. With this in mind, we describe the methods of approximating the solution of an otherwise exact FCI problem.

1.5.3. Hartree-Fock approximation

The situation of the FCI from the numerical point of view is not very favorable – expansion of the ground eigenvector may be very long i.e. too many configurations (Slater determinants) have the expansion amplitudes which cannot be neglected. On the other hand, the FCI problem is invariant to whatever transformation of the (spin-)orbital basis set. This means that any unitary transformation $\mathbf{U}$ of the (spin-)orbitals induces some other unitary transformation $\mathbf{U}^{\Lambda}$ of the set of all N-electron Slater determinants. This changes (by a similarity transformation with matrix $\mathbf{U}^{\Lambda}$) only the matrix representation of the FCI problem not affecting the eigenvalues and reduces to the transformation of the eigenvectors: i.e. of their amplitudes C_K by the same matrix $\mathbf{U}^{\Lambda}$. Operators acting in the N-electron space are transformed accordingly, so that their expectation values (i.e. the observables) do not change. This may be formulated as independence of the answer of the FCI problem on the particular choice of the orbital basis in the chosen subspace spanned by (spin-)orbitals. By contrast, the FCI does depend on the orbital subspace it is formulated in (see below).

The natural idea would be to use the freedom given by the invariance of the result with respect to the basis choice in the space of (spin-)orbitals in order to make the FCI expansion shorter. In more formal terms one can set the task as follows: to find such a transformation of the original basis of orbitals that relative to the new (transformed) basis the expansion of the N-electron ground state wave function is as short as possible. To give this a somewhat more precise meaning: we want to find such an orbital basis set so that the expansion amplitudes (or better still the overall weight – sum of the squared amplitudes) of several leading configurations take as large a fraction as possible of the exact ground state wave function.

The Hartree-Fock approximation is the first step along this way. In its original form it was proposed by V.A. Fock [36]. The idea (in its modern formulation) is to find a single Slater determinant as close as possible to the precise ground state. The expectation value of the energy operator (in agreement with the variational principle) is taken as a measure of this closeness: the lower the energy, calculated with the use of the trial wave function, the closer it approaches the true ground state i.e. the larger is the overlap integral between the trial wave function and the exact one.

The trial wave function in the Hartree-Fock approximation takes the form of a single Slater determinant:

$$\Psi(x_1,\ldots,x_N) = \frac{1}{\sqrt{N!}} \begin{vmatrix} \phi_1(x_1) & \cdots & \phi_1(x_N) \\ \vdots & \ddots & \vdots \\ \phi_N(x_1) & \cdots & \phi_N(x_N) \end{vmatrix}$$

(1.142)

$$= \frac{1}{\sqrt{N!}} \left| \phi_1(x_1) \ \cdots \ \phi_N(x_N) \right|$$

we are already familiar with. However, the spin-orbitals ϕ_k, $k = 1 \div N$ are not taken directly from the set of $2M$ basis spin-orbitals, but are selected to be linear combinations of them. Then the expansion coefficients with respect to the same original basis set $\{\phi_m\}$ become the variables of the variational procedure (and take the role of the variables ξ mentioned above).

Restricting the wave function by the form eq. (1.142) allows one to significantly reduce the calculation costs for all characteristics of a many-fermion system. Inserting eq. (1.142) into the energy expression (for the expectation value of the electronic Hamiltonian eq. (1.27)) and applying to it the variational principle with the additional condition of orthonormalization of the system of the occupied spin-orbitals ϕ_k (known in this context as molecular spin-orbitals) yields the system of integrodifferential equations of the form (see e.g. [27]):

$$\left\{ -\frac{1}{2}\nabla^2 + \hat{V}_{ne}(\mathbf{r}) + \int dx' \frac{\rho_{\mathrm{HF}}^{(1)}(x';x')}{|\mathbf{r}-\mathbf{r'}|} \right\} \phi_k(x) -$$

(1.143)

$$- \int dx' \phi_k(x') \frac{\rho_{\mathrm{HF}}^{(1)}(x;x')}{|\mathbf{r}-\mathbf{r'}|} = \epsilon_k \phi_k(x)$$

where

(1.144) $\quad \rho_{\mathrm{HF}}^{(1)}(x;x') = \displaystyle\sum_{i=1}^{N} \phi_i(x)\phi_i^*(x')$

is the one-electron density matrix (see below) in the Hartree-Fock approximation. This system must be solved self consistently since the kernels of the integral operators in its left hand part depend on the functions which are its solutions through eq. (1.144).

The Hartree-Fock equation eq. (1.143) can be rewritten using the Coulomb and exchange integral operators $\hat{J}$ and $\hat{K}$, respectively:

$$\hat{J}[\hat{\rho}_{\mathrm{HF}}^{(1)}]\phi(x) = \int dx' \frac{\rho_{\mathrm{HF}}^{(1)}(x';x')}{|\mathbf{r}-\mathbf{r'}|}\phi(x)$$

(1.145)

$$\hat{K}[\hat{\rho}_{\mathrm{HF}}^{(1)}]\phi(x) = \int dx' \phi(x') \frac{\rho_{\mathrm{HF}}^{(1)}(x;x')}{|\mathbf{r}-\mathbf{r'}|}$$

If one-electron operators in eq. (1.143) are collected into a single operator defined according to:

$$(1.146) \quad \hat{h} = -\frac{1}{2}\nabla^2 + \hat{V}_{ne}(\mathbf{r})$$

the integrodifferential operator in the left hand part of eq. (1.143) – the Fock operator – becomes:

$$(1.147) \quad \hat{F}[\hat{\rho}_{HF}^{(1)}] = \hat{h} + \hat{J}[\hat{\rho}_{HF}^{(1)}] - \hat{K}[\hat{\rho}_{HF}^{(1)}] = \hat{h} + \hat{\Sigma}[\hat{\rho}_{HF}^{(1)}]$$

With these notations the Hartree-Fock problem acquires the form of an eigenvalue/eigenvector problem:

$$(1.148) \quad \hat{F}[\hat{\rho}_{HF}^{(1)}]\phi_i = \varepsilon_i\phi_i$$

with the reservation that the operator whose eigenvalues and eigenvectors are sought depends on the solution. These solutions represent an approximate picture of a motion of a single electron in a field induced by the nuclei and by the averaged distribution of all electrons. It is usual to hear in this context that an electron moves in the field induced by *other* electrons. It is not completely true since the Hartree term represented by the operator $\hat{J}$ involves the total density of all electrons including the one which seems to be singled out. This results in self interaction which is cured by the exchange operator not having any classical analogue and coming from the averaging over the electron coordinates while taking into account the Pauli principle.

On the basis of AOs the Fock operator eq. (1.147) acquires a matrix representation of the form:

$$F_{\mu\nu} = h_{\mu\nu} + J_{\mu\nu} - K_{\mu\nu} = h_{\mu\nu} + \Sigma_{\mu\nu}$$

so that the Hatree-Fock problem written in the original AOs basis acquires the form of the generalized eigenvalue/eigenvector problem eq. (1.48):

$$(1.149) \quad \mathbf{Fu} = \varepsilon\mathbf{Su}$$

where $\mathbf{u}$ stands for the vector of molecular spin-orbitals represented by their expansion coefficients with respect to the original (nonorthogonal) basis of AOs $\{\chi_\mu\}$.

The solutions $\{\phi_i\}$ of the Hartree-Fock equation eq. (1.143) in the form of the eigenvalue/eigenvector problem eq. (1.149) are known as molecular (spin-)orbitals (MO). As in the case of the FCI problem some finite basis of M orbitals related with the particular form of the electronic Hamiltonian (one-electron potential) is chosen and the matrix elements of the operator $\hat{F}$ relative to this basis are found and the standard methods of searching for eigenvectors and eigenvalues (diagonalization) can be applied. It provides the expansion of the functions

$$(1.150) \quad \phi_i = \sum_{\mu} u_{i\mu}\varphi_\mu$$

where the quantities $u_{i\mu}$ are referred to as MO LCAO coefficients (here we have used an orthonormal basis set $\{\varphi_\mu\}$ to expand MOs).[18] Only N eigenvectors with lower eigenvalues are needed (at least at this point). Obviously the number of independent variables whose optimal values are to be determined within the Hartree-Fock procedure is already not as large as in the case of the FCI method. For N spin-orbitals entering the determinant eq. (1.142) only $N \times M$ transformation coefficients are necessary[19] which is by many orders of magnitude less than the factorial estimate for the FCI functions. The Hartree-Fock (single determinant) wave function eq. (1.142) is also invariant with respect to the orthogonal transformation of the occupied spin-orbitals only. The set of these transformations is given by the $N \times N$ orthogonal matrices. That means that the true solution of the problem of searching the wave function in the Hartree-Fock approximation requires not the specific form of the occupied MOs, but that of the N-dimensional subspace in the $2M$-dimensional original space. This suggests the formulation of the Hartree-Fock problem in terms of the operator projecting to the subspace to be found. This is done as follows. It is easy to check that the integral kernel eq. (1.144) when acting on the functions of x, behaves as a projection operator. Indeed, the action of the integral kernel $\rho_{\mathrm{HF}}^{(1)}(x;x')$ is defined in a standard way:

$$
\begin{aligned}
\hat{\rho}_{\mathrm{HF}}^{(1)} f(x) &= \int dx' \rho_{\mathrm{HF}}^{(1)}(x;x') f(x') = \\
&= \sum_{i=1}^{N} \phi_i(x) \int dx' \phi_i^*(x') f(x') \\
&= \sum_{i=1}^{N} |\phi_i\rangle \langle \phi_i | f \rangle
\end{aligned}
$$

(1.151)

Every single term in the sum in eq. (1.144) acts as the operator projecting on $\phi_i(x)$ which are mutually orthogonal and normalized, so thus $\hat{\rho}_{\mathrm{HF}}^{(1)}$ projects on the subspace spanned by the occupied spin-orbitals. The idempotency and hermiticity are checked immediately. So eq. (1.151) obviously coincides with the definition of an operator projecting to a subspace. The equation defining it reads:

$$(1.152) \quad \hat{F}\hat{\rho} = \hat{\rho}\hat{F}$$

In variance with the standard formulation of the Schrödinger equation in terms of projection operators eq. (1.95) in the above Hartree-Fock equation for the projection

---

[18]The idea to employ a finite basis set of AOs to represent the MOs as linear combinations of the former apparently belongs to Lennard-Jones [68] and had been employed by Hückel [37] and had been systematically explored by Roothaan [38]. That is why the combination of the Hartree-Fock approximation with the LCAO representation of MOs is called the Hartree-Fock-Roothaan method.

[19]The orthonormalization conditions reduce the number of independent variation variables as compared to this estimate, but do not reduce so to say the number of numbers to be calculated throughout the diagonalization procedure.

operator $\hat{\rho}$ the operator $\hat{F}$ itself depends on the projection operator to be found:

(1.153) $\quad \hat{F}[\hat{\rho}] = \hat{h} + \hat{\Sigma}[\hat{\rho}]$

It is easy to see that the self-energy operator – average electron-electron interaction – can be considered also as a linear superoperator in the space of the matrices it depends on. Indeed, from the point of view of the $2M$-dimensional space of spin-orbitals $\hat{\Sigma}[\hat{\rho}]$ acts as a $2M \times 2M$ matrix, so that $\hat{\Sigma}[\hat{\rho}]$ is a $2M \times 2M$ matrix constructed after another $2M \times 2M$ matrix $\hat{\rho}$. On the other hand, it easy to see from the definition of the Coulomb and exchange operators in eq. (1.145) that the result of calculating each of them (and thus of the sum of them) taking a sum of two functions $\rho_1(x; x') + \rho_2(x; x')$ and/or a product of this function by a number $\lambda\rho(x; x')$ as its argument yields respectively a sum of the results of the actions of $\hat{\Sigma}$ and the product in the same number as the result of action of $\hat{\Sigma}$:

(1.154) $\quad \hat{\Sigma}[\hat{\rho}_1 + \hat{\rho}_2] = \hat{\Sigma}[\hat{\rho}_1] + \hat{\Sigma}[\hat{\rho}_2]$

$$\hat{\Sigma}[\lambda\hat{\rho}] = \lambda\hat{\Sigma}[\hat{\rho}]$$

(This holds even if $\hat{\rho}$ is not a projection operator, since the property to be a projection operator is not generally conserved by linear operations, but just a matrix). Thus $\hat{\Sigma}$ can be considered a linear superoperator transforming one $2M \times 2M$ matrix to another one of the same dimension.

The projection operator formulation of the Hartree-Fock problem can be used for constructing a perturbation procedure for determining the electronic structure in terms of the latter.[20] The simplest formulation departs from the Hartree-Fock equation for the projection operator to the occupied MOs eq. (1.152). Let us assume that the bare perturbation (see below) concerns only the one-electron part of the Fock operator so that:

(1.155) $\quad \hat{h} = \hat{h}^{(0)} + \lambda\hat{W}$

Then we know that for the unperturbed Fock operator

(1.156) $\quad \hat{F}_0 = \hat{h}_0 + \hat{\Sigma}[\hat{\rho}_0]$

the following holds:

(1.157) $\quad \hat{F}_0\hat{\rho}_0 = \hat{\rho}_0\hat{F}_0$

Then we assume that the projection operator for the exact Fock operator is close to the unperturbed one so that the expansion

$$\hat{\rho} = \hat{\rho}_0 + \hat{\rho}^{(1)} + \hat{\rho}^{(2)}$$

(1.158) $\quad \hat{\rho}^{(1)} = \hat{V} + \hat{V}^\dagger$

$$\hat{\rho}^{(2)} = \hat{V}\hat{V}^\dagger - \hat{V}^\dagger\hat{V}$$

---

[20]Presentation in this section also follows the route presented in the brilliant lectures delivered by one of the author's teachers Dr. V.I. Pupyshev of the Chemistry Department of the Moscow State University [V.I. Pupyshev, Additional chapters of molecular quantum mechanics, Parts 1–3. Moscow University Publishers [in Russian], 2008] published only recently.

50   *Andrei L. Tchougréeff*

(see eq. (1.107)) can be used. Inserting this into the Hartree-Fock equation with the perturbed Fock operator one gets (keeping the terms not higher than the first order in λ and $\hat{V}$ simultaneously):

$$(1.159) \quad \left[\lambda \hat{W} + \hat{\Sigma}[\hat{V} + \hat{V}^\dagger], \hat{\rho}_0 \right] = \left[\hat{V} + \hat{V}^\dagger, \hat{h}^{(0)} + \hat{\Sigma}[\hat{\rho}_0] \right]$$

It can be formally resolved using the superoperator adjoint to the unperturbed Fock operator:

$$(1.160) \quad \hat{V} + \hat{V}^\dagger = - \operatorname{Ad}_{\hat{F}_0}^{-1} \left[\lambda \hat{W} + \hat{\Sigma}[\hat{V} + \hat{V}^\dagger], \hat{\rho}_0 \right]$$

but now $\hat{V} + \hat{V}^\dagger$ appears also on the right side which is known as renormalization ("dressing") of the original ("bare") perturbation due to average electron-electron interaction $\hat{\Sigma}$ entering the Fock operator.

The above equation can be solved iteratively, but apparently it is not a very good choice and for that reason a somewhat different approach may be useful. It is based on the variational procedure for the energy. Using the projection operator $\hat{\rho}$ the Hartree-Fock estimate for the electronic energy reads:

$$(1.161) \quad E = \operatorname{Sp}\left(\hat{h}\hat{\rho} \right) + \frac{1}{2} \operatorname{Sp}\left(\hat{\rho}\hat{\Sigma}\,[\hat{\rho}] \right)$$

which is a quadratic function of the matrix elements of $\hat{\rho}$. Now let $\hat{\rho}_0$ be the projection operator to that subspace of occupied MOs which gives the lowest possible Hartree-Fock energy for the unperturbed Fock operator in the given basis of AOs. Then using the expansion eq. (1.107) for the projection operators close to a given one, writing explicitly the corrections up to the second order in $\hat{V}$, inserting this expansion in the expression for the energy and keeping the terms of the total order not higher than two in $\hat{V}$ and taking into account that under the spur sign the argument of the self-energy part $\hat{\Sigma}$ can be interchanged with the matrix multiplier we arrive at

$$(1.162) \quad \begin{aligned} E = \underbrace{\operatorname{Sp}\left(\hat{h}\hat{\rho}_0 \right) + \frac{1}{2} \operatorname{Sp}\left(\hat{\rho}_0 \hat{\Sigma}\,[\hat{\rho}_0] \right)}_{=E_0} + \operatorname{Sp}[\hat{F}_0 \left(\hat{V} + \hat{V}^\dagger \right)] + \\ + \frac{1}{2} \operatorname{Sp}\left\{ \left(\hat{V} + \hat{V}^\dagger \right) \hat{\Sigma}\left[\hat{V} + \hat{V}^\dagger \right] \right\} + \operatorname{Sp}[\hat{F}_0 \left(\hat{V}\hat{V}^\dagger - \hat{V}^\dagger\hat{V} \right)] \end{aligned}$$

In the minimum the terms linear in $\hat{V} + \hat{V}^\dagger$ vanish so that the electronic energy becomes:

$$(1.163) \quad E = E_0 + \frac{1}{2} \operatorname{Sp}\left\{ \left(\hat{V} + \hat{V}^\dagger \right) \hat{\Sigma}\left[\hat{V} + \hat{V}^\dagger \right] \right\} + \operatorname{Sp}[\hat{F}_0 \left(\hat{V}\hat{V}^\dagger - \hat{V}^\dagger\hat{V} \right)]$$

which is a quadratic form with respect to the matrix elements of $\hat{V}$, which in its turn can be given a form of the expectation value of some quantity Λ over $\hat{V}$ considered as an element of the vector space of the $2M \times 2M$ matrices

$$E = E_0 + \frac{1}{2} \left\langle\!\left\langle \hat{V} \left| \Lambda \right| \hat{V} \right\rangle\!\right\rangle$$

This describes the quadratic response of the electronic energy to the variation $\hat{V}$ of the one-electron density matrix in the vicinity of a minimum. The quantity Λ can be considered a superoperator (supermatrix) acting in the space of variations of the density matrices taken as elements of a linear space of the $2M \times 2M$ matrices. The supermatrix Λ has four indices running through one-electron states in the ($2M$-dimensional) carrier space. To get an idea of the properties of the quantity Λ we notice that in the absence of the self-energy term $\hat{\Sigma}$ the Fock operator $\hat{F}_0$ reduces to its one-electron part. For it the Hartree-Fock approximation eq. (1.142) provides the *exact* solution: the one-electron part of the Hamiltonian $\hat{h}$ must be diagonalized and N lowest eigenstates must be taken as occupied. Let the subscript i run over the occupied MOs and the subscript j run over the empty MOs. Then the relation between the allowable matrices $\hat{V}$ and the projection operators assures that the matrix $|j\rangle\langle i|$, all filled with zeroes with only one unity in the i-th column and j-th row can be used as matrix $\hat{V}$. Then $\hat{V}^\dagger = |i\rangle\langle j|$, $\hat{V}\hat{V}^\dagger = |j\rangle\langle j|$, and $\hat{V}^\dagger\hat{V} = |i\rangle\langle i|$. On the other hand the Fock operator in the basis of its eigenvectors $|i\rangle$ and $|j\rangle$ has the form:

$$(1.164) \quad \hat{F}_0 = \sum_{i\in\mathrm{occ}} \varepsilon_i |i\rangle\langle i| + \sum_{j\in\mathrm{vac}} \varepsilon_j |j\rangle\langle j|$$

where summation separately extends to the occupied and vacant MOs. Inserting all this in to the expression eq. (1.163) one gets that for the noninteracting Hamiltonian the vectors $|j\rangle\langle i|$ (elements of the space of $2M \times 2M$ matrices) are the eigenvectors of the superoperator $\hat{\Lambda}$ with the eigenvalues $\varepsilon_j - \varepsilon_i$

$$\hat{\Lambda} |j\rangle\langle i| = (\varepsilon_j - \varepsilon_i) |j\rangle\langle i|$$

which are obviously the excitation energies corresponding to transfer of one electron from the i-th (occupied) MO to the j-th (empty) MO. It is clear (one can check) that if there is no Coulomb interaction in the Hamiltonian the self-energy term also vanishes and such excited states are *exact* ones for the Hamiltonian without interaction. If the electron-electron interaction (self-energy term) is not vanishing it couples different configurations obtained by single excitations of the ground state Slater determinant.

The projector operator formulation also allows the perturbative treatment of the Hartree-Fock problem known as the self-consistent perturbation theory. In variance with the perturbative treatment departing of the Schrödinger equation we start from the energy expression with the perturbed one-electron part of the Hamiltonian. The path, based on the perturbative treatment of the Fock equation, is more complicated as even if only the one-electron part of the Fock operator $\hat{h}^{(0)}$ gets perturbed i.e. $\hat{h}^{(0)} \rightarrow \hat{h} = \hat{h}^{(0)} + \lambda\hat{W}$ the projection operator or equivalently the one-electron density gets the correction of the order λ so that the perturbation of the Fock operator is not limited to the term $\lambda\hat{W}$ (bare perturbation), but is additionally perturbed in the same order through the self-energy (renormalized or dressed perturbation). The treatment based on the energy expression eq. (1.161) is more straightforward. Indeed, the energy of

                                          Andrei L. Tchougréeff

the ground state with the perturbed one-electron part of the Fock operator in the Hartree-Fock approximation reads:

$$(1.165) \quad E = \mathrm{Sp}\left(\left(\hat{h}^{(0)} + \lambda\hat{W}\right)\hat{\rho}\right) + \frac{1}{2}\mathrm{Sp}\left(\hat{\rho}\hat{\Sigma}\left[\hat{\rho}\right]\right)$$

whereas the unperturbed energy E_0 is as previously defined by eq. (1.162) calculated with the unperturbed one-electron operator and with the corresponding projection operator $\hat{\rho}^{(0)} = \hat{\rho}_0$. Inserting as previously the expansion of the projection operator up to second order in $\hat{V}$ results in the following:

$$E = \underbrace{\mathrm{Sp}\left(\hat{h}^{(0)}\hat{\rho}^{(0)}\right) + \frac{1}{2}\mathrm{Sp}\left(\hat{\rho}^{(0)}\hat{\Sigma}\left[\hat{\rho}^{(0)}\right]\right)}_{=E_0} + \mathrm{Sp}[\hat{F}^{(0)}\left(\hat{V} + \hat{V}^\dagger\right)] +$$

$$(1.166) \qquad + \lambda\,\mathrm{Sp}\,\hat{\rho}^{(0)}\hat{W} + \lambda\,\mathrm{Sp}\left(\hat{V} + \hat{V}^\dagger\right)\hat{W}$$

$$+ \frac{1}{2}\mathrm{Sp}\left\{\left(\hat{V} + \hat{V}^\dagger\right)\hat{\Sigma}\left[\hat{V} + \hat{V}^\dagger\right]\right\} + \mathrm{Sp}[\hat{F}^{(0)}\left(\hat{V}\hat{V}^\dagger - \hat{V}^\dagger\hat{V}\right)]$$

where we dropped the terms of presumably higher than the second order in $\hat{V}$ and λ together. As previously the term linear in $\hat{V}$ (the second one in the first row) vanishes due to the fact that $\hat{\rho}^{(0)}$ brings the minimum to the energy calculated as an expectation value of the unperturbed Hamiltonian. On the other hand the first term in the second row does not affect the subspace spanned by the occupied MOs since it does not contain the matrices of interest $\hat{V}$ and $\hat{V}^\dagger$. In fact it is merely the first order correction to the energy – the sum of the diagonal matrix elements of the perturbation operator, which can be also omitted (see above) from the problem of searching the new subspace of the occupied MOs adjusted to the perturbation. With these notions we rewrite the energy retaining only the terms relevant to the problem of searching the corrected subspace of the occupied MOs:

$$\tilde{E} = \lambda\,\mathrm{Sp}\left(\hat{V} + \hat{V}^\dagger\right)\hat{W}$$

$$(1.167) \qquad + \frac{1}{2}\mathrm{Sp}\left\{\left(\hat{V} + \hat{V}^\dagger\right)\hat{\Sigma}\left[\hat{V} + \hat{V}^\dagger\right]\right\} + \mathrm{Sp}[\hat{F}^{(0)}\left(\hat{V}\hat{V}^\dagger - \hat{V}^\dagger\hat{V}\right)]$$

Then let us take into account the form of the matrices $\hat{V}$. It represents an off-diagonal matrix block having nonvanishing matrix elements only if one of the vectors (bra) belongs to the subset of the occupied MOs and another (ket) to the subset of the vacant MOs. Then the only relevant part of the perturbation matrix $\hat{W}$ is the sum of two similar conjugate off-diagonal blocks:

$$\hat{W} = \hat{w} + \hat{w}^\dagger;$$

where

$$(1.168) \qquad (1 - \hat{P}^{(0)})\hat{w}\hat{P}^{(0)} = \hat{w};$$

$$\hat{P}^{(0)}\hat{w}^\dagger(1 - \hat{P}^{(0)}) = \hat{w}^\dagger$$

in terms of which the contribution proportional to λ rewrites as

$$(1.169) \quad \text{Sp}\left(\hat{V} + \hat{V}^\dagger\right)\hat{W} = \text{Sp}\left(\hat{V}\hat{w}^\dagger + \hat{V}^\dagger\hat{w}\right) = \langle\langle V \mid \hat{w}\rangle\rangle + \langle\langle \hat{w} \mid V\rangle\rangle$$

where an obvious notation for the scalar product in the vector space of $2M \times 2M$ matrices is introduced. With this notation the energy becomes:

$$(1.170) \quad \tilde{E} = \lambda\left[\langle\langle V \mid \hat{w}\rangle\rangle + \langle\langle \hat{w} \mid V\rangle\rangle\right] + \frac{1}{2}\left\langle\left\langle \hat{V} \mid \Lambda \mid \hat{V}\right\rangle\right\rangle$$

This expression has to be optimized with respect to $\hat{V}$ which yields the following linear relation:

$$(1.171) \quad \lambda\hat{w} + \Lambda\hat{V} = 0$$

(and an analoge for the Hermitian conjugates of $\hat{w}$ and $\hat{V}$) which can be formally solved:

$$(1.172) \quad \hat{V} = -\lambda\Lambda^{-1}\hat{w}$$

using the inverted supermatrix Λ^{-1}. It is easy to see, however, at least in the case of the interactionless Hamiltonian, that the inversion can be easily done since the supermatrix Λ is not degenerate in the case when the highest occupied MO (HOMO) is separated from the lowest unoccupied MO (LUMO) by a finite energy gap $\Delta\varepsilon$ so that:

$$\Lambda^{-1}\left|j\right\rangle\left\langle i\right| = (\varepsilon_j - \varepsilon_i)^{-1}\left|j\right\rangle\left\langle i\right|$$

If the supermatrix Λ becomes degenerate (at the point where the Hartree-Fock solution for which it is calculated loses its stability i.e. ceases to be a minimum of the energy functional) the inversion is not possible any more, but the Hartree-Fock picture of the electronic structure itself becomes invalid. In this case the above treatment obviously loses any sense.

1.5.4. Second quantization formalism

We have introduced the basic elements of the formalism to be used for describing the electronic structure of molecular systems and described two most important techniques of quantum chemistry: FCI and HF methods. The notations used so far to represent many electronic wave functions – the coordinate representation – were extremely cumbersome and in fact superfluous. There was no need to write down all the combinations of the one-electron functions and their arguments entering the determinant wave functions as the latter were completely defined by the list of entering one-electron functions. Moreover, using the coordinate representation for the Hamiltonian leads to long lasting confusion related to the so called separable Hamiltonians, which turn out, in fact, to be symmetry breaking (see below).

The notation concerns are easily overcome by the following simple construct bearing the name of second quantization formalism.[21] Let us consider the space of wave functions of all possible numbers of electrons and complement it by a wave function of no electrons and call the latter the vacuum state: $|vac\rangle$. This is obviously the direct sum of subspaces each corresponding to a specific number of electrons. It is called the Fock space. The Slater determinants eq. (1.137) entering the expansion eq. (1.138) of the exact wave function are uniquely characterized by subsets of spin-orbitals $K = \{k_1, k_2, \ldots, k_N\}$ which are occupied (filled) in each given Slater determinant. The states in the list are the vectors in the carrier space of spin-orbitals (linear combinations of the functions of the $\{\phi_k(x) = \phi_{m\sigma}(\mathbf{r}, s)\}$ basis. We can think about the linear combinations of all Slater determinants, may be of different numbers of electrons, as elements of the Fock space spanned by the basis states including the vacuum one.

The linear operators in Fock space are defined by their action on its basis elements. Action of an operator a_k^+ on the vacuum state i.e. on the empty determinant produces a 1×1 determinant formed by the function $\varphi_k(x)$. These "determinants" are obviously the basis spin-orbitals themselves. By these acts of operators a_k^+ on the vacuum state all the basis of one-electron states can be formed. It is logical to conclude that an arbitrary linear combination of the operators a_k^+ with numerical coefficients, when acting on the vacuum state, produces the function which is a linear combination of the functions $\phi_k(x)$ with the same numerical coefficients. The Hermitian conjugate a_k of the operator a_k^+ acting on the Slater determinant containing only one function $\phi_k(x)$ gets it back to the vacuum state. Acting by an operator a_k on the vacuum state yields the zero element of the vector space (not the vacuum state; the vacuum state is a state, zero is zero). The pair of operators a_k^+ and a_k are called respectively the creation and annihilation operators of an electron in the state $\phi_k(x)$ or together – Fermi operators. Further moves are necessary to extend the action of these operators to more general determinants. This is done as follows. It is declared that the operator a_k^+ acting on an N-electron determinant (that of an $N \times N$-matrix) adds a column with the values of all N functions already included in the Slater determinant for the coordinates of the $(N+1)$-electron and a row with the values of the one-electron function ϕ_k for coordinates of all electrons. Additionally the resulting determinant is multiplied by $\sqrt{N!}$ and divided by $\sqrt{(N+1)!}$. This definition is natural: if the spin-orbital ϕ_k is already in use in the N-electron determinant the result of action of the operator a_k^+ on such a Slater determinant is the zero element of the vector space. The Hermitian conjugate Fermi operator a_k acting on the Slater determinants not containing the spin-orbital ϕ_k converts them to the zero. If a row with the values of the function ϕ_k is present in the N-electron Slater determinant it is removed as well as the column with the values of all functions at the coordinates of the N-th electron. Also the normalization is adjusted by multiplying the result by $\sqrt{N!/(N-1)!}$. To complete this constructive

---

description of the creation and annihilation operators we have to add that applying the same operator two times to any Slater determinant yields the zero for sure. Applying two different operators in an arbitrary order to the vacuum operator must yield the same state; 2×2 determinants, however, obtained by the rules described above, are going to differ by the phase multiplier equal to -1. It is an automatic consequence of the fact that changing the order of rows in any determinant results in changing its sign.

All the described features of how the creation and annihilation operators act on the Slater determinants constructed from the fixed basis of spin-orbitals are condensed in the set of the anticommutation rules:

$$(1.173) \quad \{a_k^+, a_{k'}^+\} = \{a_k, a_{k'}\} = 0;$$
$$\{a_k^+, a_{k'}\} = \delta_{kk'} = \delta_{\sigma\sigma'}\delta_{\mathrm{mm'}}$$

which the Fermi operators obey. The braces above stand for the anticommutator of two operators, so that:

$$(1.174) \quad \{A, B\} = AB + BA$$

The above construct is known as second quantization formalism.

Much more important than the possibility of expressing the Slater determinants in terms of creation operators is the possibility of expressing *all the operators* acting upon the electron states in terms of the Fermi operators. We are not going to present the formal construct here (it is well described in books [39–41]), rather we are going to explain the situation.

Let us assume that we have a system of electrons in a single determinant state in which, say, the state φ_k ($k = \mathrm{m}\sigma$) is occupied (other states may be either occupied or empty). This electron propagates interacting with some external potential (for example that induced by nuclei). Under the action of this potential the electron scatters into a state $\varphi_{k'}$ ($k' = \mathrm{m}'\sigma'$). In the absence of the magnetic field the spin projection does not change so that $\sigma' = \sigma$. This process is represented by the product of the Fermi operators:

$$a_{k'}^+ a_k$$

which can be literally described as destroying an electron in the one-electron state ϕ_k and creating an electron in the one-electron state $\phi_{k'}$. The creation and annihilation operators take care of the following selection rules: the above scattering process can take place if the k'-state is not occupied and the k-state is (of course, it is possible that $k' = k$).

The corresponding contribution to the Hamiltonian reads

$$h_{k'k}^{(1)} a_{k'}^+ a_k$$

and the energy multiplier $h_{k'k}^{(1)}$ is the scattering matrix element of the operator of the scattering potential between the states ϕ_k and $\phi_{k'}$. As we mentioned previously the external potentials in the nonrelativistic case do not depend on the spin projections of the scattered electrons:

$$(1.175) \quad h^{(1)}_{k'k} = \sum_s \int d^3 r \varphi^*_{\mathbf{m}'}(\mathbf{r}) \sigma'^*(s) \left(-\frac{1}{2}\nabla^2 + \hat{V}_{ne}(\mathbf{r}) \right) \varphi_{\mathbf{m}}(\mathbf{r})\sigma(s)$$

$$= \delta_{\sigma'\sigma} h_{\mathbf{m}'\mathbf{m}} = \int d^3 r \varphi^*_{\mathbf{m}'}(\mathbf{r}) \left(-\frac{1}{2}\nabla^2 + \hat{V}_{ne}(\mathbf{r}) \right) \varphi_{\mathbf{m}}(\mathbf{r})$$

so that it is enough to calculate the integral with the *orbitals* $\varphi_{\mathbf{m}'}(\mathbf{r})$ and $\varphi_{\mathbf{m}}(\mathbf{r})$ only.

Analogously if there are two electrons in the states ϕ_k and ϕ_l their (Coulomb) interaction results in the state where one of them is scattered to $\phi_{k'}$ and the other to $\phi_{l'}$. The product of the Fermi operators

$$a^+_{k'} a^+_{l'} a_l a_k$$

corresponds to this process which is literally described as destroying two electrons in the states ϕ_k and ϕ_l, respectively, and creating two electrons in the states $\phi_{k'}$ and $\phi_{l'}$ ($l = \mathbf{n}\tau; l' = \mathbf{n}'\tau'$). The selection rules ensured by the above product of the Fermi operators are more complicated: the products of the annihilation operators take care of the elimination of "self interaction" of the electron due to the fact that $a_k a_k \equiv a^+_k a^+_k \equiv 0$; next, due to the fact that the total spin projection does not change as in the classical Hamiltonian the electron-electron interaction is spin independent the spin projections of the electrons involved satisfy the condition $s' + t' = s + t -$ total projection of spin conserves in each scattering act. The latter can be satisfied if two pairs of spin functions: $\sigma' = \sigma; \tau' = \tau$ and $\sigma' = \tau; \tau' = \sigma$ are used to construct the involved spin orbitals. As in the case of one-electron scattering the electron-electron scattering process contributes to the Hamiltonian the term:

$$h^{(2)}_{k'l'lk} a^+_{k'} a^+_{l'} a_l a_k$$

with the energy multiplier:

$$h^{(2)}_{k'l'lk} = \sum_{ss'} \int \int d^3 r d^3 r' \varphi^*_{\mathbf{m}'}(\mathbf{r})\sigma'^*(s)\varphi_{\mathbf{m}}(\mathbf{r})\sigma(s)$$

$$\times \frac{1}{|\mathbf{r}-\mathbf{r}'|} \varphi^*_{\mathbf{n}'}(\mathbf{r}')\tau'^*(s')\varphi_{\mathbf{n}}(\mathbf{r}')\tau(s') =$$

Its nontrivial part

$$(1.176) \quad (\mathbf{m}'\mathbf{m}|\mathbf{n}'\mathbf{n}) = \int \int d^3 r d^3 r' \varphi^*_{\mathbf{n}'}(\mathbf{r})\varphi_{\mathbf{n}}(\mathbf{r}) \frac{1}{|\mathbf{r}-\mathbf{r}'|} \varphi^*_{\mathbf{m}'}(\mathbf{r}')\varphi_{\mathbf{m}}(\mathbf{r}')$$

known as a two-electron integral is, as previously, calculated over the orbitals only.

These moves allow us to write the electronic Hamiltonian in the second quantized form with respect to the basis of (spin-)orbitals $\{\phi_k(x)\}$ introduced above:

$$(1.177) \quad \hat{H} = \sum_{k'k} h^{(1)}_{k'k} a^+_{k'} a_k + \frac{1}{2} \sum_{k'l'lk} h^{(2)}_{k'l'lk} a^+_{k'} a^+_{l'} a_l a_k$$

where the multiplier $\frac{1}{2}$ appears to avoid double count of the possible mutually scattering pairs of electrons; restrictions on the spin indices in the above sum appear as a result of the spin independence of either the kinetic energy, or of the electron attraction potential or of the electron-electron Coulomb interaction. Finally one can use the anticommutation rules for the Fermi creation and annihilation operators in order to calculate expectation values (matrix elements) of different operators (including the Hamiltonian) represented in the second quantized form over the pairs of many-electron functions in the same representation. As mentioned already, the Slater determinants are represented by rows of operators creating electrons in the occupied spin-orbitals acting on the vacuum state. When an operator containing an annihilation operator as a multiplier applies to such a row, the anticommutation rules are used to transpose step by step the annihilation operator so that it finally acts on the vacuum state. This latter term obviously vanishes, and the rest contributes to the matrix element of many-electron functions. The formal representation of this technique is known to physicists as the Wick theorem and to chemists as the Slater rules [30].

1.5.5. Unitary group formalism

We mentioned earlier that the dimensionality of the FCI space is significantly reduced due to spin symmetry. This can be formulated somewhat differently due to the relation existing between the spin and permutation symmetries of the many-electronic wave functions (see [30, 42]). Indeed, the wave function of two electrons in two orbitals a and b allows for six different Slater determinants

$$
\begin{aligned}
&b_\beta^+ a_\alpha^+ \, |\text{vac}\rangle = |a(1)\alpha(1), b(2)\beta(2)| \quad b_\alpha^+ a_\beta^+ \, |\text{vac}\rangle = |a(1)\beta(1), b(2)\alpha(2)| \\
&b_\alpha^+ a_\alpha^+ \, |\text{vac}\rangle = |a(1)\alpha(1), b(2)\alpha(2)| \quad a_\beta^+ a_\alpha^+ \, |\text{vac}\rangle = |a(1)\alpha(1), a(2)\beta(2)| \\
&b_\beta^+ a_\beta^+ \, |\text{vac}\rangle = |a(1)\beta(1), b(2)\beta(2)| \quad b_\beta^+ b_\alpha^+ \, |\text{vac}\rangle = |b(1)\alpha(1), b(2)\beta(2)|
\end{aligned}
$$

where $a(1)$ stands for a handy notation of $a(\mathbf{r}_1)$ and $\alpha(1)$ for that of $\alpha(s_1)$ etc. They all apparently satisfy the Pauli principle of antisymmetry of the many-electron wave function with respect to interchange of the order of electron coordinates in the wave function argument list. On the other hand, we also know that the electronic Hamiltonian does not depend on the spin coordinates of electrons so that one may think that due to this the many electron wave functions can be factorized into products of functions, one dependent on the spatial coordinates $\mathbf{r}_i$ only and the other on the spin coordinates s_i only, called respectively the spatial and the spin functions. It is easily done to the above Slater determinants which can be linearly transformed (in fact it is enough to pass to the sum and difference of the functions of the first row) to the following set of functions:

$$
\begin{array}{ll}
|a(1)b(2)| \, (\alpha(1)\beta(2) + \alpha(2)\beta(1)) & (a(1)b(2) + a(2)b(1)) \, |\alpha(1)\beta(2)| \\
|a(1)b(2)| \, \alpha(1)\alpha(2) & a(1)a(2) \, |\alpha(1)\beta(2)| \\
|a(1)b(2)| \, \beta(1)\beta(2) & b(1)b(2) \, |\alpha(1)\beta(2)|
\end{array}
$$

In this set the functions can be classified into two types: in the right column the spatial multiplier is symmetric with respect to transpositions of the spatial coordinates and the spin multiplier is antisymmetric with respect to transpositions of the spin coordinates; in the left column the spatial multiplier is antisymmetric with respect to transpositions of the spatial coordinates and the spin multipliers are symmetric with respect to transpositions of the spin coordinates. Because in the second case the spatial (antisymmetric) multiplier is the same for all three spin-functions, the energy of these three states will be the same i.e. triply degenerate – a triplet. The state with the antisymmetric spin multiplier is compatible with several different spatial wave functions, which probably produces a different value of energy when averaging the Hamiltonian, thus producing several spin-singlet states. From this example one may derive two conclusions: (i) the spin of the many electronic wave function is important not by itself (the Hamiltonian is spin-independent), but as an indicator of the symmetry properties of the wave function; (ii) the symmetry properties of the spatial and spin multipliers are complementary – if the spatial part is symmetric with respect to permutations the spin multiplier is antisymmetric and *vice versa.*

These observations are valid for the wave functions of an arbitrary number of electrons. The respective generalization is done as follows: first we notice that the permutation symmetry of a function is given by the so-called Young patterns: figures formed by boxes, e.g.:

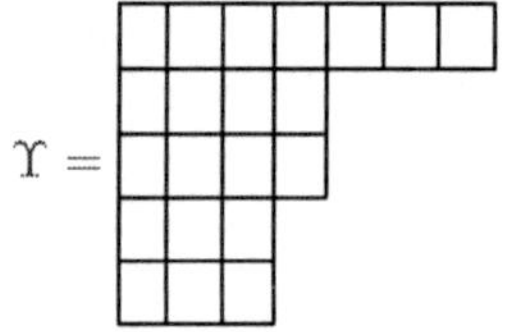

and representing the way a given number of boxes N is represented as a sum of smaller numbers of boxes (in the corresponding rows). These figures label the irreducible representations Υ of the group of permutations of N objects (the $\mathfrak{S}_N$ group) since any representation of N by the above sum uniquely corresponds to the class of conjugated elements of the permutation group which in its turn is characterized by the lengths of cycles in each permutation. The representation of many variables by the functions is achieved if one fills the boxes by functional symbols (see below) and assumes that the symbols along the rows are symmetrized and along the columns are antisymmetrized. The Young pattern Υ with the boxes filled by functional symbols is called Young tableau Υv. They represent functions of N variables possessing definite permutational symmetry. With this it is easy to understand how the spin functions of definite permutational symmetry can be constructed. Indeed, we have two functional symbols for the spin functions α and β. In order to obtain a representation of the group of permutations of N particles by products of spin functions they must be inserted in a Young pattern Υ of N boxes. It is clear however that the allowable Young patterns cannot contain more than two rows as in this case at least one of the columns contains two equal functional symbols and thus vanishes upon antisymmetrization, e.g.:

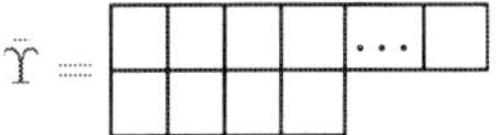

So only the patterns like:

$$\hat{\Upsilon} =$$

are allowable and their allowable fillings are:

$$\hat{\Upsilon}\bar{v} =$$

It is easy to figure out the relation of this with the spin of the many-electron state. It is clear that the above Young tableau corresponds to a many-electron state with the spin projection equal to

$$\frac{n_\alpha - n_\beta}{2}$$

It is also clear that for a given two-row Young pattern one can construct $2S + 1$ Young tableaux starting from a tableau where the first row is completely filled by αs and replacing them one by one with βs. This allows us to conclude that each two-row Young pattern corresponds not only to a state with definite permutational properties, but also to the state with definite total spin S if the first row is longer by $2S$ boxes than the second row (S may be thus half integer).

Now we can consider the symmetry properties of the spatial functions corresponding to the above spin functions. They are uniquely defined from the requirement that the product of the spatial and spin functions must be antisymmetric. In a way, what was symmetrized in the spin part (rows) must be antisymmetrized in the spatial part (columns) and *vice versa*. That means that the spin function represented by a two-row Young pattern $\hat{\Upsilon}$ with the first row longer by $2S$ boxes than the second one must be complemented by a spatial function represented by a two-column Young pattern Υ with the first column longer by $2S$ boxes than the second one, e.g.:

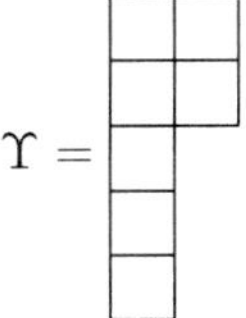

$$\Upsilon =$$

Such Young patterns have to be filled by symbols of the orbitals with the understanding that the many-electron spatial function thus obtained is formed from products of orbitals which are symmetrized along the rows and antisymmetrized along the columns. Obviously, two equal orbitals cannot appear in the same column, but can appear in different columns. As there are no more than two columns, each orbital appears no more than twice, which corresponds to the usual notion that orbitals can be no more than doubly filled. By this the permutational properties of the spatial functions complementary to the spin functions of the definite total spin are established.

Using only the spatial parts of the many electronic wave functions as the basis functions solves, to some extent, the problem of reducing the dimensionality of the FCI method. Indeed, by employing thus labeled basis set of many electron functions one uses only one basis function of a given total spin for all possible spin projections and the Hamiltonian matrix breaks down into blocks for each given value of the total spin. The problem is, however, to be able to write down the electronic Hamiltonian in the basis of the spatial parts only. This can be done by algebraic means by employing the wonderful interrelation between the irreducible representations of the permutation group $\mathfrak{S}_N$ and the representations of the group of $M \times M$ unitary matrices $U(M)$. The breathtaking fact is that the above N-box Young tableaux filled by orbital symbols taken from an orthonormal basis of an M-dimensional space of orbitals form a basis of an irreducible representation Υ of the group $U(M)$. This representation is of the N-th (tensor) power with respect to the matrix elements of an $M \times M$ matrix $u \in U(M)$ transforming the orbitals and has the permutation properties as those defined by the irreducible representation of the permutation group $\mathfrak{S}_N$ determined by the Young pattern employed. The dimensionality of this representation is given by the Weyl formula:

$$\dim(\Upsilon = \Upsilon(M, N, S)) = \frac{2S + 1}{M + 1} \binom{M + 1}{\frac{1}{2}N + S + 1} \binom{M + 1}{\frac{1}{2}N - S}$$

and the rows of this degenerate irreducible representation of $U(M)$ are numbered by the Young tableaux Υv filled by the orbital symbols. The fact that the Young tableaux Υv filled by M orbital symbols form the basis of an irreducible representation Υ of the group $U(M)$ allows one to express operators acting within the space of this representation in terms of the so-called generating operators (generators) of this group specific for this representation. The generating operators of the group $U(M)$ can be defined through the creation and annihilation operators by the relations

$$E_{ij} = \sum_\sigma \mathsf{a}_{i\sigma}^+ \mathsf{a}_{i\sigma}$$

where $i, j = 1 \div M$ run over the basis orbitals. The matrix representation $\mathbf{E}_{ij}^\Upsilon$ of the operators E_{ij} in the space of the irreducible representation spanned by the Young tableaux Υv can be found on purely algebraic grounds. In their terms the Hamiltonian matrix in the basis of the Young tableaux acquires the form

$$(1.178) \quad \mathbf{H} = \sum_{ij} h_{ij}^{(1)} \mathbf{E}_{ij}^\Upsilon + \frac{1}{2} \sum_{ijkl} (ij \mid kl) \left(\mathbf{E}_{ij}^\Upsilon \mathbf{E}_{kl}^\Upsilon - \delta_{jk} \mathbf{E}_{il}^\Upsilon \right)$$

This representation among others removes one more inconsistency: in quantum chemistry one generally deals with the systems of constant composition i.e. of the fixed number of electrons. The expression eq. (1.178) allows one to express the matrix elements of an electronic Hamiltonian without the necessity to go in a subspace with number of electrons different from the considered number N which is implied by the second quantization formalism of the Fermi creation and annihilation operators and on the other hand allows to keep the general form independent explicitly neither on the above number of electrons nor on the total spin which are both condensed in the matrix form of the generators $\mathbf{E}_{ij}^{\Upsilon}$ specific for the Young pattern Υ for which they are calculated.

1.5.6. Group function approximation

The Hartree-Fock(-Roothaan – HFR) approximation briefly described in Section 1.5.3 still remains the basis for further development of the major part of quantum chemistry methods. Even for the so-called post Hartree-Fock methods, it serves as the first step for several of them. The solutions of the Hartree-Fock equations in the fixed basis of orbitals yield a set of orthonormalized orbital functions (canonical MOs) which can be used to construct the set of the N-electron basis determinants necessary for the FCI method. This way of doing things is, however, not free from problems. Despite their appearance as a linear method, in eqs. (1.143)–(1.148), the HFR method is in fact strongly nonlinear due to the dependence of the Fock operator on its own eigenstates through the one-electron density. Thus, obtaining the solutions to the HFR problem requires a well-known iteration procedure, which does not always easily converge. On the other hand, in some situations even if the HFR approximate wave function is obtained, it turns out to be nevertheless too far from the true ground state. In this case a stepwise obtaining of further corrections to it turns out to be impractical since the expansion in terms of the MO-built determinants converges too slowly and is not easily interpretable. Thus it again goes about regrouping the terms in the FCI expansion to get better convergence. This can be done on physical grounds, which incidentally provide a good starting point for constructing hybrid methods, which is the topic of the present book. Following McWeeny [29, 30, 45] one can employ for that purpose a somewhat more general form for the zero approximate wave function, known as the group function approximation. Let us assume that the molecular system under consideration consists of "distinguishable" subsystems $A, B, C, \ldots$, each containing some definite number of electrons $N_A, N_B, N_C, \ldots N_A + N_B + N_C + \cdots = N$. The simplest choice of the wave function in the form of a product

$$\psi(x_1, \ldots, x_N) = \psi^A(x_1, \ldots, x_{N_A})\psi^B(x_{N_A+1}, \ldots, x_{N_A+N_B}) \cdots$$

was originally called the group function approximation. It is exact for so-called "separable" Hamiltonians. According to [43] in a system comprising (for the sake of simplicity) only two subsystems A and B its Hamiltonian can be written in the coordinate representation in terms of the separable Hamiltonian H_0 and the intersubsystem part (it is not, however, the interaction):

$$\hat{H} = \hat{H}_0 + \hat{H}_{AB} = \hat{H}_A + \hat{H}_B + \hat{H}_{AB};$$

$$\hat{H}_A = -\frac{1}{2} \sum_{i=1}^{N_A} \nabla_i^2 + \sum_{i=1}^{N_A} \sum_{j=1}^{i-1} \frac{1}{|\mathbf{r}_i - \mathbf{r}_j|}$$

$$(1.179) \qquad \hat{H}_B = -\frac{1}{2} \sum_{i=N_A+1}^{N_A+N_B} \nabla_i^2 + \sum_{i=N_A+1}^{N_A+N_B} \sum_{j=N_A+1}^{i-1} \frac{1}{|\mathbf{r}_i - \mathbf{r}_j|}$$

$$\hat{H}_{AB} = \sum_{i=1}^{N_A} \sum_{j=N_A+1}^{N_A+N_B} \frac{1}{|\mathbf{r}_i - \mathbf{r}_j|}$$

The true Hamiltonian H appears as a limiting case of the parametric family of the operators:

$$\hat{H}(\lambda) = \hat{H}_0 + \lambda \hat{H}_{AB};$$

$$\hat{H}(0) = \hat{H}_0; \hat{H}(1) = \hat{H}$$

For the value $\lambda = 0$ (separable Hamiltonian) any interaction between the subsystems is excluded. This, however, brings up two problems: first of all, excluding all interactions yields a very poor Hamiltonian, as even strong ones like Coulomb repulsions of electrons are dropped from the zero approximation. The second problem, however, is even more severe, as the separability condition breaks that symmetry of the system which is imposed by the fundamental requirement of equivalence of electrons. As one can easily see [44] a "separable" Hamiltonian is not invariant with respect to interchange of electronic coordinates if these latter "belong" to different groups (subsystems). It is obviously wrong, as the principle of the equivalence of electrons forbids us to make any statements concerning the relation of that or another electron to that or another subsystem. This requirement has nothing to do with the strength of the interaction between the subsystems. No matter how far the molecules are, the electrons in them do not become red or blue because of that, so the overall antisymmetry of the electronic wave function must be preserved. As the antisymmetry is not preserved, the product wave function cannot be an eigenstate even for the simplest Hamiltonian:

$$(1.180) \quad H = H_A + H_B + H_C + \dots$$

without interaction between parts $A, B, C, \dots$. Even in the absence of interactions between the subsystems, the symmetry requirement for the wave function is to belong to the fully antisymmetric representation of the symmetric group of N electrons $\mathfrak{S}_N$, and the simple product function obviously does not belong to this representation.

This problem is solved by explicitly including the antisymmetrization operator in the definition of the group function [45]:

$$(1.181) \quad \psi(x_1, \ldots, x_N) = \mathcal{A}\left[\psi^A(x_1, \ldots, x_{N_A})\psi^B(x_{N_A+1}, \ldots, x_{N_A+N_B}) \ldots\right] =$$
$$= \psi^A(x_1, \ldots, x_{N_A}) \wedge \psi^B(x_{N_A+1}, \ldots, x_{N_A+N_B}) \wedge \ldots$$

where $\mathcal{A}$ stands for the antisymmetrization operator (projection operator to the required fully antisymmetric irreducible representation of the $\mathfrak{S}_N$ group) in the coordinate representation and $\wedge$ stands for the antisymmetrized product of functions, provided each multiplier is already an antisymmetric function.

The wave function of eq. (1.181) is already an eigenfunction for the Hamiltonian with the interaction between the subsystems turned off and its eigenvalue is simply:

$$(1.182) \quad E = E_A + E_B + E_C + \ldots$$

where it is assumed that each of the group functions ψ^m is an eigenfunction for the corresponding Hamiltonian with the eigenvalue E_m.

Although upto this point we have coped with the antisymmetry of the wave function, the group function (GF) approximation still does not look too attractive, due to a strong feeling that the whole construct may be valid only in a weak interaction limit. The problem of interaction strength can be solved and the range of the systems to which the GF approximation applies can be significantly enlarged. Indeed in the second edition of his brilliant book R. McWeeny [30] includes in the list of weakly interacting subsystems the core shells of atoms, lone electronic pairs, pairs of electrons in different chemical bonds, etc. Obviously, all these subsystems of molecular systems interact very strongly at least due to the Coulomb forces. An obvious counterexample can be added for the requirement of the "strong spatial isolation": namely, the canonical MOs coming from the Hartree-Fock method. The wave function of the Hartree-Fock approximation eq. (1.142) is obviously a special case of the group expansion where each MO takes part of a function for a group formed by the single electron. In the case of the canonical MOs there is no reason to speak either about weak interaction nor about spatial isolation of these subsystems.

To find a way out of this contradiction we have to first of all elucidate the physical sense of some terms used and then find formal mathematical expression for these physical ideas. The key concept is the distinguishability of the subsystems described by the group functions. This concept has been introduced and analyzed by K. Ruedenberg, who noticed [46] that 'fragments in a molecular system can be singled out if these latter are observable, so that they manifest a reproducible and natural behavior; if for a series of molecules variations of fragments fit to that or another curve and its parameters can be found empirically by considering enough of the series members this proves that singling out the fragments makes sense'. Obviously the fragments can be singled out in organic molecules where these can be two-center two-electron bonds, or conjugate π-systems in polyunsaturated species. Analogously d-shells in transition metal complexes in many cases can be shown to be observable objects (see below). Now we have to decide what is a more formal expression for the distinguishability or observability of the fragments in molecular systems. For this we notice that the GF eq. (1.181) satisfies a specific symmetry condition: it is an eigenfunction for each of the family of the particle number operators $\hat{N}_A, \hat{N}_B, \hat{N}_C, \ldots$ with the eigenvalues $N_A, N_B, N_C, \ldots$, respectively. Each of these operators obviously commutes with the approximate Hamiltonian with no interaction, as defined above. However, the possibility of defining a set of conserved operators of numbers of particles for the system under consideration represents a much weaker requirement than that of "spatial separation" or "weak interaction" and thus can be satisfied in numerous cases. As for the general Hamiltonian, the criterion for the existence of an acceptable description in terms of the GF may be formulated as that of the existence of a set of the number of particle operators:

$$(1.183) \quad \hat{N}_m; \sum_m \hat{N}_m = \hat{N}$$

such that they all commute with a "significant" part $\hat{H}_0$ of the Hamiltonian $\hat{H}$:

$$(1.184) \quad \left[\hat{N}_m, \hat{H}_0\right] = 0; \left[\hat{N}_m, \hat{W}\right] \neq 0, \forall m$$
$$\hat{H} = \hat{H}_0 + \hat{W}$$

The noncommuting part $\hat{W}$ of the Hamiltonian takes the part of "interaction" or "perturbation" defined with respect to that part $\hat{H}_0$ of the total Hamiltonian $\hat{H}$ which is symmetric with respect to the group generators $\hat{N}_m$.[22] This perturbation must be weak in order to assure the validity of the GF approximation. The set of the operators $\hat{N}_m$ optimized with respect to this criterion yield the natural break of the system into subsystems.

Switching to a basis of orbitals allows us to proceed toward constructing the description of a molecular electronic structure based on the GF approximation under well acceptable restrictions. Indeed, let us assume that the multipliers in eq. (1.181) satisfy the additional condition of strong orthogonality:

$$(1.185) \quad \int \psi^{R*}(x_1, x_2 \ldots x_i \ldots x_{N_R})\psi^S(x_1, x_2 \ldots x_i \ldots x_{N_S})dx_i = 0$$
$$\forall i \leq \min(N_R, N_S)$$

Then the separability theorem [48, 49] states that the strong orthogonality is equivalent to dividing the complete space of one-electron functions into mutually orthogonal subspaces:

$$(1.186) \quad L = \bigoplus_m L_m; \dim L_m \geq N_m$$
$$L_m \cap L_n = \{0\}, m \neq n$$

and constructing each of the multiplier functions ψ^m as an expansion over n_m-electron Slater determinants such that the filled spin-orbitals in them belong to the

---

[22]See information on the Lie groups and its generators. For a beginning, [47] may be enough.

subspace L_m and only to it. More technically, one can think about selecting a set of orthogonal projection operators summing upto the identity operator:

$$(1.187) \quad \sum_m \hat{P}_m = \hat{I}; \, \hat{P}_m \hat{P}_n = \delta_{mn} \hat{P}_m$$

$$L_m = \text{Im} \hat{P}_m$$

and setting subspaces L_m to be the images of the corresponding projection operator. The relation with the definition of groups in terms of numbers of electron operators is reestablished by the following construct. In each of the subspaces L_m an orthogonal basis of spin-orbitals φ_{mv} can be chosen. Defining the operator $\hat{N}_m$ of the number of electrons in the carrier subspace L_m (or equivalently in the corresponding electron group) as

$$(1.188) \quad \hat{N}_m = \sum_v a^+_{mv} a_{mv}$$

we get the required expansion. From this the GF approximation is equivalently formulated as a statement about the constancy of the number of electrons in certain orthogonal subspaces of the entire space of one-electron states.

The wave function of electrons in the GF approximation can be briefly rewritten in the form:

$$(1.189) \quad \Psi_0 = \bigwedge_{A=1}^{M} \Psi^A$$

The sign $\wedge$, as previously, denotes the antisymmetrized product of the multipliers following it. The functions Ψ^A are taken as linear combinations of the N_A-electron Slater determinants such that the occupied spin-orbitals in each Slater determinant are taken from the carrier subspace L_A.

Using the second quantization formalism simplifies everything greatly: Antisymmetrization is achieved simply by putting all the operators creating electrons in the one-electron states of the A-th group to the left from those of the B-th group, provided $B < A$. The multipliers Ψ^A can be considered as linear combinations of rows of N_A creation operators a^+_{Av}.

The second quantization formalism also greatly simplifies the treatment of the Hamiltonian and allows its analysis pertinent to the GF approximation for the wave function.[23] Indeed, the total electron Hamiltonian $\hat{H}$ can be rewritten using the second quantization formalism according to the division of the orbital basis set into carrier subspace basis sets as introduced above:

$$(1.190) \quad \hat{H} = \sum_A \hat{H}_A + \sum_{A<B} \hat{W}_{AB}$$

---

where $\hat{H}_A$ contains only the products of the Fermi operators creating/annihilating electrons in the spin-orbitals ascribed to group A. The interaction operators $\hat{W}_{AB}$ are those which contain the mixed products of the creation/annihilation operators of the groups A and B. (In the general case there are, of course, the terms in the Hamiltonian, which contain products of the operators belonging to three or four different groups. We, however, omit them here for the sake of simplicity. Apart from the latter restriction the above form of the electronic Hamiltonian is quite general.) Also for the sake of simplicity we restrict ourselves to the interaction operator containing only one-electron transfer terms (resonance interaction) between the groups and restrict the two-electron interactions to those which conserve numbers of electrons in the groups. Under these conditions the interaction terms in eq. (1.190) acquire the form:

$$\hat{W}_{AB} = \hat{W}^r_{AB} + \hat{W}^c_{AB},$$

$$(1.191) \qquad \hat{W}^r_{AB} = \sum_{a \in A, b \in B} w^r_{ab}(\mathsf{a}^+\mathsf{b} + \mathsf{b}^+\mathsf{a})$$

$$\hat{W}^c_{AB} = \sum_{\substack{aa' \in A, \\ bb' \in B}} (aa' \parallel bb')\mathsf{a}^+\mathsf{b}^+\mathsf{b}'\mathsf{a}'$$

Here $(aa' \parallel bb')$ is the symmetrized two-electron matrix element of the electron-electron Coulomb repulsion.

$$(1.192) \quad (aa' \parallel bb') = (aa' \mid bb') - (ab' \mid ba')$$

and other notations are self-explanatory.

An estimate of the ground state of the system can be found by applying the variational principle to the GF trial wave function. Let us assume that the (spin-)orbital basis in each carrier subspace assigned to a group is fixed. It is not necessary, as it is enough if the carrier subspaces are themselves fixed, but this assumption eases the technical details. Under these conditions, each of the group multipliers Ψ^A can be thought to be an FCI wave function of N_A electrons in the dim L_m-dimensional space and the amplitudes of the corresponding configurations over these Slater determinants form the set of the variational variables ξ. This yields a system of interconnected eigenvalue problems with the effective Hamiltonians $\hat{H}^A_{\text{eff}}$ for the functions Ψ^A for the corresponding groups. Each of the group effective Hamiltonians depends on the ground state wave functions of all remaining groups [29, 30]:

$$\hat{H}^A_{\text{eff}}\Psi^A = E_A\Psi^A$$

$$(1.193) \qquad \hat{H}^A_{\text{eff}} = \hat{H}^A + \sum_{B \neq A} \langle \Phi^B_0 \mid \hat{W}_{BA} \mid \Phi^B_0 \rangle$$

$$\hat{W}_{BA} = \hat{W}^c_{BA}$$

The averaging of the Coulomb interaction operator can be easily performed following the prescription given in [29, 30]:

$$(1.194) \quad \langle \Phi_0^B \mid \hat{W}_{BA}^c \mid \Phi_0^B \rangle = \sum_{aa' \in A} \mathsf{a}^+ \mathsf{a}' \sum_{bb' \in B} (aa' \parallel bb') \langle\langle \mathsf{b}^+ \mathsf{b}' \rangle\rangle_B$$

where

$$\langle\langle \mathsf{b}^+ \mathsf{b}' \rangle\rangle_B = \langle \Phi_0^B \mid \mathsf{b}^+ \mathsf{b}' \mid \Phi_0^B \rangle$$

It leaves intact the fermion operators related to the A-th group itself. By virtue of this the two-electron operators $\hat{W}_{BA}^c$ result in a renormalization of one-electron terms in the Hamiltonians for each group $A = 1, ..., M$. The expectation values $\langle\langle \mathsf{b}^+ \mathsf{b}' \rangle\rangle_B$ are the one-electron densities. The Schrödinger equation eq. (1.193) can be driven close to the standard HFR form. This can be done if one defines generalized Coulomb and exchange operators for group A by their matrix elements in the carrier space of group A:

$$(1.195) \quad \hat{J}_{aa'}^{AB} = \sum_{bb' \in B} (aa' \mid bb') \langle\langle \mathsf{b}^+ \mathsf{b}' \rangle\rangle_B$$

$$\hat{K}_{aa'}^{AB} = \sum_{bb' \in B} (ab' \mid ba') \langle\langle \mathsf{b}^+ \mathsf{b}' \rangle\rangle_B$$

After this the analogy with the Hartree-Fock equation eq. (1.143) becomes complete as the effective Hamilton operator for group A acquires the form:

$$(1.196) \quad \hat{H}_{\text{eff}}^A = \hat{h}^A + \sum_{B \neq A} \hat{J}^{AB} - \sum_{B \neq A} \hat{K}^{AB} + \hat{g}^A$$

where $\hat{h}^A$ is the collection of one-electron terms involving the Fermi operators form group A; $\hat{J}^{AB}$ and $\hat{K}^{AB}$ defined just above are also the one-electron operators acting in the carrier subspace of group A; and $\hat{g}^A$ is a collection of two-electron terms formed by the Fermi operators associated with group A.

1.6. ALTERNATIVE TOOLS FOR REPRESENTING ELECTRONIC STRUCTURE

Similar to quantum mechanics, which can be formulated in terms of different quantities in addition to the traditional wave function formulation, in quantum chemistry a number of alternative tools are developed for this purpose, which may be useful in the context of the present book. We have already described different approximate models of representing the electronic structure using (many-electronic) wave functions. The coordinate and second quantization representations were employed to get this. However, the entire amount of information contained in the many-electron wave function taken in whatever representation is enormously large. In fact it is mostly excessive for the purpose of describing the properties of any molecular system due to the specific structure of the operators to be averaged to obtain physically relevant information and for the symmetry properties of the wave functions the expectation values have to be calculated over. Thus some reduced descriptions are possible, which will be presented here for reference.

1.6.1. Reduced density matrices

Closely inspecting the operator terms entering the electronic Hamiltonian eq. (1.27) one can easily see that they are sums of equivalent contributions dependent on coordinates of one or two electrons only. Analogously in the second quantization formalism only the products of two and four Fermi operators appear in the Hamiltonian. Inserting the trial N-electron wave function of the (ground) state into the expression for the electronic energy yields its expectation value in terms of the expectation values of the one- and two-electron operators:

$$E = \left\langle \hat{H} \right\rangle = \sum_{nn'\sigma} h^{(1)}_{n\sigma n'\sigma} \left\langle a^+_{n\sigma} a_{n'\sigma} \right\rangle +$$

$$(1.197) \qquad + \sum_{\substack{mm'nn' \\ \sigma\sigma'}} h^{(2)}_{m\sigma'm\sigma'n\sigma n'\sigma} \left\langle a^+_{n\sigma} a^+_{m\sigma'} a_{m'\sigma'} a_{n'\sigma} \right\rangle$$

entering the Hamiltonian expansion eq. (1.177). The averaging is performed over a function which corresponds to some specific values of the total spin S, of its projection S_z, of the irreducible representation of the point group Γ, and the row γ of the latter.

The expectation values

$$(1.198) \qquad \begin{aligned} \rho^{(1)}(k_1, k_2) &= \left\langle a^+_{k_1} a_{k_2} \right\rangle \\ \rho^{(2)}(k_1, k_4; k_2, k_3) &= \left\langle a^+_{k_1} a^+_{k_2} a_{k_3} a_{k_4} \right\rangle \end{aligned}$$

are the elements of so-called one- and two-electron reduced density matrices in the representation of the spin-orbitals ϕ_{k_i}. Habitually they are defined in the coordinate representation according to [30]:

$$\begin{aligned} \rho^{(1)}(x, x') &= N \int \Psi^*(x, x_2, \ldots x_N) \times \\ &\quad \times \Psi(x', x_2, \ldots, x_N) dx_2 \ldots dx_N \end{aligned}$$

$$(1.199)$$

$$\begin{aligned} \rho^{(2)}(x_1, x_1'; x_2, x_2') &= \frac{N(N-1)}{2} \int \Psi^*(x_1, x_2, x_3, \ldots x_N) \times \\ &\quad \times \Psi(x_1', x_2', x_3, \ldots, x_N) dx_3 \ldots dx_N \end{aligned}$$

The relation to the density matrix elements in the spin-orbital occupation numbers representation recovers from noticing that the rows of indices of spin-orbitals $\{k_1, k_2, \ldots, k_N\} = K$ (defining a row of creation operators $a^+_{k_N} \ldots a^+_{k_2} a^+_{k_1}$, forming a basis Slater determinant) can be in the same manner considered as a set of electronic coordinates in the spin-orbital representation as is the list $\{x_1, x_2, \ldots, x_N\}$, which is the list of arguments of the Slater determinant $(N!)^{-\frac{1}{2}} |\phi_{k_1}(x_1), \phi_{k_2}(x_2), \ldots, \phi_{k_N}(x_N)|$ in the coordinate representation. The value of the wave function in the point $\{k_1, k_2, \ldots, k_N\}$ of this new configuration space is nothing but C_K where $K = \{k_1, k_2, \ldots, k_N\}$. The only difference is that in the definition of the amplitudes C_K in Section 1.5.1 eq. (1.137) we assumed the N-tuple K to be ordered: $\{k_1 < k_2 < \ldots < k_N\}$. This restriction can be lifted here by noticing that according to the Pauli principle the value of the wave function on whatever unordered N-tuple $\{k_1, k_2, \ldots, k_N\}$ equals C_K times $(-1)^p$ where p is the number of the transpositions to be applied to $\{k_1, k_2, \ldots, k_N\}$ to make it ordered. From the above it is easy to understand that

$$(1.200) \quad \begin{aligned} \rho^{(1)}(k, k') &= \sum_{K,K'} C_K C_{K'} \\ \rho^{(2)}(k_1, k'_1; k_2, k'_2) &= \sum_{K,K'} C_K C_{K'} \end{aligned}$$

where in the first case the summation extends to all pairs of N-tuples such that $K = \{k, k_2, \ldots, k_N\}$, $K' = \{k', k_2, \ldots, k_N\}$ and in the second case it extends to all pairs of N-tuples such that $K = \{k_1, k_2, k_3, \ldots, k_N\}$, $K' = \{k'_1, k'_2, k_3, \ldots, k_N\}$. Using the defined elements of one- and two-electron density matrices the value of the energy can be rewritten as:

$$(1.201) \quad E(\Gamma S) = \mathrm{Sp}\rho^{(1)} h^{(1)} + \mathrm{Sp}\rho^{(2)} h^{(2)}$$

where ΓS refers to the symmetry and spin of the wave function the densities eq. (1.199) are calculated for.

The expressions eqs. (1.197), (1.199), (1.200), (1.201) are completely general. From them it is clear that the reduced density matrices are much more economical tools for representing the electronic structure than the wave functions. The two-electron density (more demanding quantity of the two) depends only on two pairs of electronic variables (either continuous or discrete) instead of N electronic variables required by the wave function representation. The one-electron density is even simpler since it depends only on one pair of such coordinates. That means that in the density matrix representation only about $(2M)^4$ numbers are necessary to describe the system (in fact – less due to antisymmetry), whereas the description in terms of the wave function requires, as we know $\frac{(2M)!}{N!(2M-N)!}$ numbers (FCI expansion amplitudes). However, the density matrices are rarely used directly in quantum chemistry procedures. The reason is the serious problem which appears when one is trying to construct the adequate representation for the left hand sides of the above definitions without addressing any wave functions in the right hand sides. This is known as the N-representability problem, unsolved until now [51] for the two-electron density matrices. The second is that the symmetry conditions for the electronic states are much easier formulated and controlled in terms of the wave functions (Density matrices are the entities of the second power with respect to the wave functions so their symmetries are described by the second tensor powers of those of the wave functions).

The density matrix description is useful when discussing the electron correlations. The statement that the motion of electrons is correlated can be given an exact sense only if the two-electron density matrices eqs. (1.199) and (1.200) are used. In terms of the wave function, the statement of the correlated character of electron motions sounds like a negative statement: the non-correlated (Hartree-Fock) wave function is one which is represented by a single Slater determinant, and the correlated one is that which cannot be represented by any single Slater determinant. This type of definition does not have strict sense as the number of determinants in the expansion of the wave function depends on the basis of spin-orbitals used for this purpose. Simply inserting the expansion of MOs over the basis of atomic spin-orbitals in each row one immediately gets a function which is a superposition of all $\frac{(2M)!}{N!(2M-N)!}$. The invariant formulation dating back to Löwdin [50] can be written as:

$$(1.202) \quad \rho^{(2)}(k_1, k_1'; k_2, k_2') = \begin{vmatrix} \rho^{(1)}(k_1, k_1') & \rho^{(1)}(k_2, k_1') \\ \rho^{(1)}(k_1, k_2') & \rho^{(1)}(k_2, k_2') \end{vmatrix} - \chi(k_1, k_1'; k_2, k_2')$$

where the first term expands as

$$(1.203) \quad \begin{vmatrix} \rho^{(1)}(k_1, k_1') & \rho^{(1)}(k_1', k_2) \\ \rho^{(1)}(k_1, k_2') & \rho^{(1)}(k_2, k_2') \end{vmatrix} = \rho^{(1)}(k_1, k_1')\rho^{(1)}(k_2, k_2')$$
$$- \rho^{(1)}(k_1, k_2')\rho^{(1)}(k_1', k_2)$$

and the second one is known as the cumulant of the two-electron density matrix (see e.g. [51, 52]). Different approximate forms of wave function produce corresponding simplification in the reduced density matrices. For example, substituting the single determinant Hartree-Fock wave function eq. (1.142) to the density matrix definitions eq. (1.199) yields a simple expression for the two-particle density matrix, which is uniquely determined by the one-electron density matrix:

$$(1.204) \quad \rho^{(2)}_{\mathrm{HF}}(k_1, k_1'; k_2, k_2') = \begin{vmatrix} \rho^{(1)}_{\mathrm{HF}}(k_1; k_1') & \rho^{(1)}_{\mathrm{HF}}(k_1; k_2') \\ \rho^{(1)}_{\mathrm{HF}}(k_2; k_1') & \rho^{(1)}_{\mathrm{HF}}(k_2; k_2'). \end{vmatrix}$$

In other words, the Hartree-Fock approximation is nothing but setting the cumulant of the two-electron density matrix to be zero:

$$(1.205) \quad \chi_{\mathrm{HF}}(k_1, k_1'; k_2, k_2') \equiv 0$$

From here it is obvious that for the Hartree-Fock approximation the parametrization of the energy not referring directly to the wave function is nevertheless possible (the Hartree-Fock density is a projection operator and it can be directly written using say eq. (1.107)), but the cost is fixing $\chi \equiv 0$ with the consequences of this. (We can say that the works devoted to foundations of DFT basically reduce to developing a more or less widely applicable form of the two-electron density.)

The GF approximation has many common features with the HF(R) approximation. According to [29, 30] the following relations for the one- and two-electron density matrices take place and allow us to calculate efficiently the matrix elements of the density with the functions of eq. (1.181):

$$\rho^{(1)}_{\mathrm{GF}}(k; k') = \sum_A \rho^{(1)}_A(k; k')$$

$$(1.206) \qquad \rho^{(2)}_{\mathrm{GF}}(k_1, k_1'; k_2, k_2') = \sum_A \rho^{(2)}_A(k_1, k_1'; k_2, k_2') +$$

$$+ \sum_{A \neq B} \begin{vmatrix} \rho^{(1)}_A(k_1; k_1') & \rho^{(1)}_A(k_1; k_2') \\ \rho^{(1)}_B(k_2; k_1') & \rho^{(1)}_B(k_2; k_2') \end{vmatrix}$$

where the density matrices $\rho_A^{(1)}$ and $\rho_A^{(2)}$ are calculated according to the formulae in eq. (1.200) by partially integrating the wave functions of the corresponding group only. The two-electron density matrix for each group $\rho_A^{(2)}$ can be expanded analogously to the total two-electron density matrix: into the determinantal HF(R) part and the cumulant χ_A. It is easy to see that in the GF case the cumulant of the two-electron density is nonvanishing only if all four spin-orbital indices k_1, k_2; k_1', k_2' belong to the same group. If at least one of them belongs to a different group the corresponding elements of the cumulant matrix vanish. By this we can see that the GF functions take into account the nontrivial part of the correlation only inside each of the groups (the trivial part of correlation – the Fermi hole is accounted for by the antisymmetry of the wave function or equivalently by either the intragroup or intergroup determinantal terms in the expansion of $\rho_{\text{GF}}^{(2)}$). The disappearance of the cumulant between the groups indicates that electrons in them behave as independent particles with no more correlation than is imposed by the antisymmetry requirement (Fermi correlation).

1.6.2. Resolvents and Green's functions

The density matrices described in the previous section allowed to significantly reduce the description of the electronic structure as compared to that provided by the wave function representation and also retain the most important features of the electron distribution. Also in the HF approximation the one-electron density matrix is tightly related to the wave function: It is an operator projecting to the subspace spanned by the occupied MOs of the single Slater determinant involved. However, the energy characteristics of this occupied manifold are missing as are the dynamic features – all of which describe the response of the system to external perturbation (including the interactions between the subsystems in complex systems) – not covered by the density matrices alone. Physicists have developed a powerful language which allows one to describe by a single quantity all aspects of the behavior of a many-particle (in special cases many-electron) system – the formalism of the Green's functions, which, despite the existence of very clear introductions to it targeted at chemists [39, 40] is not widely used so far in quantum chemistry. The reason is most probably that even in physics this language is largely used for deriving some general relations rather than for performing actual calculations. As in the present book we are more concerned with the general constructs necessary for understanding the situation in the complex molecular systems, we briefly describe this technique and use it in several occasions in our further derivations.

Probably the easiest way of introducing Green's functions in the many-electron context is to begin by using it to describe the electronic structure in the Hartree-Fock approximation. As shown in Section 1.5.3 the solution of the Hartree-Fock problem is known up to the subspace spanned by the occupied MOs of the single Slater determinant representing the trial wave function of the Hartree-Fock approximation. This subspace can be treated as an image subspace of the corresponding projection operator which is given by eq. (1.144). This projection operator is obviously a direct sum (in the sense of Section 1.5.6) of the one-dimensional orthogonal projection operators, each corresponding to an occupied MO. Due to the eigenvalue/eigenvector form of the HF problem the projection operators to the occupied MOs can be written in terms of the formal solution of this problem by resolvents:

$$(1.207) \quad \rho_i(x, x') = \phi_i^*(x)\phi_i(x') = \frac{1}{2\pi i} \oint_{C_i} \hat{R}(z)\, dz$$

$$(1.208) \quad \hat{R}(z) = \left(z - \hat{F}[\hat{\rho}]\right)^{-1}$$

The resolvent in eq. (1.208) is called the one-electron Green's function and the notation for it reads $G_{\mathrm{HF}}^{(1)}(z)$. The integration contour may be set in such a way that it encloses all the poles of the resolvent corresponding to the occupied MOs giving by this the required total projection operator. In the spin-orbital occupation number and the second quantization representations related to each other, one can write the operator projecting to the occupied (spin)-MO as an operator of the number of particles in it. Indeed, the expression

$$f_i^+ = \sum_k u_{ik} a_k^+$$

where u_{ik} are the MO LCAO expansion coefficients, is nothing but the operator creating an electron on the i-th MO; its Hermitian conjugate f_i – destroys one. Forming a product $f_i^+ f_i$ and applying it to the Slater determinants constructed in the basis of the MOs of the problem at hand, immediately shows that those determinants which have the i-th MO occupied are the eigenvectors of this operator with the eigenvalue one; all others – which have this MO empty – are eigenvectors with the eigenvalue zero: i.e. $f_i^+ f_i$ is a projection operator. The same applies to the operator $1 - f_i^+ f_i$ which in its turn projects to the Slater determinants which have the i-th MO empty. This allows to further develop the formalism by writing

$$G_{\mathrm{HF}}^{(1)}(z) = \sum_{i=1}^{2M} \left\{ \frac{\langle f_i^+ f_i \rangle_{\mathrm{HF}}}{z - \varepsilon_i - i\delta} + \frac{1 - \langle f_i^+ f_i \rangle_{\mathrm{HF}}}{z - \varepsilon_i + i\delta} \right\}$$

where both the occupied and empty MOs are included in the construction of what is called the one-electron Green's function in the Hartree-Fock approximation in the energy (frequency) representation. The expectation value is assumed over the Hartree-Fock ground state Slater determinant. The integration contour can now be taken as the entire real axis (from $-\infty$ to ∞) with the notion that it is closed in the upper half-plane. Under these conditions only the occupied orbitals whose poles $\varepsilon_i + i\delta$ turn out to be infinitesimally shifted up from the real axis enter into play. The Fermi operators f_i^+ and f_i can be replaced by their expansions over the basis operators a_k^+ and $a_{k'}$ thus giving the representation of the Green's function:

$$(1.209) \quad G_{\mathrm{HF}}^{(1)}(k, k'; z) = \sum_{i=1}^{2M} \left\{ \frac{\langle a_k^+ a_{k'} \rangle_{\mathrm{HF}}}{z - \varepsilon_i - i\delta} + \frac{\delta_{kk'} - \langle a_k^+ a_{k'} \rangle_{\mathrm{HF}}}{z - \varepsilon_i + i\delta} \right\}$$

in the basis of the original basis spin-orbitals, where from the values of the density matrix elements it is restored by integration over frequency:

$$(1.210) \quad \rho_{\mathrm{HF}}^{(1)}(k, k') = \left\langle a_k^+ a_{k'} \right\rangle_{\mathrm{HF}} = \frac{1}{2\pi i} \int d\omega \, G_{\mathrm{HF}}^{(1)}(k, k'; \omega)$$

So far nothing interesting has happened, as everything turns out to be different disguises of the same known quantities: if the solution of the Hartree-Fock problem is known, it can be rewritten in a suitable form. The situation changes if one decides to extend the above definition of the one-particle Green's function to a general ground state by dropping the subscript HF. In this situation one can hope to save the relation between this function and potentially exact one-electron density, but to do so one needs an *equation* to determine the Green's function form. Of course, one can use the perturbative techniques for the resolvent by trying to represent the difference between the exact Coulomb interaction and its average form, entering the HF approximation, as a perturbation. This gives a start to a whole variety of techniques based on the Dyson equation for the resolvent and thus for the Green's function. We are however interested in a more exciting result which is used below.

Let us assume from now on that by definition the one-electron Green's function in the energy domain is:

$$(1.211) \quad G^{(1)}(k, k'; z) = \sum_i \left\{ \frac{\left\langle a_k^+ a_{k'} \right\rangle}{z - \varepsilon_i - i\delta} + \frac{\left\langle a_{k'} a_k^+ \right\rangle}{z - \varepsilon_i + i\delta} \right\}$$

where ε_i represents, in the first term, the energies of positively ionized states of the system under study (those with one electron less) and in the second term the energies of the negatively ionized states of the system under study (those with one electron more), both counted from the ground state energy of the system which is incidentally used for calculating the expectation values in the above definition, where we also employed the anticommutation relation for the Fermi operators. This quantity when integrated over the real frequency axis produces the exact one-electron density matrix

$$(1.212) \quad \rho^{(1)}(k, k') = \left\langle a_k^+ a_{k'} \right\rangle = \frac{1}{2\pi i} \int d\omega \, G^{(1)}(k, k'; \omega)$$

By going to the time domain and back (see for details [30]) one obtains the relation for the *exact* ground state energy of the system which reads:

$$(1.213) \quad E_0 = \frac{1}{4\pi i} \int d\omega \left[\omega \operatorname{Sp} G^{(1)}(\omega) + \operatorname{Sp} \left(G^{(1)}(\omega) h^{(1)} \right) \right]$$

This is a deceivingly simple result, which is, however, useful. It says that for obtaining the exact ground state energy it is enough to know one-particle Green's function, and that there is no immediate need for two-electron quantities such as two-electron densities and cumulants. The life is of course not that easy: In fact to be on the safe side one needs to really deal with *exact* one-electron Green's function over *all* energy domain and to know exact values *all* ionization potentials and electron affinities and the corresponding matrix numerators. Namely these quantities accurately hidden in the word "exact" are dependent on *exact* two-electron densities and are not easily accessible. Nevertheless, the possibility of the above form of the exact ground state energy opens prospects for various "fitting" procedures which will be discussed in due prescription.

1.7. GENERAL SCHEME FOR SEPARATING ELECTRONIC VARIABLES

In the previous sections we gave a brief account of several approaches to molecular electronic structure. From a general point of view they can be classified as reflecting different levels of separating electronic variables and/or taking into account and/or neglecting electronic correlations. The technical implementations (specific tools used) were not very important in that context, although the ease of discussion may significantly depend on selecting an appropriate representation. The FCI wave function, being the most general form of the electronic wave function, does not assume any level of separating electronic variables: they are all correlated. By contrast, the Hartree-Fock approximation corresponds to as much uncorrelated motion of electrons as admitted by the Pauli principle. Clearly the GF approximation takes an intermediate position between these two extrema as it allows us to take into account correlations within those or other groups to the extent it is necessary for any specific problem. It can be done by selecting that form of the group function which is adequate for the specific purpose of describing characteristic physical conditions in each group. Prerequisite for such a convenient treatment is of course the validity of the GF approximation itself.

The GF form of the trial wave function is, of course, an approximation. In general, the electron transfers between the groups do take place and in general destroy the variable separation built in the structure of the GF approximation. It would be desirable to take the effect of these transfers into account without destroying the attractive features of the GF wave function: the separation of the electronic variables describing different groups. This is done using the Löwdin partition applied to derive the GF approximate form for the wave function.

1.7.1. Limitations of the GF approximations as overcome by Löwdin partition[24]

We start from a most general form [53–56] of the wave function. It differs from the GF approximation eq. (1.181) in that respect that the number of electrons in each group is not fixed, so that the generalized group function (GGF) expansion is a linear combination of functions which are antisymmetrized products of multipliers with a different number of electrons in the groups [53–56]:

$$(1.214) \quad \Psi_k = \sum_{\{n_A\}} \sum_{\{i_A\}} C^k_{\{i_A\}}(\{n_A\}) \bigwedge_A \Phi^A_{i_A}(n_A)$$

---

[24]Subsequent material is based on A.L. Tchougréeff, Group Functions, Loewdin Partition, and Hybrid QC/MM Methods for Large Molecular Systems. *Phys. Chem. Chem. Phys.* **1**, 1051, 1999. Reproduced by permission of the PCCP Owner Societies.

In other words, the entire space originally spanned by the orbitals $\{\varphi_m\}$ is represented as the direct sum of orthogonal subspaces serving as the carrier subspaces to different groups, but the numbers of residing electrons are not fixed for each carrier subspace and all their possible distributions enter the expansion. In the expansion eq. (1.214) each distribution $\{n_A\}$ of electrons among the groups satisfies the condition:

$$(1.215) \quad \sum_A n_A = N_e; \forall A, n_A \geq 0$$

Also $\Phi_{i_A}^A(n_A)$ is the i_A-th n_A-electron function where only the one-electron states of the A-th group may be occupied. Expansion coefficients $C_{\{i_A\}}^k(\{n_A\})$ are thought to be determined on the basis of the variational principle.

The GGF representation is very general. In fact whatever N_e-electron wave function in the finite basis of (spin-)orbitals can be recast into the GGF form. The latter is merely a specific regrouping of the standard FCI expansion. We however assume that we can make a justified division of the one-electron basis into the groups and that we have physical grounds to assign a specific number of electrons to each group as well. This is a usual formulation of the GF approach given in the literature [29, 30, 58]. Its exact meaning is that in the GGF expansion eq. (1.214) only the product functions with certain fixed distribution $\{\bar{n}_A\}$ of electrons among the groups give the dominating contribution.

In the GGF wave function the electronic variables describing different groups are not separated (this situation is sometimes described by saying that the states of the groups are *entangled* [57]). The separation of electronic variables is reached by projecting exact electronic wave functions (eq. (1.214)) to the subspace spanned by the functions with the fixed numbers $(\bar{n}_A)$ of electrons in the groups. Let $\hat{P}$ be the operator projecting N_e-electron functions (eq. (1.214)) to this subspace. The projection operator $\hat{P}$ when acting on the GGF type wave function cuts off all the states with the electron distribution different from that fixed above. The target states in the $\mathrm{Im}\hat{P}$ subspace have the form:

$$(1.216) \quad \Psi_k = \sum_{\{i_A\}} C_{\{i_A\}}^k(\{\bar{n}_A\}) \bigwedge_{A=1}^{M} \Phi_{i_A}^A(\bar{n}_A)$$

Provided the separation of the entire space of orbitals into carrier subspaces is performed, the Hamiltonian in the second quantized form can be rewritten accordingly, following eqs. (1.190) and (1.191). In the Hamiltonian the matrix elements of the $\hat{W}_{AB}^r$ operator contributions are vanishing for the pair of the states when both belong to the $\mathrm{Im}\hat{P}$ subspace spanned by the functions eq. (1.216). Namely these operators are responsible for electron transfers between the groups A and B. When acting upon a state from the $\mathrm{Im}\hat{P}$ subspace the $\hat{W}_{AB}^r$ operator produces the states which all have distribution of electrons among the groups different from that characteristic for the states from $\mathrm{Im}\hat{P}$. This is obviously the defect of the GF approximation as, in fact, the intergroup electron transfers do occur. The contributions of the states with the electron transfers between the groups are taken into account by applying the Löwdin partition procedure (see [29,30,32,58] and above). As mentioned above this Hamiltonian differs from the exact one by dropping the two-electron terms which may transfer electrons between different groups. They are usually omitted in any semiempirical context to which we adhere here (see below) and on the other hand they do not produce any qualitative difference from what follows. Again, in a semiempirical setting, one may think that the one-electron operators $\hat{W}^r$ collect all one-electron transfers between the groups irrespective of its physical origin: kinetic energy, electron-nuclear attraction, and electron-electron repulsion. The subsequent procedure can also be seen as perturbation theory corresponding to breaking of the Hamiltonian into unperturbed part:

$$(1.217) \quad \hat{H}^0 = \sum_A \hat{H}_A + \sum_{A<B} \hat{W}_{AB}^c$$

and the perturbation $\hat{W}^r$. Applying the Löwdin partition yields the following:

$$(1.218) \quad \begin{aligned} \hat{H}_{\text{eff}}(E) &= \hat{P}\hat{H}^0\hat{P} + \hat{P}W^{rr}(E)\hat{P} \\ \hat{W}^{rr}(E) &= \hat{W}^r\hat{Q}\hat{R}(E)\hat{Q}\hat{W}^r \end{aligned}$$

where

$$(1.219) \quad \hat{R}(E) = (E\hat{Q} - \hat{Q}\hat{H}^0\hat{Q})^{-1}$$

is the resolvent of the operator $\hat{H}^0$ in the $\text{Im}\hat{Q}$ subspace and $\hat{Q} = \hat{I} - \hat{P}$ is the complementary projection operator. By this the total Hamiltonian acting in the total functional space is projected to the subspace $\text{Im}\hat{P}$. The intergroup one-electron transfers are replaced by the virtual ones, which are included in the correction term $\hat{W}^{rr}(E)$.

After projecting to the $\text{Im}\hat{P}$ subspace with the fixed distribution of electrons among the groups, the ground state of electrons can be sought in the class of the wave functions of the GF form eq. (1.181). At this stage the GF form of the trial wave function nevertheless remains an approximation. The reason is that the functions of the $\text{Im}\hat{P}$ subspace are not the GF type functions. Although they are the some linear combinations of the GFs with the fixed electron distribution $\{\bar{n}_\alpha\}$ it is by no means guaranteed that the solution (eigenvector of the Hamiltonian $\hat{H}_{\text{eff}}$) is presented by a single GF product. A single GF function must be selected among the combinations of the latter on the basis of the variational principle. This is completely analogous to the standard HF approximation: the exact wave function is a linear combination of eventually all Slater determinants, but a single Slater determinant used in order to approximate the whole expansion is to be found from the energy minimum condition for the single-determinant class of trial functions.

The single GF function, as previously, must be found from the system of interconnected "self-consistency" eq. (1.193) for the separate groups:

$$\hat{H}_{\text{eff}}^A(E)\Phi_0^A = E^A(E)\Phi_0^A$$

$$(1.220) \quad \hat{H}_{\text{eff}}^A(E) = \hat{P}\hat{H}_A\hat{P} + \sum_{B\neq A} \langle \Phi_0^B \mid \hat{P}\hat{W}_{BA}\hat{P} \mid \Phi_0^B \rangle$$

$$\hat{P}\hat{W}_{BA}\hat{P} = \hat{P}\hat{W}_{BA}^c\hat{P} + \hat{P}\hat{W}_{BA}^{rr}(E)\hat{P}$$

The expectation values of the intergroup Coulomb operators are the same as in the original GF case (see Section 1.5.6). The situation with the expectation values $\langle \Phi_0^B \mid \hat{P}\hat{W}_{BA}^{rr}\hat{P} \mid \Phi_0^B \rangle$ is somewhat more complicated: the operator $\hat{P}\hat{W}_{BA}^{rr}\hat{P}$ has the form:

$$(1.221) \quad \hat{P}\hat{W}_{BA}^{rr}\hat{P} = \sum_{\substack{aa' \in A, \\ bb' \in B}} w_{ab}^r w_{a'b'}^r (\mathsf{a}^+\mathsf{b}\hat{R}(E)\mathsf{b}'^+\mathsf{a}' + \mathsf{b}^+\mathsf{a}\hat{R}(E)\mathsf{a}'^+\mathsf{b}')$$

Averaging it over the ground state of the B-th group Φ_0^B yields the following one-electron operator acting on the electron quantum numbers of the A-th group:

$$(1.222) \quad \left\langle\!\left\langle \hat{P}\hat{W}_{BA}^{rr}\hat{P} \right\rangle\!\right\rangle_B = \sum_{aa' \in A} \sum_{bb' \in B} \left\{ w_{ab}^r w_{a'b'}^r \left(\mathsf{a}^+ \left\langle\!\left\langle \mathsf{b}\hat{R}(E)\mathsf{b}'^+ \right\rangle\!\right\rangle_B \mathsf{a}' \right. \right.$$
$$\left. \left. + \, \mathsf{a} \left\langle\!\left\langle \mathsf{b}^+\hat{R}(E)\mathsf{b}' \right\rangle\!\right\rangle_B \mathsf{a}'^+ \right) \right\}$$

One can check that despite their asymmetric appearance the operators $\langle\!\langle \hat{P}\hat{W}_{BA}^{rr}\hat{P} \rangle\!\rangle_B$ of eq. (1.222) are Hermitian. The expectation values $\langle\!\langle \mathsf{b}\hat{R}(E)\mathsf{b}'^+ \rangle\!\rangle_B$ and $\langle\!\langle \mathsf{b}^+\hat{R}(E)\mathsf{b}' \rangle\!\rangle_B$ will be considered in more detail.

At this point we arrive at the theory with the energy dependent effective Hamiltonian. This construct reminds us of a version of the perturbation theory known as the Brillouin-Wigner perturbation theory not used in this book (for more details see [58]). The disadvantage of the above result is the energy dependence of all effective Hamiltonians, which implies an iterative solution procedure that is hardly justified in the present context so that we replace the energy dependence in the right side by estimating the required quantities at the ground state energy of the Hamiltonian $\hat{H}^0$. This will allow us to cope easily with the otherwise problematic terms in eq. (1.222).

1.7.2. Variable separation and hybrid modeling

The construct leading to the approximate separating electronic variables developed above may seem to be too cumbersome. However, it is a necessary element of the entire picture as, in a strict sense, the quantum description in terms of wave function is only possible for the entire universe, as the variables of the particles not included in the consideration explicitly, affect the result anyway. By contrast the topic of this book devoted to hybrid methods assumes the existence of a part of the complex system itself to be treated classically. The steps toward approximate separation of the electronic variables using the group functions and Löwdin partitioning undertaken in the previous section will be continued in this section where we, using the general formalism of the group functions and effective Hamiltonian, shall perform an approximate separation of electronic variables in the molecular Hamiltonian and pass to the effective Hamiltonian for its "interesting" or at least quantally treated part. Next, averaging the effective Hamiltonian over the ground state of the chemically inert (and thus possibly classically tractable) part results in formulae representing the PES of a molecular system, containing contributions from chemically active and inert parts in the form leading to more or less standard QM/MM treatment. The junction between the QM and MM parts appears as a contribution from the chemically inert system which renormalizes parameters of the electronic Hamiltonian for the chemically active part and of those of the chemically active part which renormalize parameters of the classically treated inert part of the molecular system.

To reach this, we finalize the separation of electronic variables restricting ourselves to two groups only: one explicitly treated by quantum mechanical (quantum chemical) methods and another whose explicit treatment is classical, although implicitly assuming the existence of some "underlying" quantum electronic structure.

1.7.2.1. Defining subsystems and related basic quantities

Now we pass to the formal derivations of a hybrid method. We assume that the orbitals forming the basis for the entire molecular system may be ascribed either to the chemically active part of the molecular system (reactive or R-states) or to the chemically inactive rest of the system (medium or M-states). In the present context, the orbitals are not necessarily the basis AO, but any set of their orthonormal linear combinations thought to be distributed between the subsystems. The numbers of electrons in the R-system (chemically active subsystem) N_R and in the M-system (chemically inactive subsystem) $N_M = N_e - N_R$, respectively, are good quantum numbers at least in the low energy range. We also assume that the orbital basis in both the systems is formed by the strictly local orbitals proposed in [59]. The strictly local orbitals are orthonormalized linear combinations of the AOs centered on a single atom. In that sense they are the classical hybrid orbitals (HO):

$$(1.223) \quad |t(A)\rangle = \sum_{\tau \in A} h_\tau^t(A) |\tau\rangle$$

where the expansion coefficients $h_\tau^t(A)$ are defined by a procedure discussed later. In a degenerate case, when the expansion coefficients are equal to 0s and 1s, the HOs in eq. (1.223) are the original AOs.

The electronic Hamiltonian for the whole system is now a sum of subsystem Hamiltonians and of their interaction which is taken to comprise the terms of two types – the Coulomb $\hat{W}^c$ and the resonance (electron transfer) $\hat{W}^r$ interactions:

$$(1.224) \quad \hat{H} = \hat{H}_R + \hat{H}_M + \hat{W}$$

The Hamiltonian for the M-system is a sum of the free M-system Hamiltonian $\hat{H}_M^0$ and of the attraction of electrons in the M-system to the cores of the R-system V^R. Analogous subdividing is true for the R-system. On the other hand the interaction terms further subdivide to:

$$\hat{W} = \hat{W}^r + \hat{W}^c,$$

$$\hat{W}^r = \sum_{\substack{r \in R, \\ m \in M}} w_{rm}(r^+m + m^+r)$$

$$(1.225)$$

$$\hat{W}^c = \sum_{\substack{rr' \in R, \\ mm' \in M}} (rr' \,||\, mm')r^+m^+m'r'$$

where

$$(1.226) \quad (rr' \,\|\, mm') = (rr' \mid mm') - (rm' \mid mr')$$

The Coulomb interaction matrix elements of the form $(rr' \mid r'm')$ and/or $(rm \mid m\,m')$ corresponding to one electron transfers as previously are absorbed in the one-electron term $\hat{W}^r$; those of the form $(rm \mid rm)$ corresponding to two-electron transfers between the subsystems are assumed to produce a minor effect and are omitted for the sake of simplicity. It is worth mentioning that in the case of only two electron groups (R- and M-systems) precautions concerning the Coulomb matrix elements involving more than two groups formulated in the previous section are not necessary any more and the Hamiltonian defined by eqs. (1.224), (1.225) is fairly general.

The "exact" wave function of the system is represented by a generalized group function (GGF) where numbers of electrons in subsystems are not fixed:

$$(1.227) \quad \Psi_k = \sum_n \sum_{\rho,\mu} C^k_{\rho\mu}(n)\Phi^R_\rho(n) \wedge \Phi^M_\mu(N_e - n)$$

where $\Phi^R_\rho(n)$ is the ρ-th n-electron wave function built upon the orbitals ascribed to the R-system, and $\Phi^M_\mu(N - n)$ is the μ-th $(N - n)$-electron wave function built upon the orbitals ascribed to the M-system.

The electron variables related to the R- and M-systems entering the wave functions of the complex system in eq. (1.227) are *entangled*. Separating them is reached by projecting the wave function of the eq. (1.227) of the entire system on the subspace of the GF of the eq. (1.216). This basically repeats the moves done in the previous section in a narrower context: here we are going to pay more attention to the resolvent term $\hat{W}^{rr}$ and to setting conditions on the subsystem wave functions.

The intuitive idea of molecular system, comprising the R- and M-systems of which the R-system is one whose electronic structure is strongly dependent on the geometry of molecular system, whereas the electronic structure of the M-system does not change in a wide range of the geometry variation, must be formalized in certain *Ansatz* for the electronic wave function. First of all, as previously, we assume the GF character of the function which guarantees that the numbers of electrons in the R- and M-systems must be constant (N_R and $N_M = N_e - N_R$ are good quantum numbers) at least for low energies. This assumption is generally common for the validity of the entire GF approximation. Further assumptions concern the properties of the multipliers in the GF for a complex system. So, lower energy electronic excitations touch only the R-system whereas the excitations of the M-system have much higher energy. This guarantees that under all chemical transformations happening to the R-system the electronic state of the M-system – medium – remains basically the same: no intersection of the different electronic terms of the medium happens. An additional limitation on the state of the M-system, as it can represent an inert medium, is a requirement that no unpaired electrons are located in it. By contrast no *a priori* restriction upon the electronic wave function of the R-system is set.

Defining the subsystems of the complex system by distributing the orbitals in two subsets allows us to consistently define other quantities characterizing the complex system. For any atom A we can write:

$$P_A = P_A^M + P_A^R$$

$$(1.228) \qquad P_A^R = \sum_{r \in A \cap R} \langle\langle r^+ r \rangle\rangle_R$$

$$P_A^M = \sum_{m \in A \cap M} \langle\langle m^+ m \rangle\rangle_M$$

where the expectation values

$$(1.229) \quad \langle\langle \ldots \rangle\rangle_R = \langle \Phi_0^R | \ldots | \Phi_0^R \rangle$$

$$\langle\langle \ldots \rangle\rangle_M = \langle \Phi_0^M | \ldots | \Phi_0^M \rangle$$

are calculated over the ground states of the corresponding subsystems. This treatment allows us to naturally define a *frontier* atom as one which bears orbitals belonging to different subsystems. For them the above definitions of electron densities ascribed to subsystems are nontrivial. To proceed further with the frontier atoms we also distribute the core charge of an atom A between the R- and M-systems according to [60]:

$$(1.230) \quad Z_A = Z_A^M + Z_A^R$$

Formally it applies to any atom, but it is nontrivial only for the frontier ones. The condition which specifies the distribution of the core charge Z_A between the R- and M-systems is that the cores of the R-system must be as much as possible screened by the electrons of the R-system i.e. the effective Hamiltonian $\hat{H}_M^{\text{eff}}$ must be as close as possible to the Hamiltonian of the free M-system $\hat{H}_M^0$. This reduces to the electron counting rules based on the concept of the formal oxidation state (see [60] for details). With this we arrive at the possibility of distributing not only the electronic density, but also the total effective charges between the R- and M-systems. This is done by the formulae:

$$Q_A^R = P_A^R - Z_A^R$$

$$Q_A^M = P_A^M - Z_A^M$$

$$Q_A = Q_A^R + Q_A^M$$

The requirements of the description of the electronic structure formulated above can be satisfied by using the N-electron functions in the GF approximation of the particular form:

$$(1.231) \quad \Psi_\rho = \Phi_\rho^R \wedge \Phi_0^M$$

where Φ_ρ^R is the ρ-th N_R-electron wave function of the R-system, and Φ_0^M is the ground state wave function of N_M electrons in the M-system. (We foresee here that not only the ground state of the R-system but also some low-energy excited states being incidentally – according to assumption – the low energy excitation of the entire system may be of interest to us). The difference of the wave functions eq. (1.231) from the general wave functions eqs. (1.214) and (1.227) consists not only in the elimination of superposition of the states with different distributions of electrons between the R- and M-systems (summation over n in eq. (1.227) is replaced by one term $n = N_R$) but also in replacing all possible electronic states of the M-system by one state (summation over μ is replaced by $\mu = 0$). To obtain this form from the exact wave function requires two sequential Löwdin partitioning procedures: the first one to the subspace of the states with a fixed number of electrons in the subsystems – just repeats in a simplified form of the derivation performed above; however, the second one, to the states with the ground state wave function of the free M-subsystem as the multiplier, is somewhat more involved.

The summation over n is eliminated by the projection operator $\hat{P}$ and its complementary projection operator $\hat{Q} = \hat{I} - \hat{P}$. The operator $\hat{P}$ projects to the N-electron states with N_R electrons in the R-system. By acting on the states of eq. (1.227) it cuts off all the terms with electron distributions different from the required one. The general technique with the projection operator $\hat{P}$ in the previous section leads to the effective Hamiltonian with the intersubsystem electron hopping $\hat{W}^r$ projected out:

$$
\begin{aligned}
\hat{H}^{\text{eff}}(E) &= \hat{P}\hat{H}_R\hat{P} + \hat{P}\hat{H}_M\hat{P} + \hat{P}\hat{W}^c\hat{P} + \hat{P}\hat{W}^{rr}(E)\hat{P} \\
\hat{W}^{rr}(E) &= \hat{W}^r\hat{Q}(E - \hat{Q}\hat{H}_0\hat{Q})^{-1}\hat{Q}\hat{W}^r
\end{aligned}
\tag{1.232}
$$

and the unperturbed Hamiltonian

$$
\hat{H}^0 = \hat{H}_R + \hat{H}_M + \hat{W}^c
\tag{1.233}
$$

commuting with the operator $\hat{N}_R$ of the number of particles in the R-system (and/or equivalently with $\hat{N}_M$ since $\hat{N} = \hat{N}_M + \hat{N}_R, [\hat{N}, \hat{H}^0] = 0$). By this the exact Hamiltonian acting in the entire space of N-electron functions is projected to the subspace $\text{Im}\hat{P}$ and the intersubsystem electron transfers which broke the separation of electronic variables are taken into account by the effective interaction $\hat{W}^{rr}(E)$ containing the resolvent of the operator $\hat{H}^0$ in the subspace $\text{Im}\hat{Q}$.

1.7.2.2. Effective Hamiltonian for the R-system

As mentioned earlier, it is highly desirable to get rid of the energy dependence of the effective Hamiltonians describing the subsystems. In order to do so we reconsider the general derivation of an effective Hamiltonian and specify it for the R-system.

To get the effective Hamiltonian for the R-system which is necessary to calculate Φ_ρ^R and the corresponding ground and excited state energies, we consider contributions to the effective Hamiltonian eq. (1.232). It is important from the point of view of the further separation of the Hamiltonians into unperturbed parts and perturbations. The bare Hamiltonians for the R-system $\hat{H}_R$ and for the M-system $\hat{H}_M$ defined by eq. (1.224) on the basis of attribution of the fermi-operators to the R- and M-systems turn out to be not a good starting point for developing a perturbational picture as the

Hamiltonians thus defined contain large one-electron terms describing the attraction of electrons to the unscreened atomic cores in an "alien" subsystem:

$$\hat{H}_M = \hat{H}_M^0(q) + \mathcal{V}_R(q)$$

$$\hat{H}_R = \hat{H}_R^0(q) + \mathcal{V}_M(q)$$

$$(1.234) \quad \mathcal{V}_M = -e^2 \sum_A \frac{Z_A^M}{|r - R_A|} = -\sum_{\substack{A \\ rr'}} r^+ r' W_{rr'}^A Z_A^M$$

$$\mathcal{V}_R = -e^2 \sum_A \frac{Z_A^R}{|r - R_A|} = -\sum_{\substack{A \\ mm'}} m^+ m' W_{mm'}^A Z_A^R$$

The real physical situation is much better described by the bare operators where expectation values of the Coulomb operator: $\langle\langle \hat{W}^c \rangle\rangle_R$ and $\langle\langle \hat{W}^c \rangle\rangle_M$ ensure the screening of the "alien" core charges in the effective Hamiltonians eq. (1.224):

$$\delta\hat{V}_M = \mathcal{V}_M + \left\langle\left\langle \hat{W}^c \right\rangle\right\rangle_M ,$$

$$\delta\hat{V}_R = \mathcal{V}_R + \left\langle\left\langle \hat{W}^c \right\rangle\right\rangle_R$$

$$(1.235)$$

$$\hat{H}_M^{\text{eff}} = \hat{H}_M^0 + \delta\hat{V}_R + \left\langle\left\langle \hat{W}^{rr} \right\rangle\right\rangle_R$$

$$\hat{H}_R^{\text{eff}} = \hat{H}_R^0 + \delta\hat{V}_M + \left\langle\left\langle \hat{W}^{rr} \right\rangle\right\rangle_M$$

Using the definitions of the bare Hamiltonians in eq. (1.234) and of the effective Hamiltonians for the subsystems in eq. (1.235), we get an alternative break down of the effective Hamiltonian for the R-system:

$$(1.236) \quad \hat{H}_R^{\text{eff}} = \hat{H}_R^0 + \delta\tilde{V}_M$$

and analogously for the M-system:

$$(1.237) \quad \hat{H}_M^{\text{eff}} = \hat{H}_M^0 + \delta\tilde{V}_R$$

In the above expressions:

$$(1.238) \quad \begin{aligned} \delta\tilde{V}_M &= \delta\hat{V}_M + \langle\langle \hat{W}^{rr} \rangle\rangle_R \\ \delta\tilde{V}_R &= \delta\hat{V}_R + \langle\langle \hat{W}^{rr} \rangle\rangle_M \end{aligned}$$

which must be considered as perturbations to the operators of the free subsystems.

In the frame of the target hybrid QM/MM procedure, only the electronic structure of the R-system is calculated explicitly. For this reason, we consider its effective Hamiltonian eq. (1.235) in more detail. It contains the operator terms coming from (1) the Coulomb interaction of the effective charges in the M-system with electrons in the R-system $\delta\hat{V}_M$ and (2) from the resonance interaction of the R- and M-systems.

The expectation value of the Coulomb interaction between electrons of the subsystems over the ground state of the M-system is most easy to find:

$$(1.239) \quad \langle\langle \hat{W}^c \rangle\rangle_M = \langle \Phi_0^M \mid \hat{W}^c \mid \Phi_0^M \rangle$$

Inserting the explicit form of the Coulomb interaction operator yields:

$$(1.240) \quad \begin{aligned} \langle\langle \hat{W}^c \rangle\rangle_M &= \sum_{rr'} r^+ r' \big[\sum_{mm'} (rr' \parallel mm') \langle\langle m^+ m' \rangle\rangle_M \big] \\ \delta \hat{V}_M &= \sum_{rr' \in R} \left\{ \sum_{mm' \in M} (rr' \parallel mm') \langle\langle m^+ m' \rangle\rangle_M - W_{rr'}^A Z_M^A \right\} r^+ r' \end{aligned}$$

One-electron transfers between the subsystems finally contribute to the effective Hamiltonian the following energy dependent term:

$$(1.241) \quad \hat{W}^{rr}(E) = \sum_{rmr'm'} w_{rm} w_{r'm'} \times [r^+ m \hat{R}(E) m'^+ r' + m^+ r \hat{R}(E) r'^+ m']$$

The resolvent $\hat{R}(E)$ entering it can be written as:

$$(1.242) \quad \hat{R}(E) = \sum_{i \in \mathrm{Im}\hat{Q}} \frac{|i\rangle\langle i|}{E - E_i}.$$

The above resolvent operator $\hat{R}(E)$ refers to the operator $\hat{H}^0$ including only the Coulomb interaction between the subsystems. Its poles E_i are those eigenvalues of $\hat{H}^0$ which differ from those in the subspace Im $\hat{P}$ by transfers of one electron between the M- and R-systems. We denote these states as $|\rho \rightarrow \mu\rangle$ or $|\mu \rightarrow \rho\rangle$ with respect to the direction of the transfers. The energies in the expression in eq. (1.242) are defined by the ionization potentials I_μ, I_ρ and electron affinities A_ρ, A_μ of the subsystems:

$$E_i = \begin{cases} I_\mu - A_\rho - g_{\mu\rho} \\ I_\rho - A_\mu - g_{\rho\mu} \end{cases}$$

and by the quantities $g_{\mu\rho} = g_{\rho\mu}$ which are the Coulomb interactions of an electron and a hole in the R- and M-systems. The contribution to the effective Hamiltonian for the R-system appears after averaging eq. (1.241) over the ground state of the M-system:

$$(1.243) \quad \begin{aligned} \langle\langle \hat{W}^{rr}(E) \rangle\rangle_M = &\sum_{rr'} \sum_{mm'} w_{rm} w_{r'm'} \times \\ &\times \left\{ r^+ \langle\langle m \hat{R}(E) m'^+ \rangle\rangle_M r' + r \langle\langle m' \hat{R}(E) m^+ \rangle\rangle_M r'^+ \right\} \end{aligned}$$

The idea of chemical nonactivity of the M-system assumes among other features that the energies of the states with electrons transferred between the subsystems (the poles of the resolvent eq. (1.242)) are much larger than the energies of the complex system which are of interest to us. For that reason in order to estimate the effective Hamiltonian eq. (1.232) one may set $E = 0$. By this we immediately arrive at the

Raleigh-Schrödinger perturbation theory and get rid of the energy dependence of the Hamiltonian as desired. Then the expression eq. (1.243) takes the form:

$$\langle\langle \hat{W}^{rr} \rangle\rangle_M = \sum_{rr'} \sum_{mm'} w_{rm} w_{r'm'} \times$$

$$(1.244) \qquad \times \left\{ \sum_{\rho \in \mathrm{Im}\mathcal{O}_R(N_R+1)} r^+ |\rho\rangle\langle\rho| r' G_{mm'}^{(adv)}(A_\rho) + \right.$$

$$\left. + \sum_{\rho \in \mathrm{Im}\mathcal{O}_R(N_R-1)} r |\rho\rangle\langle\rho| r'^+ G_{mm'}^{(ret)}(I_\rho) \right\}$$

where $G^{(ret)}(\epsilon)$ and $G^{(adv)}(\epsilon)$ are the advanced and retarded one-electron Green's function of the M-system, respectively, written in the basis of one-electron states of the M-system [39, 61–64]. The cases when such corrections assume crucial importance will be considered later.

The approximate ground state of the form in eq. (1.231) in the $\mathrm{Im}\hat{P}$ subspace is sought in the self consistent approximate form

$$(1.245) \quad \Psi_0 = \Phi_0^R \wedge \Phi_0^M$$

where the functions Φ_0^R and Φ_0^M satisfy the system of interconnected eigenvalue equations with the effective Hamiltonians for the respective subsystems:

$$\hat{H}_R^{\mathrm{eff}} \Phi_0^R = E^R \Phi_0^R$$

$$\hat{H}_M^{\mathrm{eff}} \Phi_0^M = E^M \Phi_0^M$$

$$(1.246) \quad \hat{H}_R^{\mathrm{eff}} = \hat{P}\hat{H}_R\hat{P} + \langle\Phi_0^M | \hat{P}\hat{W}_{\mathrm{RM}}\hat{P} | \Phi_0^M\rangle$$

$$\hat{H}_M^{\mathrm{eff}} = \hat{P}\hat{H}_M\hat{P} + \langle\Phi_0^R | \hat{P}\hat{W}_{\mathrm{RM}}\hat{P} | \Phi_0^R\rangle$$

$$\hat{P}\hat{W}_{\mathrm{RM}}\hat{P} = \hat{P}\hat{W}^c\hat{P} + \hat{P}\hat{W}^{rr}(E=0)\hat{P}$$

Averaging the interaction operators in eq. (1.246) – they are both two-electronic ones – over the ground states of each subsystem does not touch the fermi-operators of the other subsystem. The averaging of the two-electron operators $\hat{P}\hat{W}^c\hat{P}$ and $\hat{P}\hat{W}^{rr}\hat{P}$ yields the one-electron corrections to the bare subsystem Hamiltonians. The wave functions Φ_0^R and Φ_0^M are calculated in the presence of each other. The effective operator $\hat{H}_R^{\mathrm{eff}}$ describes the electronic structure of the R-system in the presence of the medium, whereas $\hat{H}_M^{\mathrm{eff}}$ describes the medium in the presence of the R-system.

1.7.2.3. Electronic structure and spectrum of R-system

Wave function of electrons in quantum R-system Φ_0^R satisfies the Schrödinger equation with the effective Hamiltonian $\hat{H}_R^{\mathrm{eff}}$ eq. (1.246), which is obtained by averaging the interaction operators in eq. (1.232) over the ground state of the M-system, i.e. over Φ_0^M, and acts on the quantum numbers (variables) of electrons in the R-system.

In the frame of the hybrid methods it must be computed by a QM method. The Schrödinger equation with the effective Hamiltonian $\hat{H}_R^{\text{eff}}$ has multiple solutions, which describe excited states of the R-system provided the M-system is frozen in its ground state. Electronic energy of the system in the state expressed by the wave function eq. (1.231), has the form [29, 30]:

$$(1.247) \quad \mathcal{E}_\rho = E_\rho^R + E^M$$

where

$$(1.248) \quad E_\rho^R = \left\langle \Phi_\rho^R \left| \hat{H}_R^{\text{eff}} \right| \Phi_\rho^R \right\rangle ; E^M = \left\langle \Phi_0^M \left| \hat{H}_M \right| \Phi_0^M \right\rangle$$

In the given expression the characteristics of the subsystems enter asymmetrically. The quantities E_ρ^R are the electronic energies of the R-system, i.e. the eigenvalues of its effective Hamiltonian. The quantity E^M is the expectation value of the bare Hamiltonian H_M, containing the nonscreened potential induced by the nuclei of the R-system although the averaging must be performed over Φ_0^M – eigenfunction of the effective Hamiltonian for the M-system, where the core's potential induced by the R-system is screened by its electrons and it cannot be interpreted as the electronic or any other energy of the M-system. These inconsistencies will be addressed later. Despite them and irrespective of the value of E^M the differences $E_\rho^R - E_0^R$ can be considered as estimates for the energies of the excited states localized in the R-system.

1.7.2.4. Electronic structure of M-system in QM/MM methods

The description of the electronic structure of the complex molecular system given by the system eq. (1.246) is perfectly sufficient when it goes about the hybrid QM/QM methods, when both the parts of the complex system are described by some QM methods. In the case of the hybrid methods in a narrow sense i.e. of the QM/MM methods, further refinements are necessary. The problem is that the description provided by eq. (1.246) suffers from the need to calculate the expectation values in these expressions over the wave function Φ_0^M i.e. over the solution of the self-consistency equations eq. (1.246) in the presence of the R-system. This result does not seem to be particularly attractive since the functions Φ_0^M are not known and are not supposed to be calculated in the frame of the MM procedure. Thus the theory must be reformulated in a spirit of the theory of intermolecular interactions [67] and to express necessary quantities in terms of the *observable* characteristics of *free* parts of the complex system.

The reformulation of the theory of interaction between the R- and M-systems in terms of observables pertinent to the M-system assumes certain procedure for evaluating either the wave functions Φ_0^M or directly the necessary expectation values taken over it. To do so, we notice that according to eq. (1.235) the effective Hamiltonian $\hat{H}_M^{\text{eff}}$ for the M-system in the presence of R-system, defining Φ_0^M is close to the Hamiltonian $\hat{H}_M^0$ for the free M-system. The assumption that the M-system is inert implies that its characteristic excitation energies are large, thus the reduced interactions $\delta\hat{V}_R$ and $\delta\tilde{V}_R$ treated as perturbations can be thought to be small ones with respect to characteristic excitation energies of the Hamiltonian H_M^0. Then we can write for the wave function Φ_0^M in the first order of the perturbation theory:

$$(1.249) \quad \left| \Phi_0^M \right\rangle = \left| \Phi_{00}^M \right\rangle - \sum_{\mu \neq 0} \frac{\left| \Phi_{0\mu}^M \right\rangle \left\langle \Phi_{0\mu}^M \left| \delta\tilde{V}_R \right| \Phi_{00}^M \right\rangle}{E_{0\mu}^M - E_{00}^M}$$

where Φ_{00}^M is the ground state eigenfunction of the Hamiltonian $\hat{H}_M^0$ for the free M-system, $\Phi_{0\mu}^M$ are the eigenfunctions of its excited states $E_{0\mu}^M$ and E_{00}^M are the corresponding eigenvalues. Employing eq. (1.249) yields, according to Section 3.8, the one-electron densities in the M-system perturbed by the R-system:

$$(1.250) \quad \langle\langle \mathrm{m}^+\mathrm{m}' \rangle\rangle_M = \underbrace{\left\langle \Phi_{00}^M \left| \mathrm{m}^+\mathrm{m}' \right| \Phi_{00}^M \right\rangle}_{=\langle\langle \mathrm{m}^+\mathrm{m}' \rangle\rangle_M^{(0)}} -$$

$$- 2 \sum_{\mu \neq 0} \frac{\left\langle \Phi_{00}^M \left| \mathrm{m}^+\mathrm{m}' \right| \Phi_{0\mu}^M \right\rangle \left\langle \Phi_{0\mu}^M \left| \delta\tilde{V}_R \right| \Phi_{00}^M \right\rangle}{E_{0\mu}^M - E_{00}^M}$$

where $\langle\langle \mathrm{m}^+\mathrm{m}' \rangle\rangle_M^{(0)}$ is the density matrix element for the free M-system. These values have to be used when estimating the effective Hamiltonian for the R-electrons eq. (1.235) and for the energy of the entire system.

1.7.2.5. Potential energy surface for combined system

Having obtained a sufficient impression of the electronic structure of the medium (M-system) we can address the PES of the complex system. From the point of view of substantiation of the hybrid methods, it is the main problem [65]. The entire PES $\mathcal{U}(q)$ of the molecular system is the sum of its electronic energy $\mathcal{E}_0(q)$ given by eq. (1.247) and of the Coulomb repulsion $U(q)$ of nuclei (or cores):

$$(1.251) \quad \mathcal{U}(q) = \mathcal{E}_0(q) + U(q)$$

Employing the distribution eq. (1.230) of the nuclear (core) charges between the subsystems eq. (1.251) the repulsive contribution to the energy can be written:

$$U(q) = U_{\mathrm{RM}} + U_{\mathrm{RR}} + U_{\mathrm{MM}},$$

$$U_{\mathrm{RM}} = \sum_{A,B} Z_A^R Z_B^M \Gamma_{AB},$$

$$(1.252) \quad U_{\mathrm{MM}} = \frac{1}{2} \sum_{A,B} Z_A^M Z_B^M \Gamma_{AB},$$

$$U_{\mathrm{RR}} = \frac{1}{2} \sum_{A,B} Z_A^R Z_B^R \Gamma_{AB}$$

In the ZDO approximation (see below) common for semiempirical methods one can set:

(1.253) $\quad \Gamma_{AB} = (1 - \delta_{AB})\gamma_{AB}$

which formally excludes the interaction of the fractions of the core (or nuclear) charges attributed to the different subsystem ($\Gamma_{AA} = 0$). The same applies to nuclear charges provided γ_{AB} are set R_{AB}^{-1} in the above definition.

To get the PES of a complex system we supply the electronic energy $\mathcal{E}_0(q)$ with the sum of the core repulsions eq. (1.252) which yields the required PES of the complex system:

$$\mathcal{U} =$$

$$= U_{\mathrm{RR}} + \qquad (1)$$

$$+ \langle \Phi_0^R | H_R^0 | \Phi_0^R \rangle + \langle\langle \hat{W}^{rr} \rangle\rangle + \quad (2)$$

(1.254)

$$+ \langle\langle \hat{W}^c \rangle\rangle + \langle \Phi_0^R | \mathcal{V}_M | \Phi_0^R \rangle + \quad (3)$$

$$+ \langle \Phi_0^M | \mathcal{V}_R | \Phi_0^M \rangle + U_{\mathrm{RM}} + \quad (4)$$

$$+ \langle \Phi_0^M | \hat{H}_M^0 | \Phi_0^M \rangle + U_{\mathrm{MM}} \quad (5)$$

This expression is the starting point for further analysis. In it the sum of rows (2) and (3) is nothing but the expectation value $\langle\langle H_R^{\mathrm{eff}} \rangle\rangle_R$ of the effective Hamiltonian for the R-system eq. (1.235) over its ground state. This is the quantity normally calculated by the QM modeling packages. It comprises a significant fraction of the intersubsystem interaction, but only a fraction: namely the interaction of *electrons* of the R-system with the entire *charge distribution* of the M-system. By contrast the sum of the rows (3) and (4) represents the total energy of the Coulomb interaction E^{coul} between the *charge distributions* of the subsystems (in the ZDO approximation the interaction of the charge distributions reduces to the interaction of the *effective charges*). Row (4) does not depend on the wave function of electrons of the R-system. It (as also row (1) – core repulsion in the R-system) can be added to the effective Hamiltonian for the R-system without changing the (ground state) electronic wave function. At the same time the row (4) represents a strong interaction of the cores of the R-system with the *charge distribution (effective charges)* of the M-system. The interactions thus distributed do not provide any basis for systematization. Below we shall regroup them in order to obtain a more physical picture of the PES of a complex system.

Now with the precision up to the second order in the small perturbations we can write the estimate for the energy of the M-system:

$$\langle \Phi_0^M | \hat{H}_M^0 | \Phi_0^M \rangle = \underbrace{\left\langle \Phi_{00}^M | \hat{H}_M^0 | \Phi_{00}^M \right\rangle}_{= E_{00}^M} +$$

(1.255)

$$+ \sum_{\mu \neq 0} \frac{\left\langle \Phi_{00}^M | \delta \tilde{V}^R | \Phi_{0\mu}^M \right\rangle \left\langle \Phi_{0\mu}^M | \delta \tilde{V}^R | \Phi_{00}^M \right\rangle}{E_{0\mu}^M - E_{00}^M}$$

Inserting the estimates for the densities $\langle\langle m^+m'\rangle\rangle_M$ obtained above using perturbation theory in the expression for the average Coulomb interaction between the electronic distributions of the subsystems we get:

$$\mathcal{U} =$$

$$U_{\mathrm{RR}} + \tag{1}$$

$$+ \langle\Phi_0^R|H_R^0|\Phi_0^R\rangle + \langle\langle\hat{W}^{rr}\rangle\rangle^{(0)} + \tag{2}$$

$$+ \langle\langle\hat{W}^c\rangle\rangle^{(0)} + \langle\Phi_0^R|\mathcal{V}_M|\Phi_0^R\rangle + \tag{3}$$

$$(1.256) \qquad + \langle\langle\mathcal{V}_R\rangle\rangle^{(0)} + U_{\mathrm{RM}} + \tag{4}$$

$$+ E_{00}^M(q) + U_{\mathrm{MM}} + \tag{5}$$

$$- \sum_{\mu\neq 0} \frac{\left\langle\Phi_{00}^M\left|\delta\tilde{V}^R\right|\Phi_{0\mu}^M\right\rangle\left\langle\Phi_{0\mu}^M\left|\delta\tilde{V}^R\right|\Phi_{00}^M\right\rangle}{E_{0\mu}^M - E_{00}^M} \tag{6}$$

This allows some interpretation. Row (1) is the core or nuclear repulsion in the R-system. It is a c-number which does not affect the wave function of the R-system. The sum of the rows (2) – (4) represents the expectation value of the effective Hamiltonian for the R-system (including other c-numbers not affecting the wave functions of the R-system) obtained using the electronic distribution of the *free* M-system. Row (5) is nothing but the PES of the free M-system (the sum of its electronic ground state energy and of the repulsion of the respective parts of the cores). One can hope that this part can be parametrized in the MM form (see below). Finally, row (6) is the second order correction to the energy of the M-system appearing due to the electronic polarization in the M-system due to interaction with the R-system. Employing for the sake of simplicity only the operator $\delta\hat{V}_R$:

$$(1.257) \qquad \delta\hat{V}_R = \sum_{mm'\in M}\left\{\sum_{rr'\in R}(rr'\|mm')\langle\langle r^+r'\rangle\rangle_R - W_{mm'}^A Z_R^A\right\}m^+m' \approx$$

$$\approx \sum_{m\in B}m^+m\sum_{r\in A}\gamma_{AB}Q_A$$

and the approximate equality (equivalent to using the ZDO scheme for the two-center Coulomb integrals – see below) we get from row (6) the correction to the energy of the Coulomb interaction of the effective charges in the R-system:

$$(1.258) \quad \sum_{AA'}Q_A^R Q_{A'}^R \sum_{BB'}\gamma_{AB}\gamma_{A'B'} \sum_{\substack{m\in B \\ m'\in B'}} \Pi_{mmm'm'}^M(0)$$

describing the weakening of their interaction due to interaction between the polarizations induced by these charges in the M-system. The polarization propagator of the free M-system entering the answer is defined as:

$$(1.259) \quad \Pi^M_{mnkl}(\omega) = \sum_{\mu \neq 0} \langle \Phi^M_0 | m^+ n | \Phi^M_\mu \rangle (\omega - \epsilon_\mu)^{-1} \langle \Phi^M_\mu | k^+ l | \Phi^M_0 \rangle$$

This correction must be included if the variation of the electronic density in the M-system of the complex system as compared to the free M-system is not considered explicitly (see below).

The above comprises the derivation of the expression for the PES of the complex system which is not only free from the necessity to recalculate the wave function of the classical subsystem in each point, but formally not requiring any wave function of the M-system at all, since the result is expressed in terms of the generalized observables – one-electron Green's functions and the polarization propagator of the free M-system. Reality is of course more harsh as the necessary quantities must be known for a system we know too little about, except the initial assumption that its orbitals do exist. Section 3.5 will be devoted to reducing this uncertainty.

The obtained explicit form will be used to analyze the strengths and weaknesses of the existing hybrid methods and pave routes to new ones.

1.7.2.6. Dispersion correction to PES of complex system

Estimates of the electronic energy of the complex system employed in the expressions eqs. (1.254), (1.256) for its PES can be further improved. For this let us notice that the solutions of the self consistent system eq. (1.246) are used as multipliers in the basis functions eq. (1.216) of the subspace $\mathrm{Im}\hat{P}$. It turns out that the effective Hamiltonian H^{eff} eq. (1.232) has nonvanishing matrix elements between the ground state of eq. (1.246) and the basis product states of the subspace $\mathrm{Im}\,\hat{P}$, differing from it by two multipliers simultaneously: by the wave function for the R-system and by that for the M-system ($\Phi^R_\rho \wedge \Phi^M_\mu$, $\rho, \mu \neq 0$). Indeed:

$$(1.260) \quad \begin{aligned} \left\langle \Phi^R_0 \wedge \Phi^M_0 \left| \hat{H}_R + \hat{H}_M + \hat{W}_{\mathrm{RM}} \right| \Phi^R_0 \wedge \Phi^M_\mu \right\rangle = \\ \left\langle \Phi^R_0 \wedge \Phi^M_0 \left| \hat{H}_R + \hat{H}_M + \hat{W}_{\mathrm{RM}} \right| \Phi^R_\rho \wedge \Phi^M_0 \right\rangle = 0 \end{aligned}$$

whereas

$$(1.261) \quad \left\langle \Phi^R_0 \wedge \Phi^M_0 \left| \hat{H}_R + \hat{H}_M + \hat{W}_{\mathrm{RM}} \right| \Phi^R_\rho \wedge \Phi^M_\mu \right\rangle \neq 0$$

These matrix elements result in an additional energy correction which can be taken into account by the moves similar to those used when we took into account the interactions of the states with the fixed electron distribution with the states with the charge transfers between the subsystems. As previously, we consider the projection operator $\mathcal{P}$ on the "single configuration" ground state of the complex system:

$$(1.262) \quad \mathcal{P} = \left| \Phi^R_0 \wedge \Phi^M_0 \right\rangle \left\langle \Phi^R_0 \wedge \Phi^M_0 \right|$$

and the complementary projection operator $Q = 1 - P$ on the subspace orthogonal to it. Inserting the projection operators P and Q into general expressions for the effective Hamiltonian acting in the one-dimensional subspace Im P, yields the expression

$$\mathcal{H}^{\text{eff}}(\omega) = P\hat{H}^{\text{eff}}P + P\hat{H}^{\text{eff}}Q\mathcal{R}(\omega)Q\hat{H}^{\text{eff}}P,$$

$$\mathcal{R}(\omega) = (\omega Q - Q\hat{H}^{\text{eff}}Q)^{-1}$$

The first term equals the energy $\mathcal{E}_0$ multiplied by the ground state projection operator P. The second gives the correction to it. Different terms in $\hat{H}^{\text{eff}}$ behave differently under this projection. Taking into account that P projects to the product of the *eigenstates* of the operators $\hat{H}_R^{\text{eff}}$ and $\hat{H}_M^{\text{eff}}$ eq. (1.246) one can see that:

$$(1.263) \quad P\hat{H}^{\text{eff}}Q = P\mathcal{W}_{\text{RM}}Q$$

where

$$(1.264) \quad \mathcal{W}_{\text{RM}} = \hat{W}_{\text{RM}} - \langle\langle\hat{W}_{\text{RM}}\rangle\rangle_R - \langle\langle\hat{W}_{\text{RM}}\rangle\rangle_M$$

is the operator of reduced interaction of the R- and M-systems where:

$$(1.265) \quad \mathcal{W}_{\text{RM}} = \mathcal{V}^c + \mathcal{V}^{rr}$$
$$\mathcal{V}^c = \hat{W}^c - \langle\langle\hat{W}^c\rangle\rangle_R - \langle\langle\hat{W}^c\rangle\rangle_M$$
$$\mathcal{V}^{rr} = \hat{W}^{rr} - \langle\langle\hat{W}^{rr}\rangle\rangle_R - \langle\langle\hat{W}^{rr}\rangle\rangle_M$$

where the expectation values of the interaction operators calculated for the ground states of the respective subsystems are subtracted (i.e. the interactions are reduced by their expectation values). With these notions the correction acquires the form:

$$(1.266) \quad P\mathcal{W}_{\text{RM}}\mathcal{R}(\omega)\mathcal{W}_{\text{RM}}P$$

As before the idea of inertness of the M-system is formalized by the assumption that the excitation energies in it are large compared to the excitation energies in the R-system, which is of interest to us. For this reason one can guess that the dependence of the resolvent on ω is weak and that the values of ω in the interesting energy range are much smaller than the resolvent poles which are all lying not lower than the first excitation energy in the M-system. These notions allow us to replace the resolvent $\mathcal{R}(\omega)$ by its value at $\omega = 0$ (by this the electronic dynamic effects in the M-system are excluded):

$$(1.267) \quad \mathcal{R}(0) = -\sum_{\rho,\mu \neq 0} \frac{\left|\Phi_\rho^R \wedge \Phi_\mu^M\right\rangle\left\langle\Phi_\rho^R \wedge \Phi_\mu^M\right|}{E_\rho^R + E_\mu^M}$$

where E_ρ^R and E_μ^M are excitation energies in the R- and M-systems. Following [30] and [67] we get

$$(1.268)$$
$$\mathcal{R}(0) = \sum_{\rho,\mu \neq 0} \frac{\left|\Phi_\rho^R \wedge \Phi_\mu^M\right\rangle\left\langle\Phi_\rho^R \wedge \Phi_\mu^M\right|}{E_\rho^R + E_\mu^M} =$$
$$= \frac{2}{\pi} \sum_{\rho,\mu \neq 0} \left|\Phi_\rho^R \wedge \Phi_\mu^M\right\rangle\left\langle\Phi_\rho^R \wedge \Phi_\mu^M\right| \int_0^\infty du \frac{E_\rho^R E_\mu^M}{\left((E_\rho^R)^2 + u^2\right)\left((E_\mu^M)^2 + u^2\right)}$$

Restricting ourselves for the simplicity in eq. (1.264) by the reduced Coulomb operator

$$(1.269) \quad \mathcal{V}^c = \sum_{\substack{mm'\in M \\ rr'\in R}} (rr'||mm')\,\{\mathsf{r}^+\mathsf{r}'\mathsf{m}^+\mathsf{m}'-$$

$$-\mathsf{r}^+\mathsf{r}'\,\big\langle\!\big\langle \mathsf{m}^+\mathsf{m}'\big\rangle\!\big\rangle_M - \mathsf{m}^+\mathsf{m}'\,\big\langle\!\big\langle \mathsf{r}^+\mathsf{r}'\big\rangle\!\big\rangle_R\}$$

we get after averaging over the ground state eq. (1.245) the following correction to the energy of the complex system:

$$(1.270) \quad \mathcal{P}\mathcal{V}^c\mathcal{R}\mathcal{V}^c\mathcal{P} = \frac{2}{\pi} \sum_{\substack{rr'\in R\ mm'\in M \\ pp'\in R\ nn'\in M}} (rr'||mm')\,(pp'||nn') \times$$

$$\times \int_0^\infty du \Pi^R_{rr'pp'}(iu)\Pi^M_{mm'nn'}(iu)$$

which is nothing but the dispersion interaction between the subsystems. Here

$$(1.271) \quad \Pi^R_{rstu}(\omega) = \sum_{\rho\neq 0}\langle\Phi^R_0\,|\mathsf{r}^+\mathsf{s}|\,\Phi^R_\rho\rangle(\omega-\epsilon_\rho)^{-1}\langle\Phi^R_\rho\,|\mathsf{t}^+\mathsf{u}|\,\Phi^R_0\rangle$$

is the polarization propagator of the R-system [30] whereas that for the M-system is defined above by eq. (1.259).

In this chapter we reviewed briefly the general theoretical techniques to be used throughout the rest of the book. Some of them have a general application, others are of relatively rare use. Using the Löwdin partition for separating electronic variables and deriving the GF form of the trial wave function is most probably original. Using it we derived general formulae for the PES of a complex molecular system comprising a chemically transforming part of a system to be treated using quantum chemistry and a chemically inert part which can be treated using molecular mechanics. Applying the general formalism of separating electron variables related to the two systems resulted in a consistent description of the PES of the combined system. It is dominated by the sum of the QM and MM contributions. The quantities usually referred to as "junctions" between the QM and MM parts of the combined system are consistently derived. It turned out that the junctions manifest themselves in renormalizations of the electronic Hamiltonian for the QC system. Respective modifications of the MM potential are at this point expected to appear only indirectly, due to variations of the one-electron density matrix of the M-system. These theoretical expressions will be used in the subsequent Chapter, where it will be employed for analysis of the meaning of the approximations and prescriptions used by different authors while constructing the hybrid QM/MM methods and paving the routes to new ones.

References

1. P.W. Atkins. *Physical Chemistry*, Oxford University Press, Oxford, 1979.
2. P.W. Atkins and J. de Paula. *Physical Chemistry*, Oxford University Press, Oxford, 2005.
3. L.D. Landau and E.M. Lifshits. *Statistical Physics. Course of Theoretical Physics*, vol. 5, Pergamon, Oxford, 1978.
4. J.M. Mayer and M. Goeppert Mayer. *Statistical Mechanics*, Wiley, New York, 1940.
5. M. Tribus. *Thermostatics and Thermodynamics.* An Introduction to Energy, Information and States of Matter, with Engineering Applications, D. Van Nostrand, Princeton, NJ, 1961.
6. L.D. Landau and E.M. Lifshits. *Mechanics. Course of Theoretical Physics*, vol. 1, Pergamon, Oxford, 1976.
7. H. Goldstein. *Classical Mechanics*, Addison-Wesley. Cambridge, MA, 1950.
8. E.B. Wilson, J.C. Decius and P.C. Cross. *Molecular Vibrations.* The Theory of Infrared and Raman Vibrational Spectra. McGraw-Hill, New York, 1955.
9. J.W. Leech. *Classical Mechanics*, Methuen, Wiley, London New York, 1964.
10. H. Eyring, S.-H. Lin and S.-M. Lin. *Basic Chemical Kinetics*, Wiley, New York, 1980.
11. P. Hanggi, P. Talkner and M. Borkovec. *Rev. Mod. Phys.*, **62**, 251, 1990.
12. B.J. Berne and M. Borkovec. *J. Chem. Soc.*, Faraday Trans, **94**, 2717, 1998.
13. H. Eyring, J. Walter and G.E. Kimball. *Quantum Chemistry*, Wiley, New York, 1946.
14. M.P. Allen and D.J. Tildesley. *Computer Simulation of Liquids.* Clarendon, Oxford, 1987.
15. A.R. Leach. *Molecular Modeling.* Principles and Applications (2 ed.). Pearson Education, 2001.
16. F. Jensen. *Introduction to Computational Chemistry.* Wiley, Chichester, 1999.
17. A. Szabo and N.S. Ostlund. *Modern Quantum Chemistry*, McGraw Hill, New York, 1989.
18. I. Mayer. *Simple Theorems, Proofs, and Derivations in Quantum Chemistry*, Kluwer/Plenum, Dordrecht/Newyork, 2003.
19. V.A. Fock. Nachala kvantovoi mekhaniki [in Russian], 1976.
20. L.D. Landau and E.M. Lifshits. *Quantum Mechanics.* Course of Theoretical Physics, vol. 3, Pergamon, London, 1977.
21. D.I. Blokhintsev. *Quantum Mechanics*, Springer, Berlin Heidelberg New York, 1964.
22. P.A.M. Dirac. *Principles of Quantum Mechanics*, Oxford University Press, London 1930.
23. J. von Neumann. *The Mathematical Foundations of Quantum Mechanics*, Princeton University Press, Princeton, 1932.
24. D. Bohm. *Quantum Theory*, Prentice Hall, New York, 1961, 1989.
25. L.C. Young. *Lectures on Calculus of Variations and Optimal Control Theory*, W.B. Saunders. Philadelphia, 1969.
26. S.T. Epstein. *The Variation Method in Quantum Chemistry*, Academic, New York, 1974.
27. L. Zülicke. *Quantenchemie. Band 1.* Deutscher Verlag der Wissenschaften, Berlin, 1973.
28. T. Kato. *Perturbation Theory for Linear Operators.* Springer, Berlin Heidelberg New York, 1966.
29. R. McWeeny and B.T. Sutcliffe. *Methods of Molecular Quantum Mechanics*, Academic, London, 1969.
30. R. McWeeny. *Methods of Molecular Quantum Mechanics*, (2 ed.) Academic, London, 1992.
31. N.F. Stepanov and V.I. Pupyshev. *Molecular Quantum Mechanics and Quantum Chemistry*, Moscow University Publishers, Moscow, 1991.
32. P.-O. Löwdin. Studies in perturbation theory. IV. Solution of eigenvalue problem by projection operator formalism, *J. Math. Phys.*, **3**, 969, 1962.
33. E.R. Davidson. Matrix Eigenvector Methods in *Methods in Computational Molecular Physics* (ed. G.H.F. Dierksen and S. Wilson) Reidel, Dordrecht, 1983.
34. I.G. Kaplan. *Int. J. Quant. Chem.*, **89**, 268, 2002.
35. D.J. Singh. *Planewaves, Pseudopotentials and the LAPW Method*, Kluwer, Norwell, MA, 1994.
36. V.A. Fock. *Z. Phys.*, **61**, 126, 1930.
37. E. Hückel. *Z. Phys.*, **70**, 204, 1931.

38. C.C.J. Roothaan. *Rev. Mod. Phys.*, **23**, 69, 1951.
39. J. Linderberg and Y. Öhrn. *Propagators in Quantum Chemistry*, Academic, New York, 1973.
40. P. Jørgensen and J. Simons. *Second Quantization-Based Methods in Quantum Chemistry*, Academic, New York, 1981.
41. P.R. Surján. *Second Quantized Approach to Quantum Chemistry*, Springer, Berlin Heidelberg New York, 1989.
42. I.G. Kaplan. Symmetry of Many-Electron Systems, Academic, New York, 1975.
43. H. Primas. Separability in many electron systems, in *Modern Quantum Chemistry. Istanbul Lectures* ed. by O. Sinanoğlu, Academic, New York, 1965.
44. I.G. Kaplan. *Molecular Interactions, Physical Picture, Computational Methods, and Model Potentials*, Wiley, Chichester, 2006.
45. R. McWeeny. *Rev. Mod. Phys.*, **35**, 335, 1960.
46. K. Ruedenberg. *Rev. Mod. Phys.*, **34**, 326, 1962.
47. J.P. Elliot and P.G. Dawber. *Symmetry in Physics*, vols. 1, 2, Macmillan, London, 1979.
48. T. Arai. *J. Chem. Phys.*, **33**, 95, 1960.
49. P.-O. Löwdin. *J. Chem. Phys.*, **35**, 78, 1961.
50. P.-O. Löwdin. *Adv. Chem. Phys.*, **2**, 207, 1959.
51. W. Kutzelnigg. *Int. J. Quant. Chem.*, **95**, 404, 2003.
52. P. Ziesche. *Int. J. Quant. Chem.*, **90**, 342, 2002.
53. A.L. Tchougréeff and I.A. Misurkin. *Dokl. AN SSSR* [in Russian], **291**, 1177, 1986.
54. R. Constanciel. In *Localization and Delocalization in Quantum Chemistry*, O. Chalvet, R. Daudel, S. Diner and J.P. Malrieu. eds., Reidel, Dordrecht, 1975.
55. P.E. Schipper. *J. Phys. Chem.*, **90**(18), 4259, 1986.
56. P.E. Schipper. *Aust. J. Chem.*, **40**, 635, 1987.
57. J.M. Jauch. *Foundations of Quantum Mechanics*, Addison-Wesley, Reading, MA, 1968.
58. S. Wilson. *Electron Correlation in Molecules*, Clarendon, Oxford, 1984.
59. G. Náray-Szabó. *Comput. & Chem.*, **24**, 287, 2000.
60. G.G. Ferenczy and J.G. Ángyán. *J. Comp. Chem.*, **22**, 1679, 2001.
61. A.B. Migdal. *Theory of Finite Fermi Systems and Properties of Atomic Nuclei* [in Russian]. Nauka, Moskva, 1965.
62. A.A. Abrikosov, L.P. Gor'kov and I.E. Dzyaloshinskii. *Methods of Quantum Field Theory in Statistical Physics* [in Russian] (2 ed.) Dobrosvet, Moscow, 1998.
63. D.J. Thouless. *The Quantum Mechanics of Many-Body System*. Academic, New York, 1972.
64. J.-P. Blaizot and G. Ripka. *Quantum Theory of Finite Systems*, The MIT Press, Cambridge, MA, 1986.
65. A.M. Tokmachev and A.L. Tchougréeff. *Int. J. Quant. Chem.*, 8. **84**, 39, 2001.
66. A.L. Tchougréeff and A.M. Tokmachev. Physical principles of constructing hybrid QM/MM procedures, In J. Maruani, R. Lefebvre, and E. Brändas, (ed), *Advanced Topics in Theoretical Chemical Physics*, vol. 12 of *Progress in Theoretical Chemistry and Physics*, Kluwer, Dordrecht, p. 207, 2003.
67. V. Magnasco and R. McWeeny. In Z.B. Maksić (ed.): *Theoretical Models of Chemical Bonding*, Part 2, Springer, Berlin Heidelberg New York, 1991.
68. J.E. Lennard-Jones. *Trans. Faraday Soc.*, **25**, 668, 1929.
69. A.L. Tchougréeff. *Phys. Chem. Chem. Phys.* 1, 1051, 1999.
70. É. Borel. *Introduction géométrique à quelques théories physiques*, Paris: Gauthier-Villars, p. 97, 1914.
71. V.I. Pupyshev. *Additional chapters of molecular quantum mechanics*, Parts 1–3. Moscow University Publishers [in Russian], 2008.

2

MODELS OF MOLECULAR STRUCTURE: HYBRID PERSPECTIVE

Abstract In this chapter we provide a hybrid perspective of the methods of molecular modeling present in the literature. The widespread viewpoint is that hybrid modeling is a rather specific, restricted field in the otherwise universal modeling realm of quantum chemistry. From this perspective, classical models of molecular potential known as Molecular Mechanics seem a completely foreign subject that has to be artificially attached to the quantum description. This presupposes the problems of the process of developing quantum-classical junctions. We take a different view of this area, based on the general scheme of variable separation as presented in the previous chapter. On that basis we analyze the entire realm of molecular modeling and arrive at a conclusion that basically all modeling methods employ – although largely implicitly – the electron variable separation. This forms the hybrid perspective of molecular modeling mentioned above. Based on it, we present a short review of the methods of quantum chemistry, including a description of the unsolved problems of semi-empirical quantum chemistry and suggest solutions to these problems, on the basis of the patterns of variable separations which are alternative to those accepted in traditional semi-empirical quantum chemistry. Finally we use the general scheme of variable separation to classify the existing methods of hybrid molecular modeling (in the narrow sense) and clarify the origins of the problems these methods face. Some ways of solving or avoiding these problems are suggested.

The theoretical tools of quantum chemistry briefly described in the previous chapter are numerously implemented, sometimes explicitly and sometimes implicitly, in ab initio, density functional (DFT), and semi-empirical theories of quantum chemistry and in the computer program suits based upon them. It is usually believed that the difference between the methods stems from different approximations used for the one- and two-electron matrix elements of the molecular Hamiltonian eq. (1.177) employed throughout the calculation. However, this type of classification is not particularly suitable in the context of hybrid methods where attention must be drawn to the way of separating the entire molecular system (eventually – the universe itself) into parts, of which some are treated explicitly on a quantum mechanical/chemical level, while others are considered classically and the rest is not addressed at all. That general formulation allows us to cover both the traditional quantum chemistry methods based on the wave functions and the DFT-based methods, which generally claim that using the wave function can be avoided due to the Hohenberg-Kohn "existence theorem". Another aspect of the description of the modeling methods given in this chapter is that of elucidating the true (minimal) set of variables involved in these methods. This helps to establish interrelations between different methods and recognize the way in which the variables characteristic of one group of methods (say QM ones) correspond to those characteristic of another group of the methods (say classical or hybrid ones). Such an approach is in line with our aim of not giving a general description of all existing computational techniques at the prescription level, but rather to present a view of both their actual hybrid nature and potential "hybridizability" i.e. the possibility to use them with other "higher" or "lower" level methods.

Modeling systems which are complex, i.e. comprise nonuniform, strongly interacting parts, are almost as old as quantum chemistry itself. The boundaries of this area may seem very uncertain and flexible. Despite its modern appearance, the true history of hybrid modeling dates back to very early times. The basic reasons for that are quite obvious as in those old times, only small parts of interesting systems could be treated even by the simplest methods of quantum chemistry. The lessons of particular importance for us to learn are the ways of physical reasoning used by the researchers of those times for singling out the subsystems and the tools employed to set the border conditions upon the parts singled out. When singling out a physically relevant subsystem, the seemingly old fashioned concept of "chromophore" is, in fact, very useful. Incidentally, this concept was widely used in the early QC methods. According to the contemporary IUPAC official definition, the chromophore is an atom, or group of atoms, in the molecule that gives color to the molecule (sic!?) [1]. Nevertheless, this unusual definition unites two important aspects. One is related to the system's response to an external perturbation (by electromagnetic field): the absorption spectrum is a "scientific" expression for color. By this, the concept of chromophore is related to the experimental behavior of molecular systems. Another aspect relates to the structure of the system, understood as a localization of the excited states controlling the tentative response to the external perturbation of certain parts of the molecule. Examples of chromophores are well known from textbooks.

In the modern theory of electronic structure, the concept of chromophore is formalized in McWeeny's concept of the electronic group. Within this theory, the approximate system's electronic wave function, as we know it, is taken as an antisymmetrized product of multipliers (group functions) which can be made rather local when physically referring to isolated elements of molecular electronic structure. These elements – electronic groups – are physically identified either as conjugated π-systems or something else. Of course, these groups are not totally isolated and describing excitations (remember the group stands for a chromophore) as localized in only one of them is an idealization. Nevertheless, the effective Hamiltonian technique described in Section 1.7.1 can be employed to reduce manifestations of the intergroup interactions to renormalization terms in the effective Hamiltonians for the local groups rather than delocalization of the excited states over the entire molecular system. This allows us to interpret the response of the system to external perturbations in terms of excitations localized in the groups. The significance of hybrid modeling is that the chromophores are treated using quantum mechanics, whereas the rest of the system, which seems to be not of great interest, is assumed to be possibly tractable with the use of classical models. The reason of course in not only in the "interest" but also in the physical possibility to single out a part of the system in such a way that in the spectrum of the entire system its lower-energy (or again "interesting") part belongs to a chromophore, whereas the rest of the system remains in its ground electronic state – the only one accessible in the considered experiment whose results are to be interpreted or predicted. Such cases were intuitively quite clear to the researchers of previous days, and the most physically powerful approximate theories of chemistry seem to have been developed in this way. In subsequent sections, we concentrate on the physical conditions that allowed us to single out chromophores/electron groups specific for different, well-known methods. By this we want to demonstrate that the basic features characteristic of the hybrid methods have been widely in use throughout the entire history of quantum chemistry. With this in mind, one can consider the various classes of contemporary methods of molecular modeling.

2.1. AB INITIO METHODS

Modeling molecular structures by ab initio QC methods is based on as complete a description of electronic structures as possible.[1] The many-electron wave functions for the ground state (eigenvector) of a system $|\Phi_0\rangle$ and its corresponding energy (eigenvalue) $E_0 = \langle\Phi_0|\hat{H}_e|\Phi_0\rangle$ are to be calculated for each nuclear configuration. For this it is necessary to specify a set of orbitals (basis functions – AOs), number of electrons in the system and nuclear charges. All subsequent modeling is computer work which effectively implies calculating the matrix components $h^{(1)}$, $h^{(2)}$ of the electronic Hamiltonian, for the set of selected basis functions. Parameters of the basis sets are orbital exponents α, and basis contraction coefficients – if any. An impressive amount of work has been done till now in developing the rules for selecting the basis sets for different atoms of the Periodic Table as adjusted for efficiency in solving specific problems.

The basic flaws of using restricted basis sets in the ab initio context are well known in the literature. Following [3] restriction of the orbital basis taking place in the ab initio is the main source of errors. They range from the fundamental ones (the Heisenberg commutation relations are broken in the finite basis set theories – see e.g. [4]) through somewhat practical concerns (the Hellmann-Feynman theorem in its electrostatic form is not valid in the finite basis – see e.g. [5]) up to very practical problems known, for example, as the basis set superposition error (BSSE), which is numerically shown to almost eliminate (at least for the HFR wave function) by extending the basis. Numerous attempts at estimating the corresponding error date back to [6,7] which reduced them basically to analysis of whether the convergence of the HFR calculation was dependent on the basis completeness. For obtaining the systematic

---

[1]Sources on ab initio methods are numerous. Probably the most uptodate for the time being is [2].

precision estimates, the schemes of the basis states construction converging eventually to the complete ones, have been proposed [8, 9].

In the G2 and G3 [10, 11] theories, the Møller-Plesset perturbation theories of the 2-nd and 4-th orders are used to estimate the consequences of extending orbital basis sets by including the diffuse and polarization functions. These attempts, however, do not allow one to eliminate a systematic error of about 6 millihartree per electronic pair, which, in the frame of the G2 and G3 theories, bears the pompous name of "higher level correlation" of unknown nature. These latter are parametrized in the form:

$$-An_\beta - B(n_\beta - n_\alpha); -Cn_\beta - D(n_\beta - n_\alpha)$$

where the expansion coefficients A, B, C, D are obtained by fitting the results of calculations to the experimentally known values of formation energy. It is characteristic of these corrections that, despite some extension of the basis sets in the G3 theory as compared to the G2 theory, the order of the "higher level correlation" remains the same, whereas diminishing of the absolute and mean square deviations of the calculated (with the higher level corrections) results from the experimental data is reached by further detalization of the corrections form (in plain words: by a larger number of fitting parameters).

Similar corrections have been considered in papers [12, 13]. They have the form:

$$a_\sigma n_\sigma + a_\pi n_\pi + a_{\mathrm{pair}}(n_\sigma + n_\pi + n_{\mathrm{pair}})$$

where $n_\sigma, n_\pi, n_{\mathrm{pair}}$ are the numbers of σ-, π-bonds and lone pairs, respectively, and $a_\sigma, a_\pi, a_{\mathrm{pair}}$ – coefficients specific for each combination of the basis set used and the level of the explicit consideration of electron correlation.

A more sequential approach to the analysis of the systematic error of ab initio methods has been proposed in [14]. The same set of molecules as in [10] has been analyzed there. For this set the series of calculations using the basis sets aug-cc-pVxZ containing both polarization and diffuse functions with the number of exponents x in their respective radial parts up to x = 6 (single zeta x = 1, double zeta – DZ – x = 2, triple zeta – TZ – x = 3, etc.) and with the account of correlation effects in the range of methods from MP2 up to CCSD(T) had been performed and then fitted to the formulae [15–18]:

$$\begin{aligned}(2.1) \qquad &E(x) = E_{\mathrm{CBS}} + b\exp(-cx) \\ &E(x) = E_{\mathrm{CBS}} + b\exp(-(x-1)) + c\exp(-(x-1)^2)\end{aligned}$$

in order to define the coefficients b and c and then to get the value E_{CBS} (CBS – complete basis set) in the "complete basis" by extrapolation (here x stands for the "richness" of the basis set: 1 for single zeta, 2 for double zeta, etc.).

Results of [14] have shown that for the test set of G2 [10] the mean absolute deviation of the CBS extrapolated atomization energies from the experimental ones is *ca.* 0.5 kcal/mole, whereas in the G2 theory itself the mean absolute deviation is 1.4 kcal/mole. However, in addition to the average deviation, the G2 theory contains also a systematic error, which does not appear in the CBS limit. Thus one may assume that the "higher level correlation" are in fact the corrections to the finiteness of the employed basis set. Nevertheless, the form of the extrapolation formulae used in [14] is absolutely arbitrary and is justified by the final result only.

A rather more justified theoretical basis exists for the estimates of the corrections for the incompleteness of the angular parts of the orbital basis. It has been shown in [19,20] that the correction to the energy of an atom in the second order perturbation theory is

$$E_{2,l} = -\frac{45}{256}(l + \frac{1}{2})^{-4}\{1 - \frac{19}{8}(l + \frac{1}{2})^{-2} + O(l^{-4})\}$$

where l is the value of the azimuthal quantum number for which the angular wave functions are not already included in the AO basis. Using these formulae the precision of 10^{-4} a.u. for the ground state of the He atom has been reached in [21] by setting $l_{\max} = 6$. Analogous formulae:

$$E_{l_{\max}} = E_{\mathrm{CBS}} + \frac{B}{(l_{\max} + 0.5)^4} + \frac{C}{(l_{\max} + 0.5)^6}$$

(2.2)

$$E_{l_{\max}} = E_{\mathrm{CBS}} + \frac{B}{(l_{\max} + d)^m} + \frac{C}{(l_{\max} + d)^{m+1}} + \frac{D}{(l_{\max} + d)^{m+2}}$$

when applied in [22, 23] for estimating the CBS of the correlation energy in the MP2 approximation at the analogous methodology (fitting the results obtained in the sequence of increasing basis sets) gave the best agreement [24] with $m = 1, d = 1$, $D = 0$. However, as it can be concluded from [14] the estimates of the CBS limit obtained by extrapolating over the sequence of angular and radial parts of the AOs are considered to be interchangeable i.e. leading to the same value of E_{CBS}, rather, by construction, these two extrapolation schemes estimate the energy contribution from different parts of the orthogonal complement to the finite-dimensional L_0 in the complete L. As for AOs, the basis spanning L_0 is the product of the functions of the radial and angular parts, respectively, neither of which is complete. The extension of either of the basis sets of the above multiplier functions does not yield a complete basis set. Thus the similar numerical estimates of E_{CBS} obtained using eqs. (2.1) and (2.2) indicate a rather similar level of underestimation of the true value of energy characteristic of the complete basis by both extrapolation procedures.

Further analysis performed in [27] allowed authors to establish the convergence specifically of the correlation energy with respect to the basis size in the form:

$$(2.3) \qquad E_{\mathrm{corr}}(x) = E_{\mathrm{CBS}}^{\mathrm{corr}} + \frac{b}{x^3}$$

which integrates the contribution of the radial and angular parts omitted from the basis set. It has been also shown that the HFR energy converges exponentially with x, thus indicating the specific effect of the basis restriction on the capacity of a calculation procedure to reproduce electron correlations.

In the ab initio setting, as one used to study the effects of the finite basis set, the HFR approximation is used to build up (may be in some effective sense) only initial estimates for $\rho^{(1)}$ and $\rho^{(2)}$ which have to be further improved by various methods of taking into account electron correlations. Among them the configuration interaction in the complete active space (CAS) [28], Møller-Plesset perturbation theory (MPn) of order n, coupled clusters' [29, 30] methods must be mentioned as being most widely used. In fact, any reasonable result within the ab initio QC requires at least minimal involvement of electron correlation. In other words in the ab initio setting the cumulant of two-electron density matrix χ is never close enough to zero to be neglected. All the technical tricks invented to go beyond the HFR calculation scheme by using different forms of the trial wave function or various perturbative procedures represent in fact attempts to estimate more or less decently the second term of eq. (1.202) – the cumulant χ of the two-particle density matrix departing from the HFR solution. This allows us to pose a question on the nature of the correlations primarily addressed in the ab initio setting since in other contexts (see below) some important part of the correlation – dynamical correlation – is absorbed either by empirical parameters (HFR-based semiempirics) or by the specific form of the energy functional taken as a function of $\rho^{(1)}$ only (DFT methods). As shown in [25, 26] the slow convergence of the QC procedures with respect to orbital basis and particularly with the maximal azimuthal quantum number of the AOs involved, is intimately related to the intrinsic weakness of the CI expansion built upon a restricted basis of AOs as a tool for treating the very short range features of the electronic wave function known as electronic cusp. According to the so-called "cusp condition" (see e.g. [5]) the wave function of two electrons in close vicinity of each other when all other interactions (with nuclei and other electrons) can be neglected and only their Coulomb repulsion is important, is one of the interelectron separation r_{12}:

$$(2.4) \qquad \Psi_l^m(r_{12}) \sim \left\{ r_{12}^l \left(1 + \frac{r_{12}}{2\,(l+1)} \right) + O(r_{12}^{l+2}) \right\} Y_{lm}(\theta, \varphi)$$

where angular variables θ, φ describe the rotational motion (with the total angular momentum l and its projection m) of two electrons under consideration around the common center of mass. The standard form of the cusp condition present in the literature (see e.g. [5]) is the simplest one for $l = 0$:

$$(2.5) \qquad \Psi_l^m(r_{12}) \sim \left(1 + \frac{1}{2} r_{12} + O(r_{12}^2) \right)$$

which is the only one representing true cusp (at higher values of l the two-electron density at $r_{12} = 0$ has only higher derivatives discontinuous). For any wave function constructed from the spin-orbitals, there is no explicit dependence on the electron-electron distances so that the cusp conditions are not generally fulfilled. On the other hand, it is more or less clear that the deviation from the exact form (one with the cusp condition) of the wave function takes place in the area which nonnegligeably contributes to the total energy since the Coulomb repulsion has the singularity at $r_{12} = 0$. Thus the total energy can be expected to improve significantly if this type of correlation is taken into account explicitly, which has basically been known since the work of Hylleraas [31] on the helium atom. Incidentally, reproducing the simplest cusp for $l = 0$ is particularly difficult in the finite basis set as it requires reproducing a non-smooth function (with the different derivatives on the left and right sides of the zero). To approach the correct cusp behavior, one needs a very large number of harmonics in the AOs basis set to simulate it. It leads to extremely high computational costs for medium and large size systems due to unpleasant scalability of the required computational resources: $M^4 \div M^7$ (where M is the dimension of a space spanned by basis orbitals). On the other hand it is difficult to ascribe to the contribution of the cusp areas any clear chemical significance except some contribution to the total (or correlation) energy: no chemical process touches this range of interelectron separations.

A general theoretical point of view on the separation of electronic variables from the first glance has not much to do with the ab initio setting as the latter is usually positioned as the "exact" one. Nevertheless the technique developed in the previous chapter allows us pose several statements concerning the true status of the ab initio theories. Indeed, as we mentioned previously, a precise description in terms of the wave function is, strictly speaking, possible for the entire universe. For this reason, when an ab initio calculation is set, it is tacitly assumed that all other electrons are somehow excluded. The techniques described in the previous chapter allow us to do so. First of all we notice that selecting an M-dimensional subspace of orbitals and setting an N-electronic calculation in it actually means that (i) some projection operator cuts out a carrier subspace of dimensionality M from the complete infinitely dimensional $L^2(\mathbb{R}^3)$ Hilbert space of orbitals. (ii) This projection operator induces a projection $\hat{P}$ also in the space of wave functions of all electrons in the world such that only those functions which correspond to N electrons residing in the selected carrier subspace are in the image $\mathrm{Im}\hat{P}$ of this second projection operator. The whole story greatly resembles the GF picture with two groups where one group is that of N electrons in the carrier space and the other is that of all other electrons in the world residing in the orthogonal complement to the selected carrier space. Obviously nobody considers the "rest of the world" group explicitly in the usual ab initio calculation, but in any case the disentanglement of the electrons included in the consideration from those excluded from it can/must be formalized in the corresponding partition procedure which results in the effective Hamiltonian of the form:

$$(2.6) \qquad \hat{H}^{\text{eff}}_{\text{ab initio}}(E) = \hat{P}\hat{H}_e\hat{P} + \hat{P}\hat{H}_e\hat{Q}(E\hat{Q} - \hat{Q}\hat{H}_e\hat{Q})^{-1}\hat{Q}\hat{H}_e\hat{P}$$

$(\hat{Q} = \hat{I} - \hat{P})$. The eigenvalues of this effective Hamiltonian coincide by construction with those of the eigenvalues of the exact Hamiltonian $\hat{H}_e$. Nevertheless all the ab initio procedures reduce to seeking the eigenvalues of the *first term* $\hat{P}\hat{H}_e\hat{P}$ effectively calculating the matrix elements involved in it *only*. It is necessary to say that under these circumstances – namely under the explicit presence of one more term in truly exact expression of the effective Hamiltonian – the possibility of getting a sensible answer by considering, only one term and not taking any precaution concerning the second one seems to be at least shortsighted: everything is supposed to be done "exactly", but as only a fraction of the whole is taken into account, the result is not supposed to be "correct" as no counterpoise is provided. As one can see, the attempts in literature to analyze and eventually overcome the restrictions imposed by using the finite basis sets are developed from the direction opposite to that suggested by the general theory of variable separation, i.e. not analyzing the general expression for the effective Hamiltonian eq. (2.6) but trying to estimate the limit "from inside". It seems, however, that an attempt to obtain estimates on the basis of the general formula would be at least interesting.

When considering ab initio methods as a part of a more general hybrid technique one has to remember that no counterpoise is built in the former to overcome its inherent limitations. Combining this – "exact" method – with any inevitably empirical classical scheme for the environment raises the question of the status of the result. We shall address this problem later when reviewing the corresponding hybrid techniques.

2.2. PSEUDOPOTENTIAL METHODS AND VALENCE APPROXIMATION

Theories in a way explicitly exploiting the concept of separating electronic variables into groups, but tending to stay in the general ab initio context, deserve our particular attention. These are the pseudopotential theory and the valence approximation closely related to it, although used far beyond the scope of the ab initio setting. The idea behind this approach is simple and natural: the number of electrons involved in any chemical process is restricted from above by that of those residing in the partially filled atomic shells of all the atoms composing the system. The physical reason for this is of course the obvious fact that on one hand the electrons occupying the deep shells of the atoms do not take part in chemical events and on the other hand, the AOs from the atomic shells with a higher energy, not occupied in atoms themselves, are not readily populated when molecules are formed. By this setting the quantum chemical problem reduces to that for the valence electrons only in a limited subspace of valence orbitals. This approach is commonly termed valence approximation. However, the electrons occupying core orbitals affect those residing in the valence subspace, which has to be taken into account. This is done with the help of the concept of the pseudopotential.

The very first formulation of the pseudopotential idea is almost as old as quantum chemistry itself. It belongs to Hans Hellmann [32–35] who proposed to use the potential energy for an electron in the valence shell of an atom in the form:

$$(2.7) \qquad V_H = -\frac{Z_c}{r} + \frac{A\exp(-\kappa r)}{r}$$

where Z_c is the quantity known as core charge (see below); A and κ are some constants and the purpose of the second (pseudopotential) term is to keep the valence orbitals orthogonal to the core orbitals although these latter are not considered explicitly. Further development along these lines brought numerous refinements to the simple picture proposed by Hellmann.

Although historically pseudopotentials appear in numerous disguises on which we do not dwell here, the modern derivation of the pseudopotential theory [36] is based on the GF technique. It starts from representing the wave function of a molecular system in the form of the antisymmetrized product:

$$(2.8) \qquad \Psi = \Phi_{\text{core}} \wedge \Phi_{\text{valence}}$$

and assuming the HFR form of the wave function for the "classical" subsystem of a complex molecular system – the core. (In a given setting even a single atom sometimes can be considered to be complex enough). The many-electron nonrelativistic Hamiltonian of a given molecule with $N_c + N_v$ electrons has the standard form of eqs. (1.27) and (1.177). An assumption is that the orbitals defining the carrier space for the atomic core and the atomic valence shell appear as a result of solving the corresponding Hartree-Fock problem. This is the reason to further assume that potentially important one-electron transfers between the core and valence shell are effectively eliminated due to the Brillouin theorem [4]. This allows one to neglect them and to restrict the treatment (at this point) to only Coulomb interactions. The one-electron density entering this equation in the Hartree-Fock approximation has the form of a sum eq. (1.144) over the occupied orbitals. Then the linearity allows one to rewrite the Coulomb and exchange operators entering the Hartree-Fock problem as sums of contributions coming separately from the core and valence states:

$$\rho^{(1)} = \rho_c^{(1)} + \rho_v^{(1)}$$

$$\rho_{\text{HF}c}^{(1)}(x; x') = \sum_{i=1}^{N_c} \phi_i^*(x)\phi_i(x')$$

$$(2.9) \qquad \hat{J}_c = \hat{J}[\rho_{\text{HF}c}^{(1)}]\phi(x) = e^2 \int dx' \frac{\rho_{\text{HF}c}^{(1)}(x'; x')}{|\mathbf{r} - \mathbf{r}'|} \phi(x)$$

$$\hat{K}_c = \hat{K}[\rho_{\text{HF}c}^{(1)}]\phi(x) = e^2 \int dx' \phi(x') \frac{\rho_{\text{HF}c}^{(1)}(x; x')}{|\mathbf{r} - \mathbf{r}'|}$$

where the summation in the expression for $\rho_{\text{HF}c}^{(1)}$ extends to the core orbitals and analogous definitions are introduced for the Coulomb and exchange operators induced by the electron density residing in the valence orbitals.

Up to this point nothing changes. The next assumption extends the above treatment of atoms to molecules. Within it the molecular orbitals – linear combinations of the atomic core orbitals with zero overlap – are taken to be the molecular core orbitals and are assumed to be filled. This allows one to write

$$(2.10) \qquad \hat{J}_c = \sum_\alpha \sum_{c_\alpha \in \alpha} \hat{J}_{c_\alpha}; \hat{K}_c = \sum_\alpha \sum_{c_\alpha \in \alpha} \hat{K}_{c_\alpha}$$

where α runs over the atoms and c_α runs over the core orbitals of the α-th atom, so that the molecular core Coulomb and exchange operators become sums of the corresponding atomic core operators. At this point, the core and valence orbitals cease to be solutions of the Hartree-Fock problem and the necessary properties (the Brillouin theorem) are postulated by assumption. This allows one to make the next move and to regroup the core part of the electron-electron interactions with the nuclear attraction potentials so that:

$$(2.11) \quad -\frac{Z_\alpha}{|\mathbf{R}_\alpha - \mathbf{r}|} + \sum_{c_\alpha \in \alpha} \left(\hat{J}_{c_\alpha} + \hat{K}_{c_\alpha} \right) \to -\frac{Z_\alpha^{\text{eff}}}{|\mathbf{R}_\alpha - \mathbf{r}|} + \hat{V}_\alpha^{\text{ECP}}$$

where V_α^{ECP} is an effective core potential of atom α, and $Z_\alpha^{\text{eff}} = Z_\alpha - N_\alpha^{\text{core}}$ ($N_c = \sum_\alpha N_\alpha^{\text{core}}$ where N_α^{core} is the number of electrons residing in the core of the α-th atom). By this the molecular Hartree-Fock equation in the effective core potential (ECP) formulation becomes:

$$(2.12) \quad \left\{ -\frac{1}{2}\nabla^2 - \sum_\alpha \frac{Z_\alpha^{\text{eff}}}{|\mathbf{R}_\alpha - \mathbf{r}|} + \sum_\alpha \hat{V}_\alpha^{\text{ECP}} \right\} \phi_k(x) +$$
$$+ \left(\hat{J}_v + \hat{K}_v \right) \phi_k(x) = \epsilon_k \phi_k(x).$$

This equation is a starting point for developing a great number of further pseudopotential methods. They can be grouped into two major sets: the pseudopotential (PP) and model potential (MP) methods, according to the basic form of the $\hat{V}_\alpha^{\text{ECP}}$ operators used. Both of them have been designed to avoid explicit treatment of the atomic core orbitals, which is achieved by shifting the energies of the core orbitals so that they do not appear as occupied solutions of the corresponding ECP Hartree-Fock equation. The two approaches are based on two different equations: the Phillips-Kleinman equation for PP and the Huzinaga-Cantu equation for MP setting respectively. In the PP picture the shift is done in such a way that the core orbitals become degenerate with the valence orbital. By contrast, in the MP picture the core orbitals are shifted to be of higher energy than the valence ones. Two corresponding contributions to pseudopotentials have the general form:

$$(2.13) \quad \begin{aligned} \text{PK(PP)} : \quad & \hat{V}_{\text{PP}}^{\text{ECP}} = \sum_c (\varepsilon_v - \varepsilon_c)|\psi_c\rangle\langle\psi_c| \\ \text{HC(MP)} : \quad & \hat{V}_{\text{MP}}^{\text{ECP}} = -2\sum_c \varepsilon_c|\psi_c\rangle\langle\psi_c| \end{aligned}$$

It can be shown that the Huzinaga-Cantu equation, and thus the MP setting, can be used for molecules, provided a reasonable approximation for the core orbitals is known. It is normally true, as the concept of the atomic core is multiply confirmed by all electron calculations on smaller molecules and there is no reason to think that this picture will change in the case of larger ones.

Since in current molecular modeling tasks the Gaussian orbitals or their linear combinations are used, one can guess that they provide the explicit form of the core states. Inserting the Gaussians in the expressions for the Coulomb and exchange superoperators yields numerous approximate forms of the pseudopotentials, which can be exemplified by the formulae employed in the ab initio model potential (AIMP) [36]:

$$(2.14) \quad \sum_{c_\alpha \in \alpha} \hat{J}_{c_\alpha} \longrightarrow \frac{1}{|\mathbf{R}_\alpha - \mathbf{r}|} \sum_{c_\alpha \in \alpha} C_k \exp(-\beta_k (\mathbf{R}_\alpha - \mathbf{r})^2)$$

The coefficients C_k and exponents β_k are fitted rather than calculated from the corresponding Gaussian AO exponents and contraction coefficients of the atomic core orbitals whose effect they are assumed to represent. By contrast, the core exchange contribution to the pseudopotential is explicitly calculated over the Gaussian AOs and represented by its spectral expansion on the basis of the primitive Gaussians.

Further development evolves around the idea of taking into account the core polarizability. In general theory this corresponds to eq. (1.258). In the pseudopotential setting, this feature is reflected by adding corresponding terms to pseudopotentials. This resembles (and in fact coincides with) by its origin, the formulae describing the polarization of the M-system, but the result is represented by the classical looking terms. For example, the pseudopotential:

$$(2.15) \quad \sum_l \left\{ -\frac{Z-N}{r} + A_l \frac{\exp(-\kappa r)}{r} \right\} \Omega_l - \frac{\alpha_d}{2(r^2 + d^2)^2} - \frac{\alpha_q}{2(r^2 + d^2)^3}$$

where Ω_l is the operator projecting to the subshell with the azimuthal quantum number l, α_d and α_q are the dipole and quadrupole polarizabilities of the core respectively, and d is the radius of the core, has been proposed in [37]. This corresponds to taking into account the expectation values of the diagonal part of eq. (1.258). The operators Ω_l take care of the fact that the pseudopotential is different for subshells of the valence shell corresponding to different values of the angular momentum. The basic reason is quite clear: the valence orbitals must be orthogonal to the core orbitals. However, the orthogonality conditions formulate differently for the orbitals with different values of l: the $2s$-function must be orthogonal to the filled $1s$-core orbitals of which the pseudopotential term takes care, whereas the $2p$-functions are automatically orthogonal to the $1s^2$-core by symmetry. Further details and many examples of pseudopotentials can be found in [38].

Further uses of pseudopotentials are numerous. The most obvious (and rather widely known ones) are to continue with the PP or MP Hamiltonians for a widely understood combination of the core and valence shells and to apply standard ab initio techniques to electrons in the valence subspace only. We do no elaborate further on this as the hybrid nature of the pseudopotential methods is rather obvious from the above and its more specific applications in a narrower QM/MM hybrid context will be described later.

2.3. HARTREE-FOCK-ROOTHAAN BASED SEMIEMPIRICAL METHODS

In the previous section we briefly described the ab initio QC methods and the problems arising when they are applied to the modeling of complex systems. These problems cannot be considered as merely technical ones: even if the computer power is sufficient and the required solution of the many electron problem can be obtained by brute force, the problem of the status of the result produced by the uncertainty introduced by poorly defined junction between the quantum and classical regions may still be important. Pragmatically, however, the resource requirements may have already become prohibitively high for using the ab initio QC techniques as a tool for massive PES modeling. In this situation, the semi-empirical methods can again come into play as they have, 40 years after the pioneering works of Pople and Beveridge [39] were published, in which the CNDO and INDO parametrization were developed. However to put the whole thing in a correct (from our point of view) perspective we must mention even earlier attempts to model properties of molecular systems on the basis of analyzing the electronic structure of their "important" chromophore: namely the π-electron subsystems treated by the Hückel [40] and Pariser-Parr-Pople (PPP) [41] methods. They are particularly remarkable from the point of view of hybrid modeling as we shall see. As these methods were developed when computing capacities were quite limited, they had to profit in full from Young's recommendation [43] that "The purpose of any mathematical theory is to reduce the amount of calculation in any specific problem". The conceptual basis for that theoretical development was the wide use of the chromophore concept introduced by organic chemists.

2.3.1. π-approximation

It was observed very early that among the colored organic compounds (which are generally not very common) the most widely found are those which contain multiple double bonds separated by single bonds. Later it was realized that such an electronic structure predominantly leads to the planar geometry of the molecule under study, which in turn allows one to classify the orbitals into the σ- and π-ones, the symmetric and the antisymmetric respectively, under reflection in the mentioned plane. The excitations responsible for the observed color touch the electrons in the (antisymmetric) π-states. This results in the π-electronic approximation, which is in many respects archetypal for the entire field of hybrid modeling. The essence of the π-approximation of the electronic structure of a polyatomic molecule can be described as the approximation considering explicitly only the electrons populating the π-states formed by the $2p_z$-AOs of carbon atoms (or heteroatoms like nitrogen, oxygen, etc.) of the molecule. Clearly such an approach allows one to reduce the dimensionality of the problem significantly. In the case of carotenoid molecules (substituted linear polyenes – an important object of studies by theoretical chemists in the 30s and incidentally an important organic pigment) of about thirty carbon atoms in the main conjugation chain going to the π-approximation significantly reduces the amount of numerical work: the overall number of valence orbitals in such a system amounts to 200 of which only 30 are treated explicitly and the reduction of numerical work as estimated from the scaling of a standard diagonalization procedure which is N^3 is quite sizable. Obviously, such a treatment must be approximate and these approximations were implemented in a series of effective Hamiltonians for the π-electronic chromophore of which the Hückel Hamiltonian was the simplest:

$$(2.16) \quad H_{\text{Hückel}} = -\sum_{a,\sigma} \alpha_a a_\sigma^+ a_\sigma - \sum_{b,\sigma} \beta_b (r_{b\sigma}^+ l_{b\sigma} + h.c.),$$

where the Fermi operators $r_{b\sigma}^{+}$ and $l_{b\sigma}$ refer to the creation and annihilation of an electron on the AOs at the right and left ends of the b-th bond, respectively. The summation in the first sum extends to all atoms bearing π-AOs and describes the attraction of electrons to the core of the sp^2-hybridized carbon (or hetero-) atom. Summation in the second sum is extended to all "bonds" i.e. pairs of atoms which are considered to be immediate neighbors in the system of conjugate bonds and describes one-electron hopping between the π-orbitals. Due to different summation schemes in the two sums, the operators in these two may create/destroy electrons in the same π-AOs. As one can see, only the one-electron terms appear here and also in a very reduced form: only the orbitals centered on the nearest neighbor atoms are involved. Even this simple picture allows one to understand and numerically reproduce numerous empirical facts known from chemistry. Further refinement was to include the Coulomb interaction of electrons in the π-system into consideration explicitly. This resulted in the Pariser-Parr-Pople (PPP) Hamiltonian which has the form:

$$(2.17) \quad H_{\mathrm{PPP}} = H_{\mathrm{Hückel}} + \gamma_0 \sum_a a_\alpha^+ a_\beta^+ a_\beta a_\alpha + \frac{1}{2} \sum_{a \neq a'} \sum_{\sigma\tau} \gamma_{aa'} a_\sigma^+ a_\tau'^+ a_\tau' a_\sigma,$$

which basically reduces to adding a model Coulomb interaction term in the ZDO approximation (the first term represents repulsion of electrons with different projection of spin occupying the same π-orbital a, the second describes repulsion of electrons located on π-AOs $a \neq a'$ – see below) to the Hückel π-electron Hamiltonian.

The approximate character of the theories based on singling out π-electrons was obvious from the very beginning. On the other hand, the methods employing π-electron Hamiltonians were enormously successful. This raised significant interest in substantiation of the π-electron theories by including them in a more general context. The sequential theory was first proposed by Lycos and Parr [44] who used the group function formalism. The formal transition to the reduced description using the π-electrons was substantiated by assuming the wave function of all electrons of the system to have the form of the antisymmetrized product of the wave function of π-electrons Π and that for electrons in the σ-core Σ:

$$(2.18) \quad \Psi = \Pi \wedge \Sigma$$

This form of the wave function fixes (among other things) the number of electrons in the π-system. In variance with the general theory, the one-electron transfers between the subsystems (π- and σ-ones taking respectively the parts of R- and M-systems of the general theory) are vanishing due to the symmetry selection rules:

$$(2.19) \quad w_{\sigma\pi} = 0$$

After that, it was suggested that it is possible to think that the wave function Π is found by applying the variational principle to the energy functional E_π:

$$(2.20) \quad E_\pi = \langle \Pi | H_\pi | \Pi \rangle$$

and H_π itself is either taken in one of the possible semiempirical forms or is somehow derived from more general principles. The latter approach would have required some detailed knowledge of the σ-core wave function Σ not available at that time. Yet it had already been realized that the GF form eq. (2.18) of the wave function is an approximation even if the electron transfers between the groups are truly absent (by symmetry). Even in this case the most general form of the wave function for the complex system comprising the σ- and π-subsystems (in agreement with a general theory) is:

$$(2.21) \quad \Psi = C_1 \Pi_1 \wedge \Sigma_1 + C_2 \Pi_2 \wedge \Sigma_2 + \cdots$$

which physically corresponds to the interaction of mutually induced polarizations i.e. to dispersion interaction between the subsystems and to renormalization of the Coulomb interaction between electrons in the π-subsystem.

The π-electron approximation, as used to develop a prototype of the hybrid QM/MM setting, can also be dated to the early period. Yet in the year 1937 in the paper by Lennard-Jones [45] a primitive hybrid QM/MM construct had been proposed. This was necessary for describing details of the relation between the electronic structure and geometry of conjugated systems. The latter was presented by the interatomic distances between nearest neighbor atoms: those between which the electron hopping is taken into account in the π-electron Hamiltonian eqs. (2.16) and (2.17). The total energy of the conjugate molecule then appears as a sum of the energies of the σ- and π-subsystems:

$$(2.22) \quad E = E_\pi + E_\sigma$$

In this case the energy of the π-subsystem described by the Hückel Hamiltonian eq. (2.16) can be presented as

$$(2.23) \quad E_\pi = -4 \sum_b \beta_b P_b$$

where as in eq. (2.16) the summation goes over the nearest neighbor bonds and their orders P_b are the simplest characteristics of the electronic structure condencing the information covered in the wave function:

$$(2.24) \quad P_b = \langle \Pi | r_{b\sigma}^+ l_{b\sigma} | \Pi \rangle$$

Here the π-system is treated with a very simple, but still quantum mechanical method: e.g. by the Hückel Hamiltonian and MO LCAO approximation (which in the particular case of the Hückel Hamiltonian gives the exact answer). No explicit interaction, i.e. junction, between the subsystems was assumed at that time; however, the effects of the geometry of the classically moving nuclei were very naturally reproduced by a linear dependence of the one-electron hopping matrix elements of the bond length:

$$(2.25) \quad \beta_b = \beta_0 + \beta'(r_b - r_0)$$

the constants β_0 and β' describe the hopping integral and its derivative with respect to interatomic distance. By this relation the dependence of the effective Hamiltonian for the quantum part of the complex system on the classical part is modeled. On the other hand, the characteristics of the σ-core also depend on molecular geometry. The natural idea was to represent its energy by a quadratic function of the bond lengths (harmonic approximation):

$$(2.26) \quad E_\sigma = \frac{K_\sigma}{2} \sum_b (r_b - r_0)^2$$

where the summation is extended to all bonds between nearest neighbor atoms. Here and above, r_0 refers to some reasonable, but *hypothetical* geometry (bond length) which would occur in a system with no π-electrons; K_σ is the second derivative of the σ-core energy at that hypothetical geometry and serves as the elasticity constant for the σ-core. These parameters were originally fit to reproduce IR-spectral and structural experiments. The latter allowed Coulson and Longe-Higgins to rationalise the entire diversity of the geometry data on conjugated molecules in terms of the famous "bond-order – bond-length" rule:

$$(2.27) \quad r_b = r_0 - \omega P_b$$

where P is the matrix element of the one-electron density matrix in the π-subsystem and

$$\omega = \frac{2\beta'}{K_\sigma}$$

The theories based upon the π-electron approximation turned out to be very successful in describing both the spectral and structural data. Additionally the description of the VIS-UV electronic spectra must be mentioned which, however, requires considering at least the PPP Hamiltonian to incorporate the effects of the Coulomb repulsion of electrons in the π-subsystem. In this context, the problem of deriving and independently estimating the parameters of the effective π-electronic Hamiltonians had been addressed by K. Freed in a series of papers [46–49]. There the general method of deriving the effective multiparticle Hamiltonian for the valence shell [50] is applied to sequential derivation for the π-subsystems. Freed's method is based on the general representation of the effective Hamiltonian for the subsystem in the form of the Löwdin partitioned Hamiltonian. Toward this end, the entire space of the orbitals in accord with the general theory is classified into three subspaces: the core ones (c) to be treated as fully occupied, the valence ones (v) to be treated explicitly, and the excited ones (e) to be assumed to be always empty. The projection operator P is taken so that the $\mathrm{Im}\,P$ subspace is one where the numbers of electrons are fixed in the above c-, v-, and e-subspaces. The orthogonal projection operator $Q = 1 - P$ is then that which projects to the subspaces where the numbers of electrons in these subspaces differ from the original one. Then the resolvent part of the exact effective Hamiltonian eq. (1.133) is estimated by expanding it in the vicinity of some arbitrary one-electron Hamiltonian (not directly related to that of the problem) commuting, however, with the operators of electron numbers in the introduced subspaces. This allows us to get numerical estimates of the parameters of the PPP Hamiltonian departing from an ab initio Hamiltonian, which are satisfactorily close to those obtained empirically.

2.3.2. All-valence semiempirical methods

The π-approximation, which allowed one to address only planar "organic" molecules with the required rigor, seemed to be very restrictive. Further evolvement of quantum chemistry turned out to be strongly dependent on such an external factor as the progress of computer power. Increasing availability of computational resources allowed one to diagonalize matrices of larger dimensionality and opened the possibility of significantly extending the carrier subspace in which the quantum chemical problem is solved to that spanned by the valence AOs of all atoms comprising the molecular system. The assumption in the all-valence semiempirical methods is that the core electrons completely screen the corresponding (integer) part of the nuclear charge so that e.g. for the second row atoms where the cores are formed by the filled $1s^2$-shells the effective core charges are:

$$Z_c = Z - 2$$

In contrast to the pseudopotential methods where the Hartree-Fock method is used to construct the subset of orbitals spanning the core and valence carrier subspaces, whereas the calculation in the valence subspace can be performed at any level of correlation accounting, for the overwhelming majority of the semi-empirical methods, the electronic structure of the valence shell is described by a single determinant (HFR) wave function eq. (1.142).

Nowadays there exists an extensive sector of semiempirical methods differing by expedients of parametrizations of the HFR approximation in the valence basis, although principles of parametrization may differ as stipulated by the need to reproduce different experimental characteristics. The general description of all these methods can be summed up as attempts to construct an acceptable parametrization for as wide a selection as possible of chemical elements and possibly to compensate by that parametrization the inherent flaws of the HFR MO LCAO paradigm used to represent the molecular electronic structure. Of course, the practical implementations of this simple idea may deviate in details from the described setting. Nevertheless, the semiempirical methods are necessary for modeling complex systems and seem to be flexible enough in terms of "hybridizability" with the classical methods mentioned above. At least no conceptual problems appear like in case when someone is trying to hybridize rigorously understood ab initio methodologies with lower level methods.

The procedure of developing a semi-empirical parametrization can be generally formalized in terms of eq. (1.197). A set of experimental energies $\mathcal{E}(Cq\Gamma S)$ (here the notation eq. (1.197) is extended to cover different chemical compositions C, and molecular geometries q) is given. When a response to an external field is to be reproduced, the latter can be included in the coordinate set q. Developing a parametrization means finding a certain (sub)set of the method parameters ω (orbital exponents, various energy parameters, expressions fitting molecular integrals etc.) which minimizes the norm of the deviation vector $\delta \mathbf{E}_\omega$ with the components $\mathcal{E}(Cq\Gamma S) - E(Cq\Gamma S|\omega)$ numbered by the tuples $Cq\Gamma S$:

$$\min_\omega(\delta \mathbf{E}_\omega | M | \delta \mathbf{E}_\omega)$$

which is calculated with some positively (semi)definite metric matrix M. In this context the parameters ω refer to the parameters of semi-empirical Fock operators (see below), since the theoretical energies and electronic structure variables (ESVs) are calculated using the HFR approximation; i.e. the electronic structure of any molecular system within the semiempirical methods is described by a single Slater determinant.

Quite a number of enterprises of this sort were very successful, leading to the whole family of semi-empirical procedures used largely for describing the ground state of "organic" molecules, [51]. The situation with "inorganic" molecules, particularly those containing "metals" (what distinguishes a metal atom from a nonmetal one from the point of view of QC?) and, even more, transition metals, is much more sophisticated and will be discussed in due prescription.

2.3.2.1. Methods without interaction

Various schemes of parametrization are traditionally organized in groups according to the subsets of the two-electron integrals taken into consideration. Taking the ab initio setting as a precise starting point for developing further approximations, one may think that semi-empirical methods develop by omitting computationally demanding two-electron integrals in the carrier subspace spanned by valence AOs of the atoms composing the molecule with the pseudopotential reducing to the complete screening of electron-nuclear attraction by the core electrons. Historically the development was just the opposite of the logical structure described above. Among the earliest attempts to extend the parametrized HFR approach developed and tested on the example of conjugated organic molecules to the systems not possessing the planar geometry (symmetry), the Mulliken-Wolfsberg-Helmholtz method (MWH) [52] and the extended Hückel theory (EHT) [53] can be mentioned. These simple methods of the semi-empirical family completely ignore the electron-electron interaction matrix elements by setting:

$$h^{(2)} \equiv 0$$

Such a setting might seem to be very poor. However, the methods of this type are capable of qualitatively correct reproducing the overall form of the MOs, and frequently their relative position on the energy scale and some chemical trends while going from one element to another in a series of similar compounds. The reasons for that success (as compared to the seeming weakness of the basic assumptions) of the "methods without interaction" can be understood if one takes a somewhat different perspective of them. In fact the MWH or EHT procedures can be considered the final diagonalization in the series of the iterative diagonalizations required to solve the Hartree-Fock-Roothaan equations eq. (1.143). The diagonalization in these methods can be thought to be performed for the Fock operator dependent on the already converged density, so that if someone is in a position to directly parametrize its matrix elements – without performing the iteration procedure which ultimately serves namely this purpose – the result may be quite acceptable. The parametrization schemes used in this context were as follows – the diagonal matrix elements of the

Fock operator in the basis of AOs were set equal to ionization potential characteristic for the AO at hand:

$$F_{\mu\mu} = h^{(1)}_{\mu\mu} = -I_\mu$$

The off-diagonal ones were defined by the relation:

$$F_{\mu\nu} = h^{(1)}_{\mu\nu} = -\frac{k}{2}(I_\mu + I_\nu)S_{\mu\nu}$$

where $S_{\mu\nu}$ is the overlap integral between the basis AOs. In the context of the "theories without interaction" some other forms were proposed for the off-diagonal matrix elements of the Fock operator [54], but they did not give any decisive improvement in performance. Directly addressing the experimental quantities such as ionization potentials was decisive for the level of success achieved by these methods.

The self-consistent nature of the Fock operator is sometimes modeled in the methods without interaction by the schemes relating the ionization potentials of each AO with the overall electronic population (or effective charge) of a given atom and sometimes with the orbital populations of the AOs centered on the considered atom. This generally leads to the expressions of the form [55]:

$$F_{\mu\mu} = A_\mu Q^2 + B_\mu Q + C_\mu$$

where the AO specific constants A_μ, B_μ, and C_μ are fitted to reproduce the experimental ionization potentials. However, despite considerable success in their time, these methods are largely only of historical interest in our days.

2.3.2.2. ZDO methods

Further development of the semiempirical methods can be described as a significant extension in terms of two-electron contribution to the energy. It has been done by the so-called zero differential overlap (ZDO) approximation. Yet in the 50s, in order to reduce the computational problems while estimating the electron-electron integrals, Mulliken suggested the replacement of the products of AOs appearing under the integral sign in the definition of the overlap integral (the differential overlap) by the following expression:

$$(2.28) \quad \chi^*_\mu(\mathbf{r})\chi_\nu(\mathbf{r}) = \frac{\chi^2_\mu(\mathbf{r}) + \chi^2_\nu(\mathbf{r})}{2} S_{\mu\nu}$$

This allowed one to approximate the four-orbital integrals of electron-electron interaction by much simpler expressions:

$$(2.29) \quad (\mu\nu|\kappa\lambda) = \frac{S_{\mu\nu}S_{\kappa\lambda}}{4}[(\mu\mu|\kappa\kappa) + (\nu\nu|\lambda\lambda) + (\nu\nu|\kappa\kappa) + (\mu\mu|\lambda\lambda)]$$

The simplification is achieved due to the fact that for an integral on the left side, which may potentially involve AOs coming from four different centers and cannot be easily calculated at least for the Slater-type AOs, representation on the right is given in terms of no more than two-center quantities, which can be easily calculated for the Slater functions.

The Mulliken approximation can be used in a slightly different way, which has actually been done. The first step when solving the Hartree-Fock equation eq. (1.149) may be going to the basis of the symmetrically orthogonalized AOs (applying the Löwdin transformation $\mathbf{S}^{-\frac{1}{2}}$ to the set of the AOs). The orthogonal basis thus obtained is commonly denoted as the λ-basis and due to the variational property of the symmetrically orthogonalized AOs these latter $\{^{\lambda}\chi_{\mu}\}$ have the largest possible overlap with the original (nonorthogonal) AOs $\{\chi_{\mu}\}$ which up to a certain point allows one to use the same subscripts for labeling AOs and OAOs. In the notation introduced previously

$$^{\lambda}\chi_{\mu} = \varphi_{\mu}$$

so we shall use them on an equal footing, depending on the convenience in that or any other situation. In the λ-basis the Hartree-Fock problem acquires the form:

$$(2.30) \quad {}^{\lambda}\mathbf{F}{}^{\lambda}\mathbf{u} = \varepsilon{}^{\lambda}\mathbf{u}$$
$$^{\lambda}\mathbf{F} = \mathbf{S}^{-\frac{1}{2}}\mathbf{F}\mathbf{S}^{-\frac{1}{2}}$$
$$^{\lambda}\mathbf{u} = \mathbf{S}^{-\frac{1}{2}}\mathbf{u}$$

In the above equalities the basis functions are orthogonal (OAOs):

$$\langle {}^{\lambda}\chi_{\mu}|{}^{\lambda}\chi_{\nu}\rangle = \langle \varphi_{\mu}|\varphi_{\nu}\rangle = \delta_{\mu\nu}$$

Now, if the *differential overlaps* of the original AO basis are approximated by the formula eq. (2.28), it turns out that applying the Löwdin transformation $\mathbf{S}^{-\frac{1}{2}}$ to the set of the AOs makes the products i.e. the *differential overlaps* of the symmetrically orthogonal OAOs vanishing:

$$^{\lambda}\chi_{\mu}(\mathbf{r}){}^{\lambda}\chi_{\nu}(\mathbf{r}) = \varphi_{\mu}(\mathbf{r})\varphi_{\nu}(\mathbf{r}) = 0$$

for $\mu \neq \nu$. If these relations for the OAOs differential overlaps are inserted in the formulae for the two-electron integrals, their values on the basis of the OAOs $\{\varphi_{\mu}\}$ shall follow the condition:

$$(\mu\nu|\kappa\lambda) = \delta_{\mu\nu}\delta_{\kappa\lambda}(\mu\mu|\kappa\kappa)$$

This makes the major part of the two-electron integrals disappear, drastically simplifying the whole setting and thus serving the basis for the subsequent development, which took about thirty years. The idea was to treat the original AOs coming with atoms when a molecule's model is constructed as already symmetrically orthogonalized. It is a serious approximation; however, it had been studied theoretically [56, 57] and tested numerically [58–60] and it had been shown that the matrix elements defined with respect to the formally OAO basis set possessed some useful characteristics. First of all [56] the two-electron matrix elements are "transferable" up to the second order with respect to interatomic overlaps in the original AO basis, which can be assumed to be small. Next it had been shown (both theoretically and numerically) [57–60] that the two-electron matrix elements which are falling out in the Mulliken approximation are indeed small in the OAO basis. On the other hand, the conceptual problem is that, taking the basis AOs as implicitly orthogonal means that the AOs themselves lose their individuality: the OAOs of the same atom are obviously different in different chemical environments in different molecules and even for different molecular geometries. This is however ignored. The idea of employing original basis AOs as implicit OAOs had been numerously implemented and we review these implementations below.

CNDO methods. The simplest among the methods developed within the ZDO paradigm was the CNDO method, which uses the Complete Neglect of Differential Overlap so that

$$(2.31) \quad \chi_\mu^*(\mathbf{r})\chi_\nu(\mathbf{r}) = 0 \text{ for all } \mu \neq \nu$$

The parameters used to construct the model Fock operator's matrix elements in the OAOs basis set are as follows:

$$(2.32) \quad F_{\mu\mu} = -\chi_\mu^0 + [Q_M - P_{\mu\mu}]\gamma_{MM} + \sum_{K \neq M} Q_K \gamma_{MK}$$

$$F_{\mu\kappa} = \beta_{MK} S_{\mu\kappa} - P_{\mu\kappa}\gamma_{MK}$$

where Q_M stands for the effective charge of the atom M:

$$(2.33) \quad Q_M = P_M - Z_M = 2 \sum_{\mu \in M} P_{\mu\mu} - Z_M$$

P_M is the electronic population of the atom M; χ_μ^0 is the Pauling's electronegativity specific for the subshell (s-, p- etc.) to which the orbital μ belongs, and β_{MK} is the strength of the resonance interaction characteristic for the pair of atoms (chemical elements) M and K. Practical implementation goes further and sets

$$(2.34) \quad \beta_{MK} = \frac{1}{2}(\beta_M^0 + \beta_K^0)$$

where β_M^0 and β_K^0 are the characteristic parameters for the respective atoms. The two-center electron-electron Coulomb interaction integrals are set to be:

$$(2.35) \quad (\mu\mu|\kappa\kappa) = \gamma_{MK} = \gamma_{MK}(R_{MK})$$

for the OAOs μ and κ centered at atoms M and K, and

$$(\mu\mu|\kappa\kappa) = \gamma_{KK}$$

if both OAOs μ and κ are centered at the atom K. The parameters γ_{KK} are set specific according to the atomic type (chemical nature) of K, whereas for γ_{MK}, numerous functional forms have been proposed. Among them the Klopman-Ohno form is one of the most widely used

$$\gamma_{MK}(R) = \frac{e^2}{\sqrt{R^2 + R_{0MK}^2}}$$

with

$$R_{0MK} = \frac{e^2}{2}\left(\frac{1}{\gamma_{MM}} + \frac{1}{\gamma_{KK}}\right)$$

although the Mataga-Nishimoto form

$$\gamma_{MK}(R) = \frac{e^2}{R + R_{0MK}}$$

with

$$R_{0MK} = \frac{e^2}{2}\left(\frac{1}{\gamma_{MM}} + \frac{1}{\gamma_{KK}}\right)$$

is also used as is the original setting by Pople and Segal [39]:

$$\gamma_{MK}(R) = (s_M s_M | s_K s_K)$$

where by s_M and s_K the s-AOs of the valence shells centered on the respective atoms are meant. In this latter case the deviation of the above integral from the Coulomb law is indirectly controlled by the Slater orbital exponents ζ_M and ζ_K.

The intraatomic electron-electron interaction integrals are usually estimated following the rule:

$$(2.36) \quad \gamma_{KK} = IP_K - EA_K$$

where IP_K and EA_K are respectively the ionization potential and electron affinity of the atom K. The same experimental quantities serve also as a source for parametrizing the diagonal matrix elements of the Fock operator in terms of the Pauling electronegativity

$$(2.37) \quad \chi_\kappa^0 = \frac{1}{2}\left(IP_K + EA_K\right)$$

where the ionization potential and the electron affinity on the right are understood as those characteristic for the subshell (s-, p- etc.) to which the orbital κ belongs. Further analysis allows us to single out more fundamental parameters of the semiempirical Fockian than the Pauling electronegativity. Indeed, assuming the CNDO approximation for an isolated atom one can write:

$$(2.38) \quad \begin{aligned} IP_K &= -U_{KK} - (Z_K - 1)\gamma_{KK} \\ EA_K &= -U_{KK} - Z_K\gamma_{KK} \end{aligned}$$

in terms of the effective quantity U_{KK} parametrize the attraction of a single electron placed to the κ-th AO to the core of the K-th atom where this AO is centered. Clearly the parameter U_{KK} is subshell specific.

Some conceptual and technical difficulties of the ZDO-based methods while defining the one-electron integrals/parameters by eq. (2.34) come as a contrast to the ease of coping with the two-electron integrals. It basically indicates that the ZDO approximation is not taken too seriously, but only as a plausible argument allowing one to reduce the number of two-electron integrals. On the other hand, using eq. (2.34) for parametrizing one-electron matrix elements allows one to reproduce the most important spatial (directional) characteristics of chemical bonds dating back to Pauling's principle of optimal (maximal) overlap, which ultimately determines the mutual orientation of chemical bonds (see below).

So far we were concerned with the electronic (quantum) part of the energy in the CNDO approximation. The classical part of the energy describing the interactions between the cores, which are not treated explicitly in the CNDO methods, takes the form:

$$(2.39) \quad Z_M Z_K \gamma_{MK}(R)$$

Combined with the electron-electron and electron-core interaction terms allows one, in this setting, to rewrite the Coulomb contribution to the molecular energy in the form:

$$(2.40) \quad \frac{1}{2} \sum_{M \neq K} Q_M Q_K \gamma_{MK}(R)$$

corresponding to the interaction of the effective atomic charges.

INDO methods. Applying the above (CNDO) treatment to the intraatomic Coulomb integrals results in the evanescence of some of them. Specifically, the intraatomic exchange integrals of the form $(\mu\kappa|\mu\kappa)$ all become zero. As a result, it is not possible to distinguish the different electronic states in the atomic valence shells on the Hartree-Fock level of theory. Also for the open shell molecules (organic radicals) the spin densities did not appear correctly, which precluded the correct description of their ESR spectra, which was important from the point of view of experiment interpretation. This promoted the further development of the HFR-based semiempirical methods along the line of releasing the strictness of the ZDO approximation and by this, allowing for a more detailed description of the Coulomb interaction integrals. In the intermediate neglect of differential overlap (INDO) group of methods all the intraatomic parameters of Coulomb interaction stipulated by the spherical symmetry are included for the sp-shells of the second row atoms. Formally they can appear if the differential overlap for AOs centered on the same atom is not set equal to zero, but maybe one simply should not take the ZDO approximation too seriously: finally the generalized Ruedenberg-Mulliken approximation which leads to eq. (2.28) by construction can be used for products of the AOs centered at different atoms. Keeping the intraatomic differential overlaps results in five independent parameters $F^0(ss)$, $F^0(sp)$, $F^0(pp)$, $G^1(sp)$, and $F^2(pp)$ which suffice to express all two-electron integrals in the sp-shells in their terms. These quantities are known as the Slater-Condon parameters [61]. It is possible in principle to get their estimates from the experimental electronic spectra of atoms and their ions. For the third row atoms the problem of whether it is worth including $3d$-orbitals into the valence basis set arises. A decisive conclusion has not been achieved; however, it is clear that in this situation the number of even one-center two-electron parameters strongly increases. The spectral data necessary to estimate all of them (to find the parameters $F^k(dd)$ one needs to know the energy of states with two electrons in the d-subshell of the valence shell which may not be easily available for an atom having in its ground state the empty d-subshell) are not always readily accessible.

Using the Slater-Condon parameters all symmetry allowable two-electron one-center integrals, which brought about a modification of the matrix elements of the Fock operator, are taken into account. This allows one to write down the matrix elements of the effective Fock operator:

$$(2.41) \quad F_{\mu\mu} = U_{\mu\mu} + \sum_{\substack{\lambda \in M \\ \lambda \neq \mu}} P_{\lambda\lambda} \left(2 \left(\lambda\lambda | \mu\mu \right) - \left(\lambda\mu | \mu\lambda \right) \right) + \sum_{K \neq M} Q_K \gamma_{MK}$$

$$F_{\mu\kappa} = \beta_{MK} S_{\mu\kappa} - P_{\mu\kappa} \gamma_{MK}; \mu \in M; \kappa \in K$$

$$F_{\mu\lambda} = 3 P_{\mu\lambda} \left(\lambda\mu | \mu\lambda \right) - P_{\mu\lambda} \left(\lambda\lambda | \mu\mu \right); \mu, \lambda \in M$$

where e.g.:

$$(2.42) \quad (ss|pp) = F^0(sp); (sp|ps) = \frac{1}{3} G^1(sp); \text{ etc.}$$

also in the INDO context. Two-center molecular integrals in the INDO setting are parametrized analogously to the CNDO method.

NDDO methods. In the NDDO (neglect of the diatomic differential overlap) family of approximations, two-center Coulomb interaction integrals are further retained in the model. Following the formulation, the differential overlap is set to zero, calculating the two-electron integrals only if the AOs involved in one are centered on different atoms. The differential overlap of the AOs centered on one atom is not excluded as in the INDO formulation, but is used to substantiate the retention of the two-electron integrals describing the interactions of the corresponding hybrid densities centered on different atoms (two-center). One can check that the transformation properties of these differential overlaps under spatial rotations coincide with those of the components of the corresponding point multipoles. It has been shown more than 50 years ago [62, 63] that the multipole expansions can adequately reproduce these electron repulsion integrals. This gave W. Thiel and M.J.S. Dewar [64] the idea of not calculating the two-center integrals of the form $(\mu\nu|\kappa\lambda)^{MK}$ where the AOs χ_μ, χ_ν are centered on the atom M, and the AOs $\chi_\kappa, \chi_\lambda$ on the atom K explicitly, but to represent them as energies of interactions of certain fictitious charge distributions mimicking necessary multipole momenta. For example, the hybrid sp-density transforming as a dipole is represented by a pair of charges $\pm\frac{e}{2}$ placed on the respective axis. The two-electron matrix elements are then set to be equal to the interaction energies of these fictitious charges, which are assumed to interact by a semiempirical potential. The most popular semiempirical potential adopted in the MNDO method is [65, 66]:

$$(2.43) \quad f_{l_1 l_2}(R) = [R^2 + (\rho_{l_1} + \rho_{l_2})^2]^{-1/2}$$

It depends on the type of interaction (indices l_1 and l_2 correspond to the 2^{l_1}- and 2^{l_2}-poles located on atoms M and K, respectively).

The drawback of this approach is that it makes the integrals non-invariant with respect to rotations of the coordinate frame. The source of invariance is that the fictitious charge configurations have non-vanishing higher multipole momenta due to the fact that the charge distributions described above are not *point* multipoles and their respective potentials contain contributions of higher multipoles. In real calculations, this non-invariance is masked by evaluating corresponding terms in the diatomic coordinate frame (rotations of the molecule induce rotation of the diatomic coordinate frame and the integrals are calculated identically for all orientations of the molecule).

In the NDDO scheme, two-center contributions involve also the "penetration" effects defined in [56] as the difference between the potential induced on the AOs of the atom M by the nucleus of atom K and by the electron distribution around atom K. The corresponding integrals are taken to be proportional to two-center Coulomb integrals:

$$V_{iiK} = -Z_K (ii|ss)^{MK}$$
$$V_{ijK} = -Z_K (ij|ss)^{MK}$$

Further enlarging the set of Coulomb integrals has been done in the Fenske-Hall method practiced more or less widely in the 1970s and 1980s. It takes into account all possible two-electron integrals, but calculates them using the Mulliken approximation eq. (2.29). Nevertheless no decisive success has been achieved in this direction.

2.3.2.3. Modified ZDO methods

Neither of the "pure" parametrizations based on both the levels of the ZDO treatment described above was particularly successful in reproducing molecular geometries or the thermochemistry even of "organic" substances. To solve this problem, M.S.J. Dewar and his coworkers adopted the idea of loading the missing information on the core-core repulsion term. In the hybrid perspective, this move corresponds to reparametrizing the classical part of the hybrid energy and reloading possible inconsistencies of the HFR/ZDO scheme upon the parameters of the classical contribution to the energy. This resulted in a series of "modified" ZDO methods. The major part of the forms of the core-core repulsion used in the modified methods developed by Dewar may be characterized as those where, at shorter interatomic separations, the repulsion is somewhat stronger than that given by the electron-electron repulsion integrals characteristic of these types of atoms at the given separations. This is logical as in the ZDO methods the two-center Coulomb integrals are parametrized to flow to the average of their one-center values i.e. to a finite quantity, although the nuclear repulsion obviously diverges. The earliest methods of this series were based on the INDO approximation and are known by the name of modified INDO (MINDO) methods. For example, in the MINDO/3 method [67] the core-core repulsion is parametrized in the form:

$$E_{AB} = Z_A Z_B \{ \gamma_{AB}(R_{AB}) + [\frac{e^2}{R_{AB}} - \gamma_{AB}(R_{AB})] \exp(-\alpha_{AB} R_{AB}) \}$$

This helped to build a robust tool for modeling thermochemistry and molecular structure of "organic" molecules yet in the 70s. The MINDO/3 parametrization covers H, B, C, N, O, F, Si, P, S, and Cl atoms, although for some atomic pairs the α_{AB} and

β_{AB} parameters had not been fitted. The reason for this is the rapid increase of the number of parameters indexed by pairs of atoms. That was one of the reasons why the idea of using the parameters indexed by atomic pairs had been abandoned in the semiempirical QC context (see, however, below). Also the Slater exponents for AOs had been included in the parametrization scheme as free adjustable parameters.

The MNDO (modified NDDO) parametrization [64] involves the core-core repulsion in the form:

$$E_{AB} = Z_A Z_B (ss|ss)^{AB} \{1 + \exp(-\alpha_A R_{AB}) + \exp(-\alpha_B R_{AB})\}$$

$$E_{XH} = Z_X Z_H (ss|ss)^{XH} \left\{1 + \frac{\exp(-\alpha_A R_{XH})}{R_{XH}} + \exp(-\alpha_H R_{XH})\right\}$$

for an arbitrary pair of atoms and for a pair involving a hydrogen atom, respectively. The MNDO method has been parametrized for the elements H, B, C, N, O, F, Al, Si, P, S, Cl, Zn, Ge, Br, Sn, I, Hg, and Pb. The parameters of the MNDO method are indexed by the atoms only so that the resonance interactions take the original CNDO form. Thus the overall potential number of parameters is significantly reduced as compared to the MINDO/3 scheme.

The methods of the NDDO family were further developed, which resulted in two quite successful parametrizations for "organic" species [68, 69] known as the Austin Model (AM1) and Parametrized Model (PM3) and further, PM5 and SAM1 (semi-ab initio model) parametrizations [74, 75].

The AM1 and PM3 parametrizations can be characterized as the MNDO ones with the core-core energy terms further modified:

$$V = V^{\mathrm{MNDO}} + \frac{Z_A Z_B}{R_{AB}} \sum_k \{a_{kA} \exp(-b_{kA}(R_{AB} - c_{kA})^2) +$$

$$+ a_{kB} \exp(-b_{kB}(R_{AB} - c_{kB})^2)\}$$

with sum over k ranging from two to four of the terms depending on the sort of atoms involved. These modifications became necessary due to incurable failures of the MNDO method to correctly reproduce certain interatomic distances. The necessary result has been achieved by explicitly adding Gaussian terms to make these distances correct. Parameters of the Gaussians entering the above expression are fit to reproduce "by brute force" the interatomic separations which do not appear correctly from the otherwise MNDO calculation. This results in certain unpleasant features of the AM1 and the PM3 methods. For example, being capable of reproducing the O-H bondlength in water molecule, it fails to do the same in the water dimer (and in higher clusters). The reason is simple: the Gaussian terms force the minimum of the potential well to appear at the characteristic intramolecular O-H distance, which distorts the intermolecular geometry in the dimer [70]. The AM1 method has been parametrized for the elements: H, B, C, N, O, F, Al, Si, P, S, Cl, Zn, Ge, Br, I, and Hg. The PM3 method evolves along the same formula as AM1, but instead of the manual tuning of the parameters performed throughout developing the AM1 scheme, PM3 uses a minimization of the weighted sum of the deviations between the calculated and experimental values of the quantities of interest. In this sense the parameters' search procedure employed to find the PM3 set is "automatic". However, the optimized penalty functional is non-linear with respect to parameters to be found and thus has multiple minima and also depends on the assignment of the weights done throughout the calculation. The PM3 method has been parametrized for the elements: H, Li, C, N, O, F, Mg, Al, Si, P, S, Cl, Zn, Ga, Ge, As, Se, Br, Cd, In, Sn, Sb, Te, I, Hg, Tl, Pb, Bi, Po, and At. However, due to the "automatism" in some cases, the values of the fitted parameters do not follow the intuitive monotonous series along the rows of the Periodic Table. E.g. the parameters for phosphorus fall out from the monotonous sequence along the row. More recent analysis indicates that this happens due to the fact that experimental values for phosphorus compounds used in the parametrization turned out to be wrong by 140 kcal/mole.

The SAM1 parametrization [74] further extends the number of two-electron integrals included in the treatment. They are calculated first for the AOs taken as in the STO-3G Gaussian basis set, but then scaled using the distance dependent functions containing adjustable parameters. The SAM1 method has been parametrized for the elements H, Li, C, N, O, F, Si, P, S, Cl, Fe, Cu, Br, and I. Unfortunately, this parametrization was never thoroughly published and studied. The same applies to the PM5 method [75] which is implemented only by a commercial software, without adequate explanation.[2] Further refinement of the system of correcting Gaussian contributions to the interatomic interaction functions has been proposed in [71].

2.3.3. Miscellanea. Further development

The schemes briefly reviewed above do not exhaust the variety of ways of developing semi-empirical schemes which are presented in the literature. First of all it must be mentioned that the schemes described so far were parametrized against the geometry-thermochemistry data and due to this are capable of reproducing only this type of experimental data. As for spectral information, semiempirical parametrizations designed to reproduce it had been developed within the ZDO approximation. Among them the spectral version of CNDO by Jaffé et al. – the CNDO/S parametrization based on even earlier spectral parametrization of the PPP method – and the INDO/S, also called ZINDO by M. Zerner [81] must be mentioned. The spectral parametrizations differ in construction because the description of the excited and ionized states involves not the single-determinant, but a many-configuration wave function. This automatically requires reparametrization, since some part of correlation, which in the thermochemical parametrizations was loaded upon parameters, explicitly appears in the spectrally oriented methods which employ restricted CI wave function to describe excited states of the systems. Specifically the CNDO/S scheme

---

[2]During the preparation of the final version of this book a comprehensive paper [295] describing development of the PM6 semiempirical parameterization for 70 atoms appeared from print.

was quite successful for organic molecules, whereas the ZINDO (INDO/S) scheme described to some extent the spectra of the transition metal complexes (see below).

The HFR-based, semiempirical methods described so far were additionally using the ZDO approximation for either all or at least a good fraction of integrals and thus were taking the AOs as implicitly orthonormalized. There exists a group of methods attempting to account for the nonorthogonality of the AO basis. The SINDO1 method [73] was designed along this line. Its parametrization employs the rules of selecting nonvanishing two-electron matrix elements, which coincide with those of the INDO scheme, but treats the overlap between AOs centered on different atoms explicitly. The Fock operator constructed in the nonorthogonal basis is transformed to the orthogonal one by applying an approximate expansion of the $S^{-\frac{1}{2}}$ matrix:

$$S^{-\frac{1}{2}} = (1+\sigma)^{-\frac{1}{2}} = 1 - \frac{1}{2}\sigma + \frac{3}{8}\sigma^2 - \ldots$$

$$\sigma_{ij} \sim (1 - \delta_{ij})$$

The resulting parametrization extends to H, Li, Be, B, C, N, O, F, Na, Mg, Al, Si, P, S, Cl, Sc, Ti, V, Cr, Mn, Fe, Co, Ni, Cu, and Zn atoms. There are also other attempts to cope with the orbital basis nonorthogonality at a more advanced and explicit level. This is realized in the CNDO-S^2 [72] and PRDDO [147] methods.

Significant progress in developing semi-empirical models with overlap have been achieved only recently in the orthogonalization model n (OMn) methods by W. Thiel's group [76, 77]. The main differences from the previous methods of treating the nonorthogonality of the AOs basis in the OMn methods are (i) significant modification of the resonance integral:

$$(2.44) \quad \beta_{\mu\nu}^{MN} = \frac{1}{2}(\beta_\mu^M + \beta_\nu^N)\sqrt{R_{MN}}\exp(-(\alpha_\mu^M + \alpha_\nu^N)R_{MN}^2)$$

where $\alpha_\mu^M, \alpha_\lambda^L, \beta_\mu^M, \beta_\lambda^L$ are atom and subshell dependent adjustable parameters and (ii) explicit inclusion of the core nonorthogonality effects into the effective Fock operator for the valence subspace by setting:

$$(2.45) \quad F_{\mu\nu}^{MN} \to F_{\mu\nu} + \sum_C V_{\mu\nu,C}(\text{ECP})$$

$$V_{\mu\nu,C}(\text{ECP}) = \sum_{\gamma \in C}(S_{\mu\gamma}G_{\gamma\nu} + G_{\mu\gamma}S_{\gamma\nu} - S_{\mu\gamma}F_{\gamma\gamma}S_{\gamma\nu})$$

where γ runs over the core orbitals of an atom $C \neq M, N$ and G's are given by eq. (2.44). Otherwise as usual for the second order (in σ) correction to the one-electron orbital-orbital hopping is introduced

$$(2.46) \quad {}^\lambda H_{\mu\nu}^{MN} = \beta_{\mu\nu}^{MN} - \sum_{\alpha \in A}\left[\frac{1}{2}(S_{\mu\alpha}M_{\alpha\nu} + M_{\mu\alpha}S_{\alpha\nu}) - \right.$$

$$\left. - \frac{1}{8}S_{\mu\alpha}(H_{\mu\mu} + H_{\nu\nu} - 2H_{\alpha\alpha})S_{\alpha\nu}\right]$$

for the OAO basis. These improvements allowed many problems of the traditional semi-empirical methods to be solved. In cases where nonorthogonality is important (stereo-discrimination) significant improvements have been achieved. However, in other important cases, additional work will be necessary (see below).

2.3.4. Unsolved problems or "Holy Grails" of the HFR-based semiempirics

The methods described so far were largely targeted toward "organic" molecules. Despite the considerable success of semi-empirical methods (the author suspects that the major part of real life modeling work performed outside the academic community is done using semiempirical methods) some important items remain inaccessible. Among them one might mention the following (as formulated in [120] with some additions):

1. rotation barriers and relative stability of conformations of organic molecules;
2. pyramidal geometry of nitrogen atom in its compounds;
3. hydrogen bonds;
4. weak van der Waal's interactions;
5. molecules containing phosphorus;
6. transition metal compounds;
7. major part of true chemical transformations involving bond breaking and forming;
8. catalytic reactions.

Some of these features have got the somewhat exaggerated name of "Holy Grails" of semiempirical theory. The author has managed to find at least two Holy Grails in the Internet: among them the linearly scaling semiempirical methods which are not mentioned in the above list and those capable of treating transition metal complexes which are. Although the topic of this book is still not listed among them, this makes the whole picture somewhat more prospective: perhaps many problems have similar solutions. Before attempting to solve these problems in the context of the theory of hybrid methods, it makes sense to begin with the analysis of the fundamental sources of the problems the HFR based semiempirics faces when applied to the mentioned objects. We address first the transition metal compounds, which are attractive to first year chemistry students because of their bright colors, attributable to the open d-shell chromophores.

The highly specific behavior of transition metal complexes has prompted numerous attempts to access this "Holy Grail" of the semi-empirical theory – the description of TMCs. From the point of view of the standard HFR-based semiempirical theory, the main obstacle is the number of integrals involving the d-AOs of the metal atoms to be taken into consideration. The attempts to cope with these problems have been documented from the early days of the development of semiempirical quantum chemistry. In the 1970s, Clack and coworkers [78–80] proposed to extend the CNDO and INDO parametrizations by Pople and Beveridge [39] to transition elements. Now this is an extensive sector of semiempirical methods, differing by expedients of parametrizations of the HFR approximation in the valence basis. These are, for example, in methods of ZINDO/1, SAM1, MNDO(d), PM3(tm), PM3* etc. [74,81–86]. From the point of view of the number of two-electron integrals involved, the modern NDDO-type methods must be quite successful. For example, PM3(tm) [87] represents an extension of the PM3 set to the transition metal atoms designed with the purpose of describing TMCs. The data to be fitted throughout the parametrization procedure are the molecular geometries as determined by X-ray technique. Similar ideas were employed while developing the MNDO/d parametrization [88]. Its main purpose was, however, to improve the description of the second row elements by including the vacant d-shells in the valence set. The resulting MNDO/d method contains parameters for the following elements (beyond those already present in MNDO): Na, Mg, Al, Si, P, S, Cl, Br, I, Zn, Cd and Hg. The calculations carried out in [89] show, however, that the method is not capable of reproducing even very simple characteristics in a series of TMCs having a similar structure, though other authors [85,90] state that in some cases, reasonable estimates of geometrical characteristics may, nevertheless be achieved. This situation can be understood by a thorough consideration of the sets of objects chosen for analysis in different works. In [89] authors study a uniform set of about 30 Ni^{2+} complexes with the ligands linked through nitrogen donor atoms. The analysis performed there clearly shows that the PM3(tm) method fails for these Ni^{2+} complexes. On the other hand, in [85,90] the authors try to explore a comparable number of complexes, although much more dispersed over the range of classes, which include compounds of the first and second transition row atom, high-spin and low-spin ones, those having "ionic" and "covalent" bonds, etc. For that reason in the test sets [85,90] the problematic *classes* are represented by a couple of examples each and seem to be completely isolated exceptions. This can serve as an example of how trying to test the method on a wide and apparently "random" selection of objects may lead to a hazy picture due to the absence of clear criteria of any adequate classification of the chosen set. Basically, the problem of the semiempirical description of TMCs has remained unsolved for decades and the most problematic feature escaping a robust modeling is the ground state spin and its dependence on geometry changes.

The 40 years of futile attempts to construct a reasonable description of the TMCs certainly calls for an explanation. We shall show that the reason is the HFR approximation built into the computation scheme of semiempirical methods. Indeed, the calculated energies $E(Cq\Gamma S|\omega)$ are the *linear* functionals of the density matrices eq. (1.199). When the cumulants of the two-electron density come into play, the energies $E(Cq\Gamma S|\omega)$ and the deviations δE_ω become quadratic functionals of the one-electron density matrices and remain the linear functionals of the cumulant (the same as the previous linear functional of the two-electron density matrix). The HFR approximation is nothing but the restriction of the corresponding functionals to their quadratic parts in the one-electron density matrix and dropping the cumulant - dependent contribution completely. By this, two states having wave functions yielding the same one-electron density matrices but different two-electron density matrices are deemed to have the same energy. Despite the somewhat difficult terminology it is a very simple thing. Let us assume that we want to fit some experimental data to the model

$$f(x, y) = ax + by$$
$$(2.47) \qquad y = x^2 + z$$
$$f(x, z) = ax + bx^2 + bz$$

linear in x and y. Quantitatively a simplified model

$$(2.48) \qquad f_0(x) = ax + bx^2$$

may not be that bad, if z is small. But *qualitatively* the approximate model eq. (2.48) cannot distinguish experimental points which have the same value of x and differ by the value of z only. The capacity of a theoretical method to reproduce such features is intimately related to the (*grammatically*) correct treatment of the cumulant of the two-electron density matrix. If there is no z as in eq. (2.48) nothing can help. This is precisely the situation one might face while treating the electronic structure of the TMC's most important characteristic, which is the sophisticated structure of the low energy spectrum of their partially filled d-shell. It can be easily understood that the cumulant of the two-particle density matrix serves to distinguish the different many-electron states in the d-shells.

Let us consider a two-orbital two-electron model system with the orbitals a and b which can be understood as notation for one-dimensional irreducible representations of the point group of a TMC. In this case, it is easy to see that the corresponding singlet and triplet states ^1B and ^3B ($\Gamma = B, S = 0, 1$) are given respectively by:

$$(2.49)$$
$$\Psi_{1_B}(x_1, x_2) = \frac{1}{2}(\alpha(s_1)\beta(s_2) - \beta(s_1)\alpha(s_2))(a(\mathbf{r}_1)b(\mathbf{r}_2) + b(\mathbf{r}_1)a(\mathbf{r}_2))$$
$$\Psi_{3_B}(x_1, x_2) = \frac{1}{2}(\alpha(s_1)\beta(s_2) + \beta(s_1)\alpha(s_2))(a(\mathbf{r}_1)b(\mathbf{r}_2) - b(\mathbf{r}_1)a(\mathbf{r}_2))$$

Performing the integration according to eq. (1.199) we immediately see that irrespective of the total spin of these states, the *exact* one-electron density matrices are:

$$(2.50) \qquad \rho_{BS}^{(1)}(x, x') = \frac{1}{2}(\alpha^*(s)\alpha(s') + \beta^*(s)\beta(s'))(a^*(\mathbf{r})a(\mathbf{r}') + b^*(\mathbf{r})b(\mathbf{r}'))$$

and do not depend on the total spin. This result has been well-known for decades and appears in textbooks [91, 92]. Obviously, the HFR approximate two-electron density matrices coming from the one-electron densities eq. (2.50) give a wrong result since the *exact* two electron density matrices calculated according to their definition eq. (1.199) from the wave functions eq. (2.49) *are* really different:

$$\rho_{B\binom{0}{1}}^{(2)}(x_1 x_2, x_1' x_2') =$$
$$(2.51) \quad \frac{1}{4}(\alpha^*(s_1)\beta^*(s_2) \mp \beta^*(s_1)\alpha^*(s_2))(\alpha(s_1')\beta(s_2') \mp \beta(s_1')\alpha(s_2')) \times$$
$$(a^*(\mathbf{r}_1)b^*(\mathbf{r}_2) \pm b^*(\mathbf{r}_1)a^*(\mathbf{r}_2))(a(\mathbf{r}_1')b(\mathbf{r}_2') \pm b(\mathbf{r}_1')a(\mathbf{r}_2'))$$

with the upper sign corresponding to $S = 0$ and the lower one to $S = 1$. The physical consequences of this difference are well known: it is responsible for the validity of the first Hund's rule, stating that in an atom the term of a higher spin has a lower energy (under other equal conditions). In a more complex situation than that of two electrons each occupying its orbital, one can expect much more sophisticated interconnections between the total spin and two-electron densities than those demonstrated above. The general statement which follows from the theorem given in [93,94] is that no one-electron density can depend on the permutation symmetry properties and thus on the total spin of the wave function. For that reason, the difference between states of different total spin is concentrated in the cumulant. If there is no cumulant in a theory, there is no possibility of describing this difference.

This simple example shows clearly that in the case of TMCs, the data $\mathcal{E}(Cq\Gamma S)$ related to a set of states of different spin with the same number of d-electrons having to be reproduced in different ligand environments, is precisely the situation one faces in the model eq. (2.48). The HFR theory in its simplest form is, however, the case in eq. (2.48) and it does not provide any quantity to which this difference between these energies can be somehow ascribed. The problem is not in that or another type of the Coulomb exchange integrals whether appearing or not in the parameterization scheme, but in their density matrix cumulant counterpart. Even in the case of the HFR the density matrix provides a multiplier to be combined with that or another integral ultimately responsible for the energy difference between the states of the different total spin, but in the absence of the component of the two-electron cumulant dual to this integral, this difference remains zero. This explains to some extent the failure of almost 40 years of attempts to squeeze the TMCs into the semiempirical HFR theory by extending the variety of the two-electron integrals included in the parametrization. The real situation is somewhat more complex. In fact, during the period when the Hartree-Fock theory was the only available tool, it was realized that in some cases one can reproduce correctly the energies of different spin states staying within the single-determinant picture. For example the $S_z = 1$ component of the above triplet ^3B state can be presented as a single determinant function:

$$\Psi_{^3B1}(x_1, x_2) = \frac{1}{\sqrt{2}}\alpha(s_1)\alpha(s_2)(a(\mathbf{r}_1)b(\mathbf{r}_2) - b(\mathbf{r}_1)a(\mathbf{r}_2))$$

The Coulomb interaction energy in this state can be calculated according to general rules and it is obviously equal to that of the $S_z = 0$ component of the same triplet. The one-electron density matrices for the two spin components are of course different, but only in the case of the full one-electron density matrices in the basis of four involved spin-orbitals. If one restricts oneself to the spatial part of the density matrix it again turns out that two equal (spatial!) density matrices produce different energies if they are obtained from different multiplets and produce the same energy if they are obtained from the different components of the same multiplet state. It is worth noticing that the Coulomb integrals coming from the "Hartree" part of the self energy eqs. (1.145) and (1.147) appear in the answer with the same numerical coefficients independent either of the total spin or the total spin projection and these are the exchange integrals whose numerical coefficients distinguish the states of the different total spin. Surely, it was the integration over the spin variables which took care of the correct numerical coefficient at the exchange integral in these two situations. They absorbed the necessary features of the exact two-electron density matrix cumulants which are different for different values of the total spin. One may ask to what extent analogous moves can be generalized. How far can we get by avoiding the explicit consideration of the cumulant or equivalently of considering truly many configuration wave functions. The answer is twofold and in a way contradictory. On the one hand, for any number of spatial orbitals M and for any number of electrons N and for any value of the total spin S conforming previous two values, it is possible to construct states described by the Young tableaux i.e. belonging to the row v of the representation Υ of the group $U(M)$ of the unitary $M \times M$ matrices. This representation has the rank N i.e. the corresponding tableau contains N boxes arranged in no more than two columns of the length not larger than M, such that the first column is by $2S$ boxes longer than the second one. This shape is called the Young pattern and defines an irreducible representation Υ of the group $U(M)$. This representation is degenerate and its rows are numbered by distributing integers – in the present context the subscripts distinguishing OAOs in the basis – from 1 to M in the above N boxes in such a way that they do not decrease along the rows of each tableau and increase in each column. Under this rule, some numbers may appear twice in a two-column tableau representing a doubly occupied spatial orbital, while those appearing once represent singly occupied orbitals. Such tableaux represent states transforming according to the row v of the representation Υ. They are some linear combinations of the N-electron Slater determinants possessing the total spin specified by the Young pattern. For each of the states $|\Upsilon v\rangle$ represented by the Young tableau with the Young pattern Υ and the filling v the expectation value of the Coulomb interaction of electrons is expressed through the Coulomb and exchange integrals with respect to the orbitals involved in the construction of the states represented by the Young tableaux:

$$
\begin{aligned}
&\text{Coulomb/Hartree} \sum_{ij} (ii|jj) \left\langle \mathbf{E}_{ii}^{\Upsilon} \mathbf{E}_{jj}^{\Upsilon} - \delta_{ij} \mathbf{E}_{ii}^{\Upsilon} \right\rangle_{\Upsilon v} + \\
&\text{exchange} \qquad \sum_{ij} (ij|ji) \left\langle \mathbf{E}_{ij}^{\Upsilon} \mathbf{E}_{ji}^{\Upsilon} - \mathbf{E}_{ii}^{\Upsilon} \right\rangle_{\Upsilon v}
\end{aligned}
\tag{2.52}
$$

The operators $\mathbf{E}_{ij}^{\Upsilon}$ ($ij = 1 \div M$) are the generators of the group $U(M)$ in the space of the irreducible representation Υ whose matrix elements between the tableaux v and v' can be calculated on purely algebraic grounds. The diagonal generators $\mathbf{E}_{ii}^{\Upsilon}$ are also diagonal in the matrix representation and their matrix elements $\langle \mathbf{E}_{ii}^{\Upsilon} \rangle_{\Upsilon v} = \langle \Upsilon v | \mathbf{E}_{ii}^{\Upsilon} | \Upsilon v \rangle$ are equal to the occupation number (2, 1 or 0) of the i-th orbital in the Young tableau Υv. From this we see that the Hartree part of the Coulomb energy is uniquely defined by the occupation numbers of the spatial orbitals i.e. by the spatial density only.

For a given set of orbital occupation numbers there may exist sets of Young tableaux differing only by the positions of the orbital indices in the tableaux. The generators $\mathbf{E}_{ij}^{\Upsilon}$ (raising ones if $i > j$ and lowering ones if $j > i$) are *by definition* off-diagonal elements of the spatial one-electron density matrix. The expectation values of their products entering eq. (2.52) are uniquely defined by the occupation numbers of the orbitals in the tableau, by their mutual positions in the tableau, and by the total spin. In this case one can say that for given M, N, S uniquely defining the representation Υ of the group $U(M)$ and for the row v of the latter defined by a specific location of the orbital indices in the tableau, a "Hartree-Fock-like" energy functional can be written, whose electron-electron interaction part is given by eq. (2.52). It can be optimized with respect to the expansion coefficients of the involved orbitals over the AO's basis. It is easy to check that the positions of the orbital indices in the tableaux really matter. For example, for two Young tableaux states:

$$(2.53) \quad |\Upsilon v\rangle = \left|\begin{array}{|c|c|}\hline 1 & 2 \\\hline 3 & 4 \\\hline\end{array}\right\rangle \ ; |\Upsilon v'\rangle = \left|\begin{array}{|c|c|}\hline 1 & 3 \\\hline 2 & 4 \\\hline\end{array}\right\rangle$$

both representing (*different*) singlet states of four electrons in four orbitals, the exchange parts of the corresponding energy functionals, respectively, are [95]:

$$(2.54) \quad \begin{aligned} \Upsilon v : \quad & (12|21) + (34|43) - \frac{1}{2}((13|31) + (14|41) + (23|32) + (24|42)) \\ \Upsilon v' : \quad & -(12|21) - (34|43) + \frac{1}{2}((13|31) + (14|41) + (23|32) + (24|42)) \end{aligned}$$

On the other hand, one can easily conclude that for a pair of Young tableaux Υv and $\Upsilon v'$ for which $n_i = n_j = 1$ and the difference is only the positions of the orbitals i and j in the tableaux the operators $\mathbf{E}^{\Upsilon}_{ij}\mathbf{E}^{\Upsilon}_{ji}$ entering as multipliers of the $(ij|ji)$ exchange integrals in the exact Hamiltonian yield also an off-diagonal matrix element in the FCI:

$$(2.55) \quad \langle \Upsilon v|H|\Upsilon v'\rangle = \frac{\sqrt{3}}{2}(-(13|31) + (14|41) + (23|32) - (24|42)) \neq 0,$$

which requires at least a 2×2 diagonalization for obtaining the energy correctly. Unless there is an additional symmetry relation (in the above example it suffices that exchange integrals $(13|31), (14|41)$ and $(23|32), (24|42)$ are pair-wisely equal) which makes some of the integrals entering the above expressions equal and by this allows the diagonalization to be feasible on the purely symmetry grounds yielding the energy expression which is linear in the Coulomb and exchange integrals, there is no way to avoid at least a square root irrationality in the answer. Sometimes, in highly symmetric systems like atoms, it is possible: for example, in a half-filled p-shell (configuration p^3) the energies of all states can be expressed through the Coulomb and exchange integrals as these latter themselves are the linear combinations of only two Slater-Condon parameters F_0 and F_2 of which only F_2 contributes to the exchange integrals between the atomic p-functions. However, as one can make sure by considering the tables of the d^n state energies in [96, 97] at $n \geq 3$ there appear pairs of states having the same total spin (Young pattern) and also the same total angular momentum. Thus true 2×2 diagonalization yielding the square root irrationality in the expressions for the respective energies is required. From another angle it means that the cumulant of the two-electron density matrix cannot be recovered by simple, symmetry-based manipulations.

The situation clearly becomes less favorable in lower symmetries where the terms of the same spin and symmetry span the subspaces of dimensionalities higher than two. For example, in an octahedral environment, the LS states of d^4- (d^6-) configuration span up to seven-dimensional spaces of many-electronic states [98]. Clearly, at an arbitrarily low symmetry, the problem of linearly expressing the exact energy of many-electronic terms through the Coulomb and exchange integrals cannot be solved.

Very similar reasoning applies to the attempts to treat the TMCs with open d-shells, based on density functional theory (DFT), whatever the champions of this otherwise decent theory say. Methods of DFT originate from the X_α method originally proposed by Slater [99] on the basis of a statistical description of atomic electron structure within the Thomas-Fermi theory [100, 101]. The fundamental idea of the DFT-based methods consists first of all in approximate treatment of the electron-electron interaction energy which is represented as:

$$\langle V_{ee} \rangle = E_H + E_{xc}$$
$$E_{xc} = E_x + E_c$$

The "classical" part of the interaction energy – the Hartree energy:

$$(2.56) \quad E_H = \frac{1}{2} \sum_{\sigma\sigma'} \int \frac{\rho^{(1)}(\mathbf{r}\sigma, \mathbf{r}\sigma)\rho^{(1)}(\mathbf{r}'\sigma', \mathbf{r}'\sigma')}{|\mathbf{r} - \mathbf{r}'|} d\mathbf{r} d\mathbf{r}'$$

is taken exactly, whereas the exchange and correlation parts:

$$(2.57) \quad E_x = -\frac{1}{2} \sum_{\sigma} \int \frac{\rho^{(1)}(\mathbf{r}\sigma, \mathbf{r}'\sigma)\rho^{(1)}(\mathbf{r}'\sigma, \mathbf{r}\sigma)}{|\mathbf{r} - \mathbf{r}'|} d\mathbf{r} d\mathbf{r}'$$

$$(2.58) \quad E_c = -\frac{1}{2} \sum_{\sigma\sigma'} \int \frac{\chi(\mathbf{r}\sigma, \mathbf{r}\sigma; \mathbf{r}'\sigma', \mathbf{r}'\sigma')}{|\mathbf{r} - \mathbf{r}'|} d\mathbf{r} d\mathbf{r}'$$

whose precise definitions eqs. (2.57) and (2.58) consistent with the theoretical setting given by eqs. (1.197) and (1.202), are assumed to be functionals of the one-electron density only (diagonal of the one-electron density matrix in the coordinate representation).

The main aim of the DFT paradigm is to reduce the whole electronic structure theory to a single quantity: one-electron density – the diagonal part of the one-electron density matrix. If it had been possible, it would have considerably simplify the life. Pragmatic methods pertaining to the DFT realm are based on the use of the Hohenberg-Kohn "existence theorems" [102, 103] which state, first, the existence of a universal one-to-one correspondence between one-electron external potential and the one-electron density in the sense that not only does the one-electron potential acting upon a given number of electrons uniquely define the ground state of such a system i.e. its wave function and thus the one-electron density – which is trivial – but also that, for each given density, integrating to a given number of electrons a one-electron potential yielding that given density is uniquely defined.

Two further *hypotheses* are an important complement to the *theorems* cited above. One is the locality hypothesis, and the other is the Kohn-Sham representation of the single determinant reference state in terms of orbitals. The locality has been seriously questioned by Nesbet in recent papers [104, 105]; however, it remains the only practically implemented solution for the DFT. The single determinant form of the reference state in its turn guarantees that all the expectation values of the electron-electron interaction appearing in this context are in fact calculated with the two-electron density given by the determinant term in eq. (1.202) with no cumulant. It particularly applies to the "hybrid" (the word is used in a sense different from its use in the present book) functionals, calculating a certain part of the exchange energy with the formulae of the single determinant Hartree-Fock method eq. (2.57). In that respect the situation is analogous to that in the HFR-based semi-empirical methods. In terms of the "data-fit" model of eq. (2.48) the DFT methods can be understood as those with the fitting model of the form:

$$\tilde{f}_0(x) = ax + g(x)$$

using maybe a very sophisticated function $g(x)$ instead of bx^2 to mimic the independent variable y. Irrespective to how much refined $g(x)$ is used the resulting model will not be able to distinguish the data which differ only by the value of the independent variable z (see above) and have the same values of x. However, the relative energies of the states differing by the cumulant of the two-electron density matrix must be correctly reproduced to obtain a satisfactory description of the spectra (relative energies of the states) of the TMCs' d-shells. In this context it is possible to say that the DFT-based methods take into account electron correlations in the same sense, as all (even the elementary) semiempirical QC methods do. If these latter are parametrized to reproduce some experimental characteristics of molecules, the parameters of these methods implicitly take correlation into account. By this it may be possible to achieve quantitative agreement with a narrow segment of experimental data, but not with those which require reproducing qualitative manifestations of correlations. The latter can be simulated neither by semiempirical methods nor by the DFT-based methods. Therefore, the advantages of the DFT-based methods are are primarily observed for trivial TMCs where the correlations in the open d-shell representing a problem for single determinant methods are actually absent (as in d^0- or d^{10}-complexes or in the complexes of the second and third transition row or in carbonyls or other organometallic compounds cited in abundance in [106]). Despite that, during the past decades, DFT based methods have received a wide circulation in calculations of TMCs' electronic structure [106–110]. It is, first of all, due to the widespread use of extended basis sets, allowing one to improve the quality of the calculated electronic density, and, second, due to the development of successful (so called – hybrid) parametrizations for the exchange-correlation functionals (see above). It is generally believed that the DFT-based methods give, in the case of TMCs, more reliable results than the HFR non-empirical methods and that their accuracy is comparable to that which can be achieved after taking into account perturbation theory corrections to the HFR at the MP2 or some limited CI level [110–112].

Remarkably enough, the counter-example eq. (2.49) is well known in the DFT context, and it brought the author [106, 113] to the conclusion that the theory employing the local spin density approximation for the exchange energy is valid only for the single determinant wave function. That is precisely what other people meant by saying that the DFT (at least in its original form) does not apply to TMCs at all, which also may be an exaggeration.

The prescription proposed in [113] to solve this problem is to apply the Slater sum rules, allowing one to express the energy of a singlet state with two electrons occupying two different orbitals by linearly combining the diagonal matrix elements of the energy over the single-determinant states not having specific spin. These rules are of course a specific case of the old Roothaan prescription for the Hartree-Fock treatment of open shells [114] which in turn is a special case of the above formula using the expectation values of the generator products over the Young tableaux states for small values of M, N, S. The problem is that in the case of TMCs the Roothaan solution is expected to be capable of yielding the energy values of the states of open d-shells. It is not, however, the case. For the p-shell the solution [114] worked well – all operators projecting the Young tableau states to the rows of the irreducible representations of the $SO(3)$ group – the states with definite angular momentum produced a new state and by this the Hamiltonian matrix turned out to be diagonalized purely by symmetry based moves. It turned out, however, that for the d-shells even in free ions neither the solution [114] nor its extension [115] of constructing the two-electron density matrix work for a major part of the atomic electronic terms of the transition metal ions and multiple states appear requiring the explicit diagonalization of the Hamiltonian matrix blocks. This moment is crucial – it is not possible to get rid of the irrationality (square root) in the expression for the energy by linearly combining the parameters of the Hamiltonian. The example of such a possibility, given in [109], applies only to the case explicitly considered in that paper: that of the d^2 configuration, for which as one can see from [96, 97] the energies of all terms can be linearly expressed through the Racah parameters. Obviously the situation is going to deteriorate when the symmetry is not high enough: in this case the number of different symmetry labels simply does not suffice to distinguish all the states coming from the Young tableaux with the given set of orbital occupation numbers.

In this context another suggestion by Gunnarsson and Lundqvist [116] and von Barth [117], known also at an early stage of the development of the DFT technique of employing different functionals to describe different spin or symmetry states, deserves attention. In other words, the simplified model for the data fit eq. (2.48) changes to:

$$\tilde{f}_0^{\Gamma S}(x) = ax + g_{\Gamma S}(x)$$

where $g_{\Gamma S}(x)$ represent exchange-correlation functionals specific for each ΓS. Like the model $f_0(x)$ the model $\tilde{f}_0^{\Gamma S}(x)$ cannot distinguish experimental points with equal values of x differing by the values of z if they belong to the same spin and symmetry, but the difference in z which distinguishes one set of ΓS from another one is implicitly built into the functional. One can expect, however, that using the Young tableaux Υv to label the permutation symmetry, predefined relations between the orbital occupation numbers and total spin on the one hand and the two-electron density cumulant on the other hand can allow one to construct in the future an exhaustive classification of the allowable exchange functionals of the one-electron density.

Turning back to the traditional HFR-based semiempirics, it may be mentioned that the considerations similar to those given above apply when it concerns problems which hardly fit into the traditional HFR-based semiempirical context. It concerns the formation and cleavage of chemical bonds (H_2 dissociation being the archetypal example), catalytic action, etc. In all these cases the very structure of the HFR approximation precludes the correct description since the correlations become not only "strong", but also "nontrivial": the cases with similar x's, but different z's require to be uniformly reproduced to correctly describe these interesting situations.

That does not mean that a valid semiempirical parametrization based on the HFR MO LCAO scheme cannot be built for a certain narrow class of compounds or even for a specific purpose. It is done for example in [86] even for iron(II) porphyrins. But in a more general case there is no way to arrive at any definite conclusion [118] about the validity of a semi-empirical parametrization relying on the HFR approximation. On the other hand we have to mention that the semiempirical method ZINDO/1 [119] which allows for some true correlation by taking into account the configuration interaction may be considered a prospective setting for further parametrization, provided the HFR solution required by this method as a zero approximation can be obtained. This will be discussed in more detail later.

After that long explanation of why the HFR-based methods do not work, some people might be interested in why they nevertheless do work in many cases. The reason can be understood by inspecting the *exact* expression for the energy eq. (1.213) given in terms of the *exact* and thus unknown, but still one-electron Green's function for the system under consideration. It means that the form of the ground state energy accepted in the HFR-based semiempirics is in fact *grammatically* correct. If one manages to figure out a good approximation to the exact one-electron Green's function, then, to the extent of the range where the adopted approximation is valid, the corresponding parametrization will work. The problems must be expected at the borders between the areas where that or another approximation is stable: good candidates to the rôles of such borderlines are the situations when the ground state of the system – that over which the Green's function is by definition an expectation value – changes. Then an approximation good enough to represent a Green's function being an expectation value over one ground state is not a good one for doing similar work with another ground state and explicit addressing of nontrivial correlations becomes necessary.

Finalizing the brief review of the HFR-based semiempirical methods of quantum chemistry, we notice first of all that being themselves examples of general hybrid paradigm of separate treatment of different electron groups, they provide enough flexibility for being used in the hybrid schemes in a narrow sense, in which they are of special interest to us. Their parameters can be modified if necessary and other variations can possibly be introduced. The HFR-based methods of quantum chemistry are rather successful although not universal. In this context a program of developing a next generation NDDO-based semiempirical MO technique has been put forward in [120]. Among the lines tentatively bringing this "generation next" methods into reality are:

1. including orthogonalization corrections tentatively improving performance of the methods for rotation barriers, conformational equilibria, structural details and many other aspects.
2. extending the basis set with d orbitals for elements heavier than silicon, and maybe even for some first-row elements.
3. paying more attention to the magnitudes of the one-electron energies of the atomic orbitals, and perhaps even fixing them at the spectroscopic values.
4. using effective core potentials should improve performance for heavier elements.
5. including dispersion forces.

In our opinion, although development along these lines is highly desirable, the key problems not solved in the semi-empirical context for decades will not be solved. Among these problems we mention a topic addressed in this book as an example: developing a technique to access one of the "Holy Grails" of semi-empirical theory [120] – transition metal complexes – and eventually include it in a classical context, which will require, as we shall see, much more elaborate treatment of the wave function than simply including nonorthogonality into the otherwise HFR treatment or anything of this sort.

2.4. NON-HARTREE-FOCK SEMIEMPIRICAL QUANTUM CHEMISTRY

In the previous section we described existing HFR-based semiempirical methods and demonstrated their hybrid nature in a wide sense. We have also shown that for certain physical situations the semiempirical methods may become invalid due to the necessity to explicitly address nontrivial electronic correlations manifesting themselves either in numerous Slater determinants to be included in the consideration or in nonvanishing matrix elements of the cumulant of two-electron density matrix whose presence must be somehow reproduced in the calculation.

The standard prescription in this situation would be to apply the ab initio methods and to attempt to take into account missing correlations in their framework. This is, however, possible only for the systems of very modest size due to the $M^5 \div M^7$ scalability of the correlated ab initio methods already mentioned. Using DFT methods in the situation when explicit correlations are necessary may be completely wrong, due to the structural deficiency of this class of methods, which precludes any treatment of nontrivial parts of correlations, requiring multiconfigurational wave function.

Our account of the cumulant properties indicates, however, another possibility: to develop a local correlated theory such that the correlations are taken into account only for some specific electronic groups of the entire system (chromophores) while the rest is treated on a noncorrelated level. The cumulant being a local quantity restricted to only one group can be then economically estimated by some appropriate method. Of course such an approach would require parametrization of the semiempirical method to avoid double counting of correlations in parameters (dynamic i.e. trivial correlation), but it is a reasonable price to pay.

A semiempirical method can be developed for the arbitrary form of the trial wave function of electrons, which is predefined by the specific class of molecules to be described and by the physical properties and/or effects which have to be reproduced within its framework. Two characteristic examples will be considered in this section. One is the strictly local geminal (SLG) wave function; the other is the somewhat less specified wave function of the GF form selected to describe transition metal complexes.

Strictly local geminals are two-electron wave functions of full CI constructed in the basis of two one-electron states centered on the atoms between which the bond is formed. This wave function has the correct asymptotic behavior under the bond cleavage. In this relation it has to be noticed that the dominance of the HFR approximation in constructing the computational methods is largely accidental: a priori it is not clear what is preferable: to take into account first the one-electron transfers between the AOs of the different atoms which lead to delocalization of the one-electron states and to the incorrect description of the homolytic cleavage of the σ-bonds, or to concentrate first on taking into account pair correlations of electrons at least within the bonds and by this to assure the physically correct asymptotic behavior under the homolytic cleavage and then to take into account the missing one-electron transfers as corrections.

One more achievement along the geminal way may be that the HFR based semiempirics scales as N^3 with the increase of the size of the system. Working in the basis of strictly local HOs in the SLG framework opens up the prospect of obtaining the linearly scaling semiempirical method.

We have already mentioned that the HFR lacks the cumulant of the two-electron density matrix. As we have shown above, it is indispensable for describing the multiplet structure of central transition metal ion. The specific form of the wave function allowing for it will be used in the semiempirical context for constructing a method targeted at the transition metal complexes (TMCs). It will be described in Section 2.4.2.

2.4.1. Linear scaling semiempirics for "organic" molecules

2.4.1.1. Historical precedent. PCILO method

An early example implementing the general approach: to take into account first the intrabond correlation, is presented by the PCILO – perturbational configuration interaction of localized orbitals method [121, 122]. As one of its authors, J.-P. Malrieu mentions in [122], the PCILO method opposes the majority of the QC methods in all the fundamental concepts. In contrast to the majority of the methods based on the variational principle, the PCILO method is based on estimating the energy by perturbation theory. Also, the majority of the QC methods use one-electron HFR approximation, at least as an intermediate construct, whereas the PCILO is claimed to addresses directly the N-electron wave function and takes into account all surviving matrix elements of the electron-electron interactions. In contrast with other QC methods based largely on delocalized one-electron states (MO) the PCILO employs exclusively local one-electron states.

All these statements, although correct in principle, are not precise from the technical point of view. For example, the zero approximate wave function in the PCILO method is a one-electron approximate function constructed from the bond wave functions determined by an a posteriori localization procedure from an HFR function. Thus the bond orbitals appear after a unitary transformation of the canonical MOs, which correspond to some more or less arbitrary localization criteria [123–125].

The basic general conclusion to be derived from the experience of applying the PCILO method is the demonstration of the fact that a major part of the theoretical intuition based on the picture of the electronic structure, as it appears from the HFR approximation, does not constitute mandatory element of the theory. The PCILO approach constructively proves the existence of a method adequately describing the relative energetics of organic molecules but not leading to concepts counter-intuitive for chemistry, like the idea of the obligatory and exclusive rôle of delocalized canonical MOs. The key feature here is of course the return to the concept of chemical bond intuitively clear to all chemists. A shortcoming is the nonvariational character of the energy estimate, which reduces the proving power of the results. Using the perturbation theory for electron correlation energy is not well-founded as the matrix elements of Coulomb interaction responsible for admixing the excited configurations in the basis of local one-electron states are not small, compared to the corresponding excitation energy. In this context it would be interesting to construct a method which would not be opposed to the traditional QC methods in that many positions simultaneously.

2.4.1.2. The SLG approximation

The PCILO method bears all the characteristic features of the GF approach described by eq. (1.181). The main difference (incidentally, not improving but deteriorating the quality of description of electronic structure) is that the configuration interaction in the bond functions is taken into account as a perturbation. Meanwhile, it can be taken into account variationally in the zero order. To do so it has been proposed to consider the wave function of the form of the antisymmetrized product of strictly localized geminals [126]. In the framework of this method, the set of strictly local orbitals is divided into subsets, each ascribed to a separate chemical bond or electron lone pair. For each such subset a two-electron wave function – geminal—is formed and expressed as a linear combination of two-electron Slater determinants written with respect to the basis of orbitals ascribed to the considered subset. The amplitudes of the geminal expansion with respect to these determinants are found from the variational principle. This construct is obviously a specific case of the GF approximation. Each group corresponds to the two-electron bond or to a lone pair (LP). Such a choice is supported in [126] by the following arguments: (1) chemical bonds have a relatively small number of internal degrees of freedom as compared to the entire molecule, (2) global molecular properties (energy, dipole moment, etc.) for certain classes of molecules can be represented with acceptable accuracy as sums of bond increments (additive scheme [127, 128]) and moreover for certain properties, the increments of separate bonds possess the property of transferability from one molecule to another in the scope of a wide enough class of molecules.

The limitations of the picture, based exclusively on two-electron groups while addressing a general N-electron system, are obvious. In some systems, the collective character of behavior of electrons is a significant part of the general picture. The simplest example of such a behavior is provided by benzene molecule, where the system of π-electrons requires considering a six-electron group in addition to two-electron ones, which can be singled out in the σ-core. Another example is provided by the EHCF theory described in Section 2.4.2 which employs the d-shell containing n_d electrons as one of the groups. With these caveats, the idea of SLG is very reasonable.

Implementations of the SLG approach at the ab initio level are reviewed in [126]. This approach had been tested on a relatively narrow range of the simplest molecules. The results obtained do not permit to make a definitive conclusion concerning the applicability of this method to larger molecules. As for the orbitals used in the implementation [126] they are obtained by a posteriori localization of canonical MOs. This then requires applying the hardly formalizable procedure of 'tail' cutting, which also cannot give any hope on the transferability of such states. A significant amount of work on geminal-based ab initio models has been performed also by I. Røeggen [129] who developed a series of extended geminal models EXGEMn with n = 0 ÷ 7 with the ultimate aim of applying this approach to the treatment of intermolecular interactions.

2.4.1.3. Semiempirical implementations of SLG wave function

In this section, we consider a family of semiempirical implementations of the antisymmetrized product of the strictly local geminals (SLG). Quite naturally, this approach applies only to compounds (largely organic) with well localized two-center two-electron bonds. It had been originally developed for an old-fashioned MINDO/3 type of parametrization of the molecular Hamiltonian and then extended to the more contemporary NDDO family of parametrizations. First, the description of the wave function is given in detail and then the energy functional is described and analyzed. Its variation provides the equilibrium values of the electronic structure variables (ESVs) relevant for this method.

Constructing the SLG trial wave function according to [130–133] requires the following moves. First, the one-electron basis of the strictly local hybrid orbitals (HOs) must be constructed [134]. These orbitals are obtained by an orthogonal transformation of the s and p valence AOs for each "heavy" (non-hydrogen) atom. These transformations are represented by 4×4 orthogonal matrices $h^A \in O(4)$ for each heavy atom A. For each pair of atoms connected by a single bond with number m, two such HOs are selected $|r_m\rangle, |l_m\rangle$ referring to the right and left ends of this bond, respectively. The expression for the corresponding electron annihilation operators is written in terms of the similar operators for AOs:

$$(2.59) \quad \mathsf{t}_{m\sigma} = \sum_{i \in T_m} h^t_{mi}(T_m)\mathsf{a}_{i\sigma}$$

where notation T_m (R_m or L_m) refers to the "right" and "left" atoms of the m-th bond. The superscript can be ascribed a numerical value, t ($=\pm1$) used below to refer to r and l. It is assumed that in variance with other methods using the HOs, those of the present method are determined on the basis of the variational principle for the electronic energy, like other variables determining the wave function, i.e. from the energy minimum condition.

Chemical bonds and lone pairs are described by singlet two-electron functions – geminals [135] taken in the form originally proposed by Weinbaum [136]. Using the second quantization notation they are written as:

$$(2.60) \qquad g_m^+ = u_m r_{m\alpha}^+ r_{m\beta}^+ + v_m l_{m\alpha}^+ l_{m\beta}^+ + w_m (r_{m\alpha}^+ l_{m\beta}^+ + l_{m\alpha}^+ r_{m\beta}^+)$$

for chemical bonds and:

$$(2.61) \qquad g_m^+ = r_{m\alpha}^+ r_{m\beta}^+$$

for lone pairs. The normalization condition imposed on the amplitudes of thus defined geminals reads:

$$(2.62) \qquad u_m^2 + v_m^2 + 2w_m^2 = 1$$

The geminals defined in the carrier space spanned by HOs constructed by the above formulae are termed to be strictly local geminals (SLG).

The wave function of electrons in the molecule is then taken as the antisymmetrized product of the geminals given by eqs. (2.60), (2.61):

$$(2.63) \qquad |\Phi\rangle = \prod_m g_m^+ |0\rangle$$

The Hamiltonian for a molecular system in a general semiempirical approximation can be represented as a sum of one- and two-center contributions:

$$(2.64) \qquad H = \sum_A H_A + \frac{1}{2} \sum_{A \neq B} H_{AB}$$

In the second quantization representation related to the HOs eq. (2.59) we get the one-center contributions:

$$
\begin{aligned}
(2.65) \qquad H_A =\ & \sum_{m \in A, \sigma} \left(\delta_{t_m t_{m'}} U_m^t - \sum_{B \neq A} V_{t_m t_{m'} B}^A \right) t_{m\sigma}^+ t_{m'\sigma} - \\
& - \sum_{m_1 < m_2} \sum_\sigma \beta_{m_1 m_2}^A \left(t_{m_1\sigma}^+ t_{m_2\sigma} + h.c. \right) + \\
& + \frac{1}{2} \sum_{\substack{m_1, m_2, \\ m_3, m_4 \in A}} \sum_{\sigma\tau} (t_{m_1} t_{m_2} \mid t_{m_3} t_{m_4})^A\, t_{m_1\sigma}^+ t_{m_3\tau}^+ t_{m_4\tau} t_{m_2\sigma}
\end{aligned}
$$

and the two-center ones:

$$
\begin{aligned}
(2.66) \qquad H_{AB} =\ & - \sum_{m_1 \in A, m_2 \in B} \sum_\sigma \beta_{m_1 m_2}^{AB} \left(t_{m_1\sigma}^+ t_{m_2\sigma} + h.c. \right) + \\
& + \sum_{\substack{m_1, m_2 \in A, \\ m_3, m_4 \in B}} \sum ((t_{m_1} t_{m_2} \mid t_{m_3} t_{m_4})^{AB} \sum_{\sigma\tau} t_{m_1\sigma}^+ t_{m_3\tau}^+ t_{m_4\tau} t_{m_2\sigma}
\end{aligned}
$$

where $h.c.$ stands for the hermitean conjugated terms and it is assumed in each sum that the HOs $|t_{m_i}\rangle$ belong pairwise to atoms A and B.

Using the expansions of the HOs over the basis of AOs in eq. (2.59), we get the molecular integrals in the HOs basis as linear combinations of those in the AOs basis. The parameter U_m^t, describes the attraction of an electron placed at the HO $|t_m\rangle$ to the core of that atom where the HO is centered:

$$(2.67) \quad U_m^t = \sum_{i \in A} (h(A)_{mi}^t)^2 U_{ii}^A$$

The subscript i enumerates the s- and p-AOs of a heavy atom A. Using the $O(4)$ matrix h^A it can be expressed in terms of the weight of the s-orbital only:

$$(2.68) \quad U_{t_m t_m}^A = U_{pp}^A + (U_{ss}^A - U_{pp}^A)(h_{ms}^A)^2$$

The SLG form of the wave function significantly reduces the number of necessary integrals describing the electron repulsion on one atom. We express them through the integrals in the AO basis. These are the repulsion of two electrons occupying one HO $|t_m\rangle$:

$$(2.69) \quad \begin{aligned} (t_m t_m \mid t_m t_m) &= \sum_i (h_{mi}^t)^4 (ii \mid ii) + \\ &\quad + 2 \sum_{i<j} (h_{mi}^t h_{mj}^t)^2 [(ii \mid jj) + 2(ij \mid ij)] \end{aligned}$$

The integrals describing the Coulomb repulsion of electrons in two HOs centered at the same atom appear only in the form of the reduced repulsion integrals for pairs of HOs $|t_m\rangle$ and $|t_m'\rangle$ centered at the same atom:

$$(2.70) \quad g_{t_k t_m'}^{T_k} = 2(t_k t_k | t_m' t_m')^{T_k} - (t_k t_m' | t_m' t_k)^{T_k}$$

$(T_m = T_m')$. In the sp-shell they can be expressed using only the s-weights of the HOs:

$$(2.71) \quad \begin{aligned} (t_m t_m \mid t_m t_m)^A &= C_1^A + C_2^A (h_{ms}^A)^2 + C_3^A (h_{ms}^A)^4 \\ g_{t_k t_m'}^A &= 2(t_k t_k \mid t_m' t_m')^A - (t_k t_m' \mid t_m' t_k)^A = \\ &= C_4^A + C_5^A [(h_{ms}^A)^2 + (h_{ks}^A)^2] + C_3^A (h_{ms}^A h_{ks}^A)^2 \end{aligned}$$

where C_n are the linear combinations [137] of the five independent Slater-Condon parameters (see [61]) characterizing the Coulomb interactions in the valence sp-shell:

$$(2.72) \quad \begin{aligned} C_1^A &= F_0^A(pp) + 4F_2^A(pp) \\ C_2^A &= 2F_0^A(sp) + 4G_1^A(sp) - 2F_0^A(pp) - 8F_2^A(pp) \\ C_3^A &= F_0^A(ss) - 2F_0^A(sp) - 4G_1^A(sp) + C_1^A \\ C_4^A &= 2F_0^A(pp) - 7F_2^A(pp) \\ C_5^A &= 2F_0^A(sp) - G_1^A(sp) - 2F_0^A(pp) + 7F_2^A(pp) \end{aligned}$$

From eqs. (2.68) and (2.71) one can see that the one-center molecular integrals do not depend on the orientations of the HOs.

Now we turn to the two-center contributions to the molecular Hamiltonian. The resonance integral describing the one-electron transfer between the HOs can be written through the resonance integrals for the AOs:

$$(2.73) \quad \beta^{AB}_{t_1 m_1 t_2 m_2} = \sum_{i \in A} \sum_{j \in B} h^{t_1}_{m_1 i}(A) h^{t_2}_{mj}(B) \beta^{AB}_{ij}$$

The attraction of an electron occupying an HO centered on atom A to the cores of atom B takes the form:

$$(2.74) \quad V^A_{t_m t_m, B} = \sum_{i,j \in A} V^A_{ij,B} h^t_{mi}(A) h^t_{mj}(A)$$

The Coulomb matrix elements for the HOs localized on different atoms A and B have the form:

$$(2.75) \quad (t_{m_1} t_{m_1} \mid t'_{m_2} t'_{m_2})^{AB} = \sum_{\substack{i,j \in A \\ k,l \in B}} (ij \mid kl)^{AB} \times$$

$$\times \, h^t_{m_1 i}(A) h^t_{m_1 j}(A) h^{t'}_{m_2 k}(B) h^{t'}_{m_2 l}(B)$$

If the atoms A and B are connected by a multiple bond, additional two-electron matrix elements are necessary:

$$(2.76) \quad (t_{m_1} t'_{m_2} \mid \bar{t}'_{m_2} \bar{t}_{m_1})^{AB} = \sum_{\substack{i,j \in A \\ k,l \in B}} (ij \mid kl)^{AB} \times$$

$$\times \, h^{t_{m_1}}_{m_1 i}(A) h^{t'_{m_2}}_{m_2 j}(A) h^{\bar{t}'_{m_2}}_{m_2 k}(B) h^{\bar{t}_{m_1}}_{m_1 l}(B)$$

where $\bar{t} = l$ for $t = r$ and $\bar{t} = r$ for $t = l$.

Due to the fact that the SLG wave function belongs to the GF approximation (Section 1.7), it is subject to numerous selection rules characteristic of GF. Their explicit form can be easily obtained using the second quantization formalism. Since the operators of electron creation on the right and left HOs satisfy usual anticommutation relations for orthogonal basis and the number of particle operators have the usual form:

$$(2.77) \quad \hat{n}^t_m = \sum_\sigma \mathrm{t}^+_{m\sigma} \mathrm{t}_{m\sigma}$$

the expectation values of the products of Fermi operators defining one-electron densities over the SLG wave function have the form:

$$P^{tt'}_m = \langle 0 | g_m \mathrm{t}^+_{m\sigma} \mathrm{t}'_{m\sigma} g^+_m | 0 \rangle$$

$$(2.78) \quad P^{rr}_m = u^2_m + w^2_m, \ P^{ll}_m = v^2_m + w^2_m$$

$$P^{rl}_m = P^{lr}_m = (u_m + v_m) w_m$$

where t and t' correspond to the right and left ends of the bond and to the Fermi operators r and l, respectively. Applying the Coulson definition of effective charges to the SLG wave function yields them in the form:

$$(2.79) \quad Q_A = 2 \sum_{t_m \in A} P_m^{tt} - Z_A$$

For the expectation values of the Fermi operator products defining matrix elements of the two-electron density we have:

$$\langle \Phi | t_{m_1\sigma}^+ t_{m_2\tau}'^+ t_{m_3\tau}' t_{m_4\sigma} | \Phi \rangle = \delta_{m_1 m_2}\delta_{m_3 m_4}\delta_{m_1 m_3}(1 - \delta_{\sigma\tau})\Gamma_{m_1}^{tt'} +$$

$$(2.80) \qquad + (1 - \delta_{m_1 m_2})[\delta_{m_1 m_4}\delta_{m_2 m_3} - \delta_{m_1 m_3}\delta_{m_2 m_4}\delta_{\sigma\tau}]P_{m_1}^{tt} P_{m_2}^{t't'}$$

$$\Gamma_m^{tt'} = \langle 0 | g_m t_{m\beta}^+ t_{m\alpha}'^+ t_{m\alpha}' t_{m\beta} g_m^+ | 0 \rangle$$

Intrageminal elements of the two-electron density matrix easily write through the amplitudes of the corresponding geminal:

$$\Gamma_m^{rr} = \langle 0 | g_m r_{m\beta}^+ r_{m\alpha}^+ r_{m\alpha} r_{m\beta} g_m^+ | 0 \rangle = u_m^2$$

$$(2.81) \qquad \Gamma_m^{ll} = \langle 0 | g_m l_{m\beta}^+ l_{m\alpha}^+ l_{m\alpha} l_{m\beta} g_m^+ | 0 \rangle = v_m^2$$

$$\Gamma_m^{lr} = \Gamma_m^{rl} = \langle 0 | g_m r_{m\alpha}^+ l_{m\beta}^+ l_{m\beta} r_{m\alpha} g_m^+ | 0 \rangle = w_m^2$$

whereas the intergeminal elements of the two-electron density matrix are the products of the matrix elements of the one-electron densities.

The SLG energy can be rewritten as a function of the intrabond matrix elements of spinless one- and two-electron density matrices. However, some regrouping of terms makes the picture more clear. Using the above expressions for the density matrix elements, one can easily write the equation for electronic energy. The contributions from the one-center terms $\langle \Phi | H_A | \Phi \rangle$ have the form:

$$(2.82) \quad E_{\text{attr}} = 2 \sum_A \sum_m (U_m^t - \sum_{B \neq A} V_{t_m t_m B}^A) P_m^{tt}$$

– attraction to the cores;

$$(2.83) \quad E_{\text{Coul}} = \sum_A \sum_{m \in A} \left[(t_m t_m \mid t_m t_m) \Gamma_m^{tt} + \right.$$

$$\left. + 2 \sum_{m_1 < m_2} g_{t_{m_1} t_{m_2}}^A P_{m_1}^{t_{m_1} t_{m_1}} P_{m_2}^{t_{m_2} t_{m_2}} \right]$$

– one-center electron-electron repulsion.

The resonance contribution to the energy of each bond is proportional to the off-diagonal element of the one-electron density matrix known as the Coulson bond order between the HOs of the m-th geminal P_m^{rl} eq. (2.78):

$$(2.84) \quad E_{\text{res}} = -4 \sum_m \beta_{r_m l_m}^{R_m L_m} P_m^{rl}$$

From this we see that the structure of the wave function allows the transfer of electrons between the one-electron states only within a geminal and the possible delocalization of electrons between the geminals is not taken into account.

Now we turn to an analysis of the two-center Coulomb contributions to the energy. They can be written as follows:

$$(2.85) \quad \begin{aligned} E_{\text{rep}} &= 2 \sum_{A<B} \sum_{t_{m_1}\in A} \sum_{t'_{m_2}\in B} (t_{m_1}t_{m_1} \mid t'_{m_2}t'_{m_2})^{AB} \times \\ &\times [2(1-\delta_{m_1 m_2})P^{tt}_{m_1}P^{t't'}_{m_2} + \delta_{m_1 m_2}\Gamma^{rl}_{m_1}] \end{aligned}$$

In the NDDO Hamiltonian the exchange interaction between electrons of two geminals involved in one multiple bond contributes to the electronic energy:

$$(2.86) \quad E_{\text{mb}} = -4 \sum_{t_{m_1}<t'_{m_2}\in A} (t_{m_1}t'_{m_2} \mid \bar{t}'_{m_2}\bar{t}_{m_1})^{AB} P^{rl}_{m_1}P^{rl}_{m_2}$$

Electronic energy in the semiempirical SLG approximation is thus represented as a sum of five contributions:

$$(2.87) \quad E^{\text{SLG}}_{\text{el}} = E_{\text{attr}} + E_{\text{coul}} + E_{\text{res}} + E_{\text{rep}} + E_{\text{mb}}$$

To simplify the interpretation of the energy in the SLG approximation, we further regroup the individual terms and rewrite them as:

$$(2.88) \quad \begin{aligned} E_{\text{total}} &= \sum_A E_A + \sum_{A<B} E_{AB} \\ E_A &= \sum_{t_m\in A} [2U^t_m P^{tt}_m + (t_m t_m \mid t_m t_m)^{T_m}\Gamma^{tt}_m] + 2 \sum_{\substack{t_k t'_m\in A \\ k<m}} g^{T_k}_{t_k t'_m} P^{tt}_k P^{t't'}_m \\ E^{\text{bond}}_{R_m L_m} &= 2\gamma_{R_m L_m}[\Gamma^{rl}_m - 2P^{rr}_m P^{ll}_m] - 4\beta^{R_m L_m}_{r_m l_m} P^{rl}_m \\ E^{\text{nonbond}}_{AB} &= Q_A Q_B \gamma_{AB} + Z_A Z_B D_{AB} \end{aligned}$$

(for the bonded atoms the contribution of nonbonding interactions $E^{\text{nonbond}}_{R_m L_m}$ also must be included). We see that in this form certain energy contributions further reduce to Coulomb interaction of effective atomic charges residing on the atoms. In addition to the molecular integrals the function D_{AB} enters into the expression for the total energy eq. (2.88), which describes the difference of the core-core repulsion of the atoms A and B from the corresponding Coulomb repulsion in the MINDO/3 and NDDO approximations.

The above form of the total energy, which is somewhat close to the MM energy (see below) with interactions between bonded and non-bonded atoms treated differently, closely relates it to that given in [138] in the context of analysis of a variety of additive schemes of molecular energy. The electronic (and total) energy of the molecular system in the SLG approximation therefore depends on the electronic structure variables (ESVs) of two types: (i) on M triples of amplitudes u_m, v_m and w_m defining the bond geminals according to eq. (2.60) through elements of density matrices eqs. (2.78), (2.81) and (ii) on the $O(4)$ matrices h^A defining hybridized orbitals. The latter enter the theory indirectly – through the molecular integrals. The total number of independent variables defining the amplitudes equals $2M$ (M is the number of chemical bonds) due to normalization condition eq. (2.62) imposed on the geminal amplitudes. The total number of variables defining the hybridization equals $6L$

(L is the number of heavy atoms). This result appears as follows. The h^A matrices, as we noticed, belong to the group of 4×4 orthogonal matrices $O(4)$. This group however, has a rather sophisticated structure, which is not of particular interest to us. However, two close hybridization matrices h and h' can be obtained by multiplying any of these two matrices by an $SO(4)$ matrix, i.e. an orthogonal matrix with unit determinant. For the minimum search it is enough to be able to locally vary the systems of the hybrids. To describe the variation it suffice to have the matrices of the $SO(4)$ group which is a six-parametric one. We use parametric representation of the $SO(4)$ group based on six subsequent Jacobi rotations in two-dimensional subspaces of a four-dimensional space spanned by the AOs residing at each heavy atom (see below). Therefore, six parameters are the corresponding angles of Jacobi rotations. The determination of the ESVs is performed by using a variational principle by a series of iterations. The first step is calculation of geminal amplitudes by diagonalizing 3×3 effective bond Hamiltonians for each geminal representing a chemical bond. The next step is a series of energy minimizations with respect to sextuples of parameters defining $SO(4)$ transformations for each heavy atom. These minimizations are performed using analytical gradients of the energy with respect to the Jacobi angles. The alternating diagonalizations and minimizations are performed until convergence. The number of iterations remains approximately constant with the increase of the size of the molecular system. By this only the time per iteration changes with the size of the system and thus the linear scalability of the entire procedure is achieved.

The above scheme of determining the optimal wave function of the SLG approximation, together with optimizing the corresponding energy functional with respect to the nuclear coordinates, has been implemented and tested for a range of semiempirical Hamiltonians on a series of examples: largely organic molecules containing the atoms of the second row: carbon, nitrogen, oxygen, fluorine (and, of course, hydrogen). The results are briefly discussed in the subsequent sections.

Method SLG-MINDO/3. The first semiempirical implementation of the SLG method was undertaken using the well known MINDO/3 [67] parametrization, which is known to satisfactorily reproduce energy characteristics (thermochemistry) and geometry parameters with the HFR (SCF) trial wave function. In the frame of the MINDO/3 approximation, the matrix elements of attraction to "other" cores and two-center matrix elements of Coulomb interaction do not depend on the orbital quantum number l, i.e. they coincide for the s- and p-AOs and can be expressed through the single two-center integral γ_{AB} and the core charges Z_B for all HOs. Incidentally the energy contribution eq. (2.86) of the two-center exchange interactions in multiple bonds also vanishes. This simplifies the formulae for the energy to some extent.

The reparameterization reduces to a minor variation of the resonance parameters of the MINDO/3 set. Using the adjusted parameters, we calculated the characteristics of electronic structure (effective charges and bond orders), molecular geometries, and heats of formation of the test set of "organic" compounds containing hydrogen, carbon, nitrogen, and oxygen atoms. The test set has been borrowed from the papers devoted to parametrization of the HFR-based semiempirical methods MNDO [64],

Table 2.1. Parameters β_{AB} of the SLG- and SCF-MINDO/3 methods.

A	B	β_{AB} (SLG)	β_{AB} (SCF)
H	H	0.243007	0.244770
H	C	0.316049	0.315011
H	N	0.356416	0.360776
H	O	0.414559	0.417759
C	C	0.428097	0.419907
C	N	0.426086	0.410886
C	O	0.486514	0.464514
N	N	0.379342	0.377342
O	O	0.659407	0.659407

AM1 [68], and PM3 [69]. The obtained parameters β_{AB} of the SLG-MINDO/3 method as compared to analogous parameters of the SCF-MINDO/3 method are given in Table 2.1.

In order not to overload the text we do not give here the extensive tables [130, 140] of the calculated and experimental heats of formation and present the results of statistical analysis of these data. It is performed using the empirical distribution function for the errors [141]. The latter is constructed in the assumption that the deviation of the heat of formation of some molecules, calculated by the SLG-MINDO/3 method from its experimental value – the error – is a random variable. Constructing the graph of the empirical distribution function for this random quantity in the normal scale allows us to check the hypothesis of normality of the distribution of the errors (it is commonly believed that the random errors must be normally distributed) and to find the parameters of this distribution. It has been shown that for the test set (40 heats of formation for molecules containing H, C, N, and O atoms and single and multiple bonds) the errors of the SCF-MINDO/3 method are normally distributed with acceptable accuracy (the empirical distribution function is linear in the normal scale). At the same time, for the SLG-MINDO/3 method, the same test set seems to be less uniform (the linearity of the empirical distribution function in the normal scale is much worse). Nevertheless it turns out that the statistical parameters of this function (the average a which is the measure of the systematic error of the method and mean square deviation σ) in the case of the SLG-MINDO/3 are somewhat smaller ($\sigma = 10.98$ kcal/mole and $a = -1.57$ kcal/mole), than the corresponding values for the SCF-MINDO/3 method ($\sigma = 12.01$ kcal/mole and $a = -4.45$ kcal/mole).

Using the adjusted parameters β_{AB} given in Table 2.1, we studied the optimal geometry structures of several simplest molecules. It turned out that the H-H bond length is larger by 0.01 Å than the experimental value. The internuclear separations C-H in methane, calculated by the SLG-MINDO/3 method, coincide with the experiment up to 0.001 Å. Meanwhile the calculated C-C bondlength in ethane turns out to be 1.512 Å. This is much smaller than the experimental value of 1.536 Å. It has to be noticed, however, that the SCF-MINDO/3 method yields for the C-C bond in ethane the value of 1.474 Å. For propane and higher homologues, the deviation between the calculated and the experimental values is smaller.

As one can see, the SLG-MINDO/3 in general improves the description of molecular geometry as compared to the SCF-MINDO/3 method. Obvious deterioration takes place only for the ammonia molecule (particularly for the valence angle). However, in other cases, transition to the SLG wave function improves the values of the valence angles. This is seen in the example of the hydrogen peroxide molecule.

The SCF-MINDO/3 method significantly shortens the lengths of the single bonds between nonhydrogen atoms. In the case of the C-C bond, switching to the SLG wave function does not rectify this shortcoming completely, although it noticeably improves the situation. In the case of the bonds containing the heteroatoms, this leads to the improvement of the agreement with the experiment. This is related to the fact that for these bonds the contribution of the covalent configuration into geminal (w_m^2) turns out to be rather large and the account of the intrabond correlation becomes important for correct description of the electronic structure at the semi-empirical level.

Methods of the SLG-NDDO family. Despite some success attained by the concerted use of the MINDO/3 parametrization and the SLG form of the trial wave function, some problems have not been solved by this approach. The heats of formation for unsaturated organic compounds turn out to be strongly underestimated, whereas those for the branched compounds are strongly overestimated; the lengths of the bonds between the atoms bearing LPs are systematically underestimated and the valence angles are reproduced unsatisfactorily. These shortcomings (inherent as it will be seen for the MINDO/3 *parametrization*) have been partially lifted in the frame of the HFR based methods with the NDDO parametrization. So it is interesting to consider semiempirical schemes based on the SLG trial wave function using the three most widespread NDDO parametrizations of the Hamiltonian – MNDO [64], AM1 [68], and PM3 [69]. It has been done in our work [142].

The SLG-MINDO/3 and SCF-MINDO/3 methods have approximately the same accuracy while calculating the heats of formation of organic compounds. Significant deviations from the experiment (for both wave functions) are observed for the branched compounds. The heats of formation are significantly overestimated for both types of wave functions. It is clear that the intrabond correlation has not much to do with this defect. The NDDO parametrization partially rectifies this by a more detailed account of two-center integrals.

In our paper [133] we have performed calculations of the heats of formation using all three parametrizations (MNDO, AM1, PM3) and both types of the variation wave function (SLG and SCF). Empirical functions of distribution of errors in the heats of formation [141] for the SLG-MNDO and SCF-MNDO methods are remarkably close to the normal one. That means that the errors of these two methods, at least in the considered data set, are random. In the case of the SLG-MNDO method, the systematic error practically disappears for the most probable value of the error

$a = -0.5$ kcal/mole, whereas for the SCF-MNDO method this value amounts to *ca.* -3 kcal/mole.

An example of the performance of the SLG-NDDO scheme on a qualitative level is provided by the cyclobutane molecule. The experimental structure is nonplanar with one of the carbon atoms going out by $27°$ of the plane formed by three other atoms. Reproducing its correct structure is believed to be a complex problem. For example, analysis performed in [139] shows that in the ab initio setting, the correct structure is accessible only with large basis sets containing the polarization functions. In the SCF-NDDO methods, the nonplanar geometry of the cyclobutane molecule is not reproduced either. Switching to one of the SLG-NDDO methods yields the CCCC angle close to the experimental. Analogously the SLG function allows us to reproduce the torsional angle in the hydrogen peroxide molecule – a problem not solved in the frame of the SCF-NDDO methods [140].

Linear scaling of SLG based semiempirical methods.   As mentioned above, semiempirical quantum chemistry of large molecules faces the important problem of constructing calculation procedures with the growth of computational costs linear in N (N characterizes the size of the system). Solving this problem requires applying that or some other approach to the separation of electronic variables. The standard way of doing this assumed in quantum chemistry is the HFR approximation for the wave function of the ground state of electrons. The requirements for computational resources of the HFR procedure grow as N^3 and hence the latter cannot be considered a basis for constructing methods linear in N. Moreover, the HFR approximation requires an additional account of correlation to become useful for describing bond cleavage. In the literature, different means are proposed to significantly reduce computational costs without reducing the quality of the obtained results. The first way is to smooth the dependence of computational costs on the size of the system. These approaches usually exploit the localization of electronic degree of freedom, based on the "principle of nearsightedness" [143]. It should be stressed that the use of local one-electron basis states can significantly reduce computational costs [144]. In this context direct determination of localized Hartree-Fock orbitals can be especially useful and viable [145]. Significant acceleration of computation can be achieved by using pseudodiagonalization procedures [146] or by special choice of the trial electronic wave function [132, 133, 147, 148] alternative to the standard HFR form.

To demonstrate the computational capacities of the SLG-MINDO/3 method we carried out calculations (for the fixed geometry) for a series of normal, saturated hydrocarbons ranging from CH_4 to $C_{20}H_{42}$ by the SLG-MINDO/3 and SCF-MINDO/3 methods. It has been shown that the dependence of computation time on the system size is essentially non-linear in the case of the HFR approximation and is practically linear for the geminal approach. The SLG-MINDO/3 procedure is faster than the SCF-MINDO/3 one even for the simplest hydrocarbons. In the case of the normal hydrocarbon $C_{20}H_{42}$ (its semiempirical calculation uses 122 basis functions) the computation time for two methods differs 30 times in favor of the SLG approach.

Even greater advance has been reached by the SLG-NDDO procedures, using the multipole representation of the Coulomb interaction between the atoms. To make the scheme truly linearly scaling, it is necessary to neglect interactions between very distant atoms. Cut-off procedures of that sort are justified only for local states. In the SLG method it is particularly well substantiated because one-electron states forming carrier spaces are atom-centered. The dependence of the required computation time on the system's size n for the multipole SLG-NDDO method, where all interactions between atoms separated by more than 20 Å are totally neglected, is unequivocally linear for the systems for up to 9000 atoms. It is important that the cut-off procedure leads to a very small modification of the calculated heat of formation (less than 0.03 kcal/mol per CH_2 fragment). Of course, in the case of more polar molecules with significant effective atomic charges, the charge-charge interactions beyond 20 Å should be explicitly considered to obtain the same accuracy.

In this section we have considered a family of semiempirical methods of analysis of the electronic structure of molecules, using the trial wave function in the form of the antisymmetrized product of strictly local geminals. The studies performed on these methods allow us to conclude that:

- wave function of the SLG approximation is comparable in quality to the SCF wave function for the characteristic intramolecular interatomic separations, but in variance with the SCF, it has a correct asymptotic behavior under bond cleavage;
- minor modification of the pair resonance parameters β_{AB} for the SLG-MINDO/3 and of β_s^A, β_p^A for the SLG-NDDO methods allowed us to reach a better agreement on the heat of formation calculated by these methods with experiment than it is possible for the corresponding HFR-based methods;
- the SLG function allowed us to set the calculation in terms of the intuitive concepts of bonds, lone pairs and their respective polarities, hybridizations and other intuitively clear concepts common for other areas of computational chemistry, e.g. for MM, with the hope of using this method as a starting point for the hybrid QM/MM methods.
- using the SLG approximation for the wave function together with a parametrization of the NDDO type, but with truly point atomic multipoles and reasonable cut-off of the long-range two-electron Coulomb terms, allowed us to reach $O(N)$ scaling of the entire computational scheme.

2.4.2. Semiempirical method for transition metal complexes with open d-shells

Transition metal complexes (TMCs) represent another, somewhat better known, "Holy Grail" of the semiempirical theory. The HFR-based semiempirical methods and the DFT-based methods suffer from structure deficiency, which does not allow it to reproduce relative energies of electronic states of different spin multiplicity within their respective frameworks without serious ad hoc assumptions.

The difficulties arise precisely when modeling is to be applied to molecules involving transition metal atoms, mainly of the second half of the first transition row. Even among the TMCs formed by these atoms, the problems seem not to be uniformly distributed; the reason is that the standard chemical nomenclature does not provide an adequate classification. In the case of metal carbonyls or metals of the second or even third transition row, the DFT methods are very effective. However, turning to open d-shell compounds of the first transition row metals raises many problems. On the other hand, intuitive distinction in the behavior of two types of the metal compounds is clear to any chemist. In a row of isoelectronic species $Ni(CO)_4$, $Co(CO)_4^-$, $Ni(CN)_4^{2-}$, $Fe(CO)_4^{2-}$ they readily recognize "not a family member", but probably fail to give a reason.

2.4.2.1. Physical picture of TMCs electronic structure

Further analysis is based on the idea that the characteristic experimental behavior of different classes of compounds and the suitability of those or other models used to describe this behavior is ultimately related to the extent to which chromophores responsible for the observed behavior and physically present in the molecular system are reflected in these models by adequate electron groups. The TMCs of interest, can be physically characterized as those bearing the d-shell chromophores. (The analogy between the chromophore concept and McWeeny's theory for the special case of TMCs has also been noticed early in a remarkable work [149]). The basic features in the electronic structure of TMCs of interest, distinguishing these compounds from others, are the following:

1. These molecules contain strongly correlated electrons in the partially filled valence d-shell of the transition metal central atom.
2. The overall charge transfer between the d-shell of transition metal atom and its ligand environment is small.
3. The low-energy spectrum is spanned by excited states of the partially filled d-shell (d-d-spectrum) and it is rather dense.

These properties of the d-shell chromophore (group) prove the necessity for a special description of d-electrons of transition metal atoms in TMCs with an explicit account for the effects of electron correlations in it. Therefore, during the time of QC development (more than three quarters of a century) there was a period when two directions based on two different zero approximations of electronic structure of molecular systems coexisted. This reproduced the division of chemistry into organic and inorganic and took into account the specificity of the molecules related to these classical fields. The organic QC was then limited to the Hückel method [40], the elementary version of the HFR MO LCAO approximation. As it is discussed above, the HFR is not very reliable when it describes TMCs with open d-shells. The description of inorganic compounds – mainly TMCs – within the QC of that time was based on the crystal field theory (CFT) [96, 150, 151]. The latter provided a qualitatively correct description of electronic structure, magnetism and optical absorption spectra of TMCs by explicitly addressing the d-shell chromophore. So the CFT is used by spectroscopists and allows one to interpret and systematize experimental data related to the spectra of the d-shell chromophores. This can be considered a strong experimental support to use the CFT as a theoretical construct for describing TMCs. Our goal here is to include it in the general hybrid context and that is what we shall do in the next sections.

The crystal field theory. The basics of the CFT were introduced in the classical work by Bethe [150] devoted to the description of splitting atomic terms in crystal environments of various symmetry. The splitting pattern itself is established by considering the reduction in the symmetry of atomic wave functions while the spatial symmetry of the system goes down from the spherical (in the case of a free atom) to that of a point group of the crystal environment. It is widely described in inorganic chemistry textbooks (see e.g. [152]).

In quantum mechanics the splitting of electronic terms is described by using the degenerate version of the perturbation theory. The Hamiltonian for electrons in the atom in the crystal environment acquires the form:

$$(2.89) \quad H = H_0 + V_{cf}$$

where H_0 is the Hamiltonian for electrons in the d-shell of a free transition metal ion, which includes the kinetic energy of electrons and potential energy of the interaction of these electrons among themselves and with the metal nucleus, and V_{cf} is the *effective* potential of the crystal environment. In the basis of the atomic functions with the angular parts taken in the form of spherical harmonics, the matrix elements of this operator can be estimated with an assumption concerning the nature of this potential. The simplest possibility is to use the model of an isolated transition metal ion surrounded by point charges – the ionic model. As a rule, in this case only the charges in the first coordination sphere are taken into account. In this case the operator V_{cf} has the form [96, 97]:

$$(2.90) \quad V_{cf}(\mathbf{r}) = \sum_{k=1}^{N} \frac{q_k}{|\mathbf{r} - \mathbf{R}_k|}$$

where $\mathbf{r}$ is the electron radius vector and $\mathbf{R}_k$ is the radius vector of the point charge eq_k. The matrix elements eq. (2.90) in the basis of the one-electron wave functions can be expressed in terms of the radial integrals $F_k(R)$ [96, 97]

$$(2.91) \quad F_k(R) = R^{-(k+1)} \int_0^R r^k R_{nl}^2(r) r^2 dr + R^k \int_R^\infty r^{-(k+1)} R_{nl}^2(r) r^2 dr$$

dependent on the radial parts $R_{nl}(r)$ of the atomic d-functions:

$$(2.92) \quad V_{mm'} = \sum_L Q_L V_{mm'}^L; \quad V_{mm'}^L = F_k(R_L) Y_k^{m-m'}(\theta_L, \phi_L) A_k^{mm'}$$

Here Q_L is the effective charge of the atom L of the ligand; (R_L, θ_L, ϕ_L) are the spherical coordinates of the ligand atom L (the transition metal atom is located in the center of the coordinate frame). Functions $F_k(R_L)$ depend on the distance R_L from the metal atom to the atom L; $Y_k^{m-m'}(\theta_L, \phi_L)$ are the spherical functions with the phases defined following Condon and Shortley [61], which are also given in Section 1.5.1.3; quantities $A_k^{mm'}$, $k = 0, 2, 4$, are numerical coefficients tabulated in [97]. The matrix elements $V_{\mu\nu}^L$ relative to the cubic harmonics can be obtained from $V_{mm'}^L$ by a unitary transformation from the spherical harmonics basis $|m\rangle$ to that of the cubic harmonics $|\mu\rangle$.

In the simplest and widespread case of the octahedral environment there is only one parameter $\Delta \equiv 10Dq$, equal to the difference of the energies of the t_{2g}- and e_g-orbitals (the classification goes along the irreducible representations of the O_h – octahedron – group to which the $L = 2 - D$ – representation of the $SO(3)$ group splits under symmetry reduction) of the metal ion:

$$(2.93) \quad \Delta = \epsilon(e_g) - \epsilon(t_{2g}) = \frac{5}{3}qF_4(R)$$

where R is the distance between the central atom and the point charge located in any of the vertices of the octahedron.

The energies of the d-d-excitations in this model are obtained by diagonalizing the matrix of the Hamiltonian constructed in the basis of n_d-electronic wave functions (n_d is the number of d-electrons). Matrix elements of the Hamiltonian are expressed through the parameters describing the crystal field and those of the Coulomb repulsion of d-electrons, which are Slater-Condon parameters F^k, $k = 0, 2, 4$, or the Racah parameters A, B, and C. In the simplest version of the CFT these quantities are considered empirical parameters and determined by fitting the calculated excitation energies to the experimental ones.

Although the CFT gives a description of the characteristic properties of TMCs at the phenomenological level since the fundamental features of their electronic structure are fixed within the structure of this theory, this approach does not provide any predictive force due to the presence of empirical parameters, which are specific for each compound. Obtaining independent estimates of its parameters (strength of the crystal field) remains its constant problem. All subsequent development of the CFT was centered around this [153]. Within the standard CFT it, however, has no solution, due to the oversimplified description of the transition metal ion's environment (ligands): the CFT employs the ionic model of the environment and calculates the splitting of the initial term of the free metal ion as if it were a pure electrostatic effect. The symmetry is perfectly reproduced even by this simple scheme, but the chemical specific of the environment is completely lost. It is therefore not surprising that the heaviest strike upon the CFT from the (semi)quantitative side was given by TMC spectroscopy in the 1930s. Spectroscopic experiments allowed one to range different ligands according to the strengths of the crystal fields induced by them (parameter Δ or $10Dq$) to the so-called spectrochemical series [96, 97, 151, 159]:

$$(2.94) \quad \mathrm{F}^- < \mathrm{OH}^- < \mathrm{Cl}^- < \mathrm{Br}^- < \frac{1}{2}\mathrm{Ox}^{2-} < \mathrm{H_2O} < \mathrm{SCN}^- <$$

$$< \mathrm{NH_3, py} < \frac{1}{2}\mathrm{En} < \mathrm{CN}^- < \mathrm{CO}.$$

From it one can see that the crystal fields are systematically weaker for charged species than for the uncharged ones with the extremal example of CO inducing the strongest crystal field, but bearing neither charge nor even noticeable dipole moment. Therefore the relative strengths observed in the experiment cannot be explained by the ionic model of the environment. These observations clearly indicate that purely electrostatic effects may be only of minor significance in determining the strength of the effective crystal field felt by the d-shell. Early attempts to get better estimates led to the ligand field theory (LFT) [96, 151]. In its simplest version it is assumed that it is enough to consider the valence shell of the metal ion, containing $3d$-, $4s$-, and $4p$-orbitals and to include one lone pair (LP) orbital per donor atom:

$$\psi^a(e_{gc}) = -x_{e_g}\phi(d_{z^2}) + \frac{y_{e_g}}{\sqrt{12}}(2\chi_z + 2\chi_{-z} - \chi_x - \chi_{-x} - \chi_y - \chi_{-y})$$

$$\psi^b(e_{gc}) = y_{e_g}\phi(d_{z^2}) + \frac{x_{e_g}}{\sqrt{12}}(2\chi_z + 2\chi_{-z} - \chi_x - \chi_{-x} - \chi_y - \chi_{-y})$$

$$\psi^a(e_{gy}) = -x_{e_g}\phi(d_{x^2-y^2}) + \frac{y_{e_g}}{2}(\chi_x + \chi_{-x} - \chi_y - \chi_{-y})$$

$$\psi^b(e_{gy}) = y_{e_g}\phi(d_{x^2-y^2}) + \frac{x_{e_g}}{2}(\chi_x + \chi_{-x} - \chi_y - \chi_{-y})$$

$$\psi^a(a_{1g}) = -x_{a_{1g}}\phi(4s) + \frac{y_{a_{1g}}}{\sqrt{6}}(\chi_x + \chi_y + \chi_z + \chi_{-x} + \chi_{-y} + \chi_{-z})$$

$$\psi^b(a_{1g}) = y_{a_{1g}}\phi(4s) + \frac{x_{a_{1g}}}{\sqrt{6}}(\chi_x + \chi_y + \chi_z + \chi_{-x} + \chi_{-y} + \chi_{-z})$$

$$\psi^a(t_{1u\gamma}) = -x_{t_{1u}}\phi(4p_\gamma) + \frac{y_{t_{1u}}}{\sqrt{2}}(\chi_\gamma - \chi_{-\gamma})$$

$$\psi^b(t_{1u\gamma}) = y_{t_{1u}}\phi(4p_\gamma) + \frac{x_{t_{1u}}}{\sqrt{2}}(\chi_\gamma - \chi_{-\gamma})$$

$$(2.95)$$

By this the environment is considered more realistically: the one electron states of the surrounding atoms are explicitly taken into consideration. In the symmetric environment assumed in eq. (2.95) the d-orbitals of the t_{2g}-symmetry do not get any admixture, whereas those of the e_g-symmetry are shifted upwards due to the said admixture:

$$(2.96) \quad E_d^* = H_{dd} + \frac{|H_{d\lambda}|^2}{H_{dd} - H_{\lambda\lambda}}$$

where H_{dd}, $H_{\lambda\lambda}$, $H_{d\lambda}$ are matrix elements of the one-electron Hamiltonian of the two-level model. Within such a setting, only qualitative explanations can be obtained. First of all, it is not clear where to get the values of $H_{\lambda\lambda}$, which are presumably affected by the details of the composition and structure of the ligands. Next, applying a general HFR-based picture opens a Pandora's box: why not to apply the same approximation to the entire complex, which is better substantiated, but, as we know, can bring a disastrous result.

The LFT has been further formalized within the angular overlap model (AOM) [153, 154] with the additional observation that different ligands (or more precisely – donor atoms) contribute to the effective crystal ligand field almost independently of each other, although this is in strong contradiction with the generally delocalized HFR picture and that each ligand when interacting with a given transition metal ion can be characterized by a relatively small number of parameters (AOM parameters) describing its contribution to the total effective field felt by the d-shell. The angular overlap model of the LFT is briefly described in the following section.

Angular overlap model. The angular overlap model (AOM) as a semiempirical method for estimating the parameters of the CFT going beyond the simple ionic model has been developed by Schäffer and Jørgensen [154–158] (see also a review in [159]). The difference between AOM and the classical CFT is how the potential of the effective field induced by ligands is parametrized. In the AOM it is assumed that the energy of the d-orbitals in the complex changes, as compared to their atomic values, due to overlap with the σ- and π-orbitals of the ligands (the last option is not included in eq. (2.95)). Due to this variance with the classical CFT, it is assumed that there is a certain set of one-electron states on the ligands. In addition, instead of the multipolar expansion of the crystal field over the spherical harmonics [159], a cellular expansion [157, 158, 160] is used in the AOM for the potential of the crystal environment V_{cf} and its matrix elements are given by:

$$(2.97) \quad V_{\mu\nu} = \sum_l v_{\mu\nu}^l = \sum_l (v_{\mu\nu}^l)_{stat} + \sum_l (v_{\mu\nu}^l)_{dyn}$$

where μ, ν are the d-functions of the metal. Summation is extended to all "cells" [160], which are not, however, clearly defined. The cellular contributions to the matrix element of the crystal field $v_{\mu\nu}^l$ are related to the cellular parameters $e_{\lambda\lambda'}^l$ by the geometry-based formula:

$$(2.98) \quad v_{\mu\nu}^l = \sum_{\lambda\lambda'} R_{\mu\lambda}^{l+} e_{\lambda\lambda'}^l R_{\lambda'\nu}^l$$

Coefficients $R_{\lambda\mu}^l$ form an orthogonal matrix $\mathbf{R}^l$ transforming the d-orbitals under rotation of the laboratory coordinate frame (LCF) to the local coordinate frame related to the ligand l and constructed such that its Oz axis is going through the metal atom and the ligand atom (DCF – diatomic coordinate frame). The perturbation caused by the ligand has a matrix representation $e_{\lambda\lambda'}^l$ in the DCF with $\lambda = \sigma, \pi(x), \pi(y), \delta(xy), \delta(x^2 - y^2)$. These quantities are considered parameters of the AOM.

Simple estimates for the cellular potentials can be extracted from the MO LCAO method in the two-level model eq. (2.96). If one considers the interaction of the type λ ($\lambda = \sigma, \pi, \delta$) between the d-orbital of the metal ϕ_d and the orbital ϕ_λ, representing a mixture of the orbitals of the ligand and corresponding by local symmetry s- and p-orbitals of the metal and taking into account that according to the definition of the cellular potentials $E_d^* = \langle \phi_d \mid \nu_\lambda^l \mid \phi_d \rangle$, one gets the following expression for the energy of the antibonding MO of predominantly d-character and thus for ν_λ^l:

$$(2.99) \quad E_d^* = \nu_\lambda^l = H_{dd} + \frac{\mid H_{d\lambda} \mid^2}{H_{dd} - H_{\lambda\lambda}}$$

Successful application of the AOM parametrization scheme for interpretation of the electronic spectroscopy data based on the values extracted from experiment [159] demonstrates that the general parametrization scheme eq. (2.99) implied by the AOM, most probably reflects some general features of the electronic structure of the good fraction of TMCs. However, numerical estimates of its parameters according to formula eq. (2.99) were not particularly successful. As a result the AOM requires for its application large parameter sets (for the cells) specific for each pair of metal - ligand, which makes the parametrization boundless. The AOM parameters remain empirical quantities just as the $10Dqs$ were in the original CFT.

The problem of estimating crystal field parameters can be solved by considering the CFT/LFT as a special case of the effective Hamiltonian theory for one group of electrons of the entire N-electronic system in the presence of other groups of electrons. The standard CFT ignores all electrons outside the d-shell and takes into account only the symmetry of the external field and the electron-electron interaction inside the d-shell. The LFT acts in a similar way: only the d-shell and electrons in it are considered, although the orbitals used for it are not precisely the atomic d-states, but some orbitals of the same symmetry and of predominantly d-character having an uncontrollable contribution from some poorly defined states of the environment. This of course allows one to improve the overall description by giving additional degrees of freedom, but in fact this move is somewhat similar to the recommendation to extend the quantum subsystem: the boundary conditions and/or parameters for the LP states involved are not set in the LFT and it is not clear how this can be done. The problem is solved by sequential deduction of the effective Hamiltonian for the d-shell, carried out in [161]. It is based on the representation of the wave function of a TMC in the form of an antisymmetrized product of group functions of d-electrons and other (valence) electrons of a complex. As a result, the rest of the system from the point of view of the CFT is explicitly taken into consideration and allows one to express the CFT (LFTs or AOMs) parameters through the characteristics of the electronic structure of the environment of the metal ion. We shall characterize the effective Hamiltonian of the crystal field (EHCF) method and the numerical results obtained within its framework.

2.4.2.2. *Effective Hamiltonian of the crystal field (EHCF)*

In this section we construct a semiempirical method for describing the electronic structure of the TMCs, which allows us to calculate the d-d-spectra for a wide variety of TMCs of different types. The TMCs' electronic wave function formalizing the CFT ionic model has a fixed number of electrons in the d-shell. In the EHCF method it is used as a zero approximation and the electron transfers between the d-shell and the ligands are treated as perturbations. Following the standard semiempirical setting we restrict the AO basis for all atoms of the TMC by the valence orbitals. All the AOs of the TMC are then separated into two subsets of which one (the d-system) contains $3d$-orbitals of the transition metal atom, and the other (the "ligand subsystem", or the l-system) contains the $4s$- and $4p$-orbitals of the transition metal atom and the valence AOs of all ligand atoms.

Formally the theory evolves as follows. The low-energy d-d-spectrum of the TMC can be obtained if the Hamiltonian is rewritten in the form:

$$(2.100) \quad H = H_d + H_l + W^c + W^r$$

where H_d is the Hamiltonian for d-electrons, H_l is the Hamiltonian for the ligand system, W^c is the Coulomb interaction, and W^r is the resonance interaction operators. Exact wave functions of the system described by the Hamiltonian eq. (2.100) can be represented as a superposition of the functions, corresponding to different distributions of N valence electrons over two subsystems with natural identification of the d-shell with the R-system and of the rest with the M-system

$$(2.101) \quad \Psi = \sum_{n_1 n_2} \sum_i C_i(n_1 n_2) \Phi_i(n_1 n_2)$$
$$n_1 + n_2 = N$$

(n_1 and n_2 are the numbers of electrons in the subsystems). On the basis of the physical concepts described above, one can assume that the main contribution to the ground state wave function is provided by the functions with the number of d-electrons corresponding to the valence state of the metal atom in the TMC. As an approximate form of the trial wave function for the description of the TMCs, it is reasonable to take a function from the subspace with the fixed number of the d-electrons n_d (by this the specific type of the TMCs subject to our consideration is formalized; these are the so-called Werner complexes). Contributions of the functions with different numbers of d-electrons will be taken into account as described in Chapter 1 i.e. using Löwdin partitioning.

As in the general theory the operator W^c commutes with the operators of numbers of particles in the d- and l-systems. In the Hamiltonian eq. (2.100) only the term W^r mixes the states with a different distribution of electrons between subsystems. The $H^{\mathrm{eff}}(E)$ for the TMC can be written in the form:

$$
\begin{aligned}
H^{\mathrm{eff}} &= P H^0 P + W^{rr} \\
(2.102) \quad H^0 &= H_d + H_l + W^c \\
W^{rr} &= P W^r Q (EQ - Q H^0 Q)^{-1} Q W^r P
\end{aligned}
$$

The electronic wave function for the n-th state of the complex is written then as the antisymmetrized product of the wave functions of the electron groups introduced above:

$$(2.103) \quad \Psi_n = \Phi_n^d \wedge \Phi^l$$

The EHCF theory can be applied by assuming Φ_n^d to be a full CI function for n_d electrons in the d-shell, and Φ^l to be a HFR single determinant ground state for the

l-system. This reflects the main feature of the electronic structure of the TMC, that is the presence of the strongly correlated d-shell with low energy excitations localized in it and of relatively inert (i.e. having rather high excitation energies) ligands.

We have restricted ourselves to the case of complexes for which the excitation energies in the l-system are much higher than the excitation energy in the d-shell of the metal. In these complexes the number of valence electrons in the ligand subsystem is even and thus the ground state of the l-system can be approximated by a single Slater determinant Φ^l with zero total spin. Then the wave function Ψ_n acquires the form of eq. (1.245):

$$(2.104) \quad \Phi_n = \sum_k c_k^n \, |n_d S\sigma\Gamma\gamma k\rangle \wedge \Phi_l = \Phi_n^d \wedge \Phi^l$$

In this case the spin and symmetry of the function eq. (2.104) coincide with the spin and symmetry of the wave functions of the d-system Φ_n^d. An assumption that the functions Φ_n^d and Φ^l satisfy the strong orthogonality condition of eq. (1.185) together with the variational principle yields a pair of the coupled equations for the functional multipliers:

$$(2.105) \quad \begin{aligned} H_d^{\text{eff}} \Phi_n^d &= E_n^d \Phi_n^d \\ H_l^{\text{eff}} \Phi^l &= E^l \Phi^l \end{aligned}$$

which repeats eq. (1.246) with a correction for notation. Effective Hamiltonians for the subsystems have the form:

$$(2.106) \quad \begin{aligned} H_d^{\text{eff}} &= H_d + \langle \Phi^l \, |W^c + W^{rr}| \, \Phi^l\rangle \\ H_l^{\text{eff}} &= H_l + \langle \Phi_n^d \, |W^c + W^{rr}| \, \Phi_n^d\rangle \end{aligned}$$

According to the previous notes for obtaining the estimates of the parameters of the effective crystal field, we need a rather precise description of the electronic structure of the l-system. A simple algorithm for solving the system of eq. (2.105) proposed in [161] reduces to solving the equation for Φ^l in a semiempirical HFR approximation and in calculating the corresponding one-electron density matrix, orbital energies, and MO LCAO coefficients. They are used for constructing H_d^{eff} as shown below.

In the described model, in agreement with eq. (1.247), the energies of the low lying excited states of the entire complex are identified with the energies of the excited states of the d-system as it should be in the CFT. However, in variance with the CFT (and the LFT), this approach explicitly takes into account the electronic structure of the ligand environment. The excited states of the l-system do not appear in this setting explicitly. According to the general theory, projecting them out leads to reducing (effective screening) Coulomb interactions between the electrons in the R-system. In this case the part of the R-system is taken by the d-shell. Reduction of the effective electron-electron repulsion in it is widely known from inorganic chemistry textbooks under the name of nephelauxetic effect. This beautiful Greek word means that the cloud (nephelon) of d-electrons extends when the ion becomes a central one in a complex rather than a free one. Then the average electron-electron separation in it increases whereas the average Coulomb repulsion decreases. Numerically the magnitude of this effect is usually expressed by a ratio $B/B_0 < 1$, where B is one of the Racah parameters in the complex and B_0 is its value in the free ion. The mechanism of extension is believed to be the formation of delocalized MOs of predominantly d-character, but having some contribution from the ligand states (LFT). In this theory, the delocalization of the d-states due to mixing with the ligand states is included in the effective crystal field. Thus one may expect that the nephelauxetic series (that of the ligands with decreasing B/B_0) has to be similar to the spectrochemical series (that of increasing $10Dq$ – see above). This however does not happen. On the other hand the general theory allows for reducing the electron-electron interaction in the d-shell due to interference between induced polarizations in the d- and l-systems. The latter is not included explicitly in the EHCF treatment, but is present in the general theory. Incidentally the nephelauxetic series is known to go parallel to the polarizability of the donor atoms [159].

Effective Hamiltonian for the l-system.   Further formal development of the theory evolves as follows. Expression for H_l^{eff} has the form:

$$(2.107) \quad H_l^{\mathrm{eff}} = H_l + \langle\langle W^c \rangle\rangle_d + \langle\langle W^{rr} \rangle\rangle_d$$

where symbol $\langle\langle \ldots \rangle\rangle_d$ stands for averaging over the state of the d-shell. We assume that H_l is written in one of the approximations of the ZDO family. Following the remarks given in Chapter 1 for calculating the effective Hamiltonian for the d-shell, it is necessary to know the spectrum of the operator H^0 in the subspace $\mathrm{Im}\,Q$ which is defined by the orbital energies of the l-system. These energies must be calculated without the d-system, i.e. with the complete screening of the part of the core charge of the metal atom by d-electrons but without effective resonance interaction. For that reason when calculating the electronic structure variables of the l-system for using them in a calculation of the effective Hamiltonian for the d-system in eq. (2.107) it is necessary to drop the term $\langle\langle V^{rr} \rangle\rangle_d$. Then in the second quantization form the Hamiltonian for the l-system reads:

$$
\begin{aligned}
H_l = {} & \sum_{m,\sigma} (U_{mm} - \sum_L V_{ML})\mathsf{m}_\sigma^+ \mathsf{m}_\sigma + \\
& + \sum_L \sum_{l\in L,\sigma} (U_{ll} - \sum_{L'\neq L} V_{LL'} - V_{LM})\mathsf{l}_\sigma^+ \mathsf{l}_\sigma + \\
& + \sum_{ml,\sigma} \beta_{ml}(\mathsf{m}_\sigma^+ \mathsf{l}_\sigma + h.c.) + \sum_{ll',\sigma} \beta_{ll'} \mathsf{l}_\sigma^+ \mathsf{l}_\sigma' + \\
& + \frac{1}{2} \sum_{\substack{ll',\sigma+\\ l''l''',\tau}} (ll' \mid l''l''')\mathsf{l}_\sigma^+ \mathsf{l}_\tau''^+ \mathsf{l}_\tau''' \mathsf{l}_\sigma';
\end{aligned}
$$

$$(2.108)$$

here l_σ^+ (l_σ) are the creation (annihilation) operators of an electron with the spin projection σ on an l-AO. The first term in eq. (2.108) describes the interaction of the 4s- and 4p-electrons of the metal ($m = 4s, 4p_x, 4p_y, 4p_z$) with the metal core (parameters $U_{mm} < 0$) and the ligand atom cores (parameters $V_{ML} > 0$). The second term describes the interaction of the ligand electrons with the ligand cores (parameters $U_{ll} < 0$), with the cores of the other ligand atoms (parameters $V_{LL'} > 0$) and with the metal core (parameter $V_{LM} > 0$). The third and fourth terms describe the resonance interactions in the ligand subsystem (parameters $\beta_{ml} < 0$ and $\beta_{ll'} < 0$). The last term describes the Coulomb interactions between electrons ($(ll' \mid l''l''')$ are the corresponding two-electron integrals).

The term $\langle\langle V^c \rangle\rangle_d$ describes the Coulomb interaction of electrons of the l-system with the electron density in the d-shell. It reduces to renormalization of the one-electron parameters of the bare Hamiltonian H_l according to the formulae [161]:

$$(2.109) \quad U_{ii}^{\text{eff}} = U_{ii} + \frac{n_d}{5} \sum_\mu g_{\mu i}$$

$$V_{LM}^{\text{eff}} = V_{LM} - e^2 n_d F_0(R_L)$$

This expression has a simple physical meaning, being in agreement with the solutions for separating the core charge of the frontier atoms in the hybrid methods. In agreement with eq. (2.109) the attraction of electrons on the 4s- and 4p-orbitals of the metal atom to its core is screened by the Coulomb interaction with d-electrons. The second term in the right hand part of eq. (2.109) in both cases describes the screening of the core charge of the metal by d-electrons, as a result of which the attraction of electrons on the orbitals of the ligands to the metal core also weakens. We see that for the theory of the TMC, constructing such a separation of the entire system into subsystems is characteristic and the strong Coulomb intersubsystem interactions are minimized. Due to the screening of a part of the metal core charge by the d-electrons the total charge of the metal atom from the point of view of the l-system turns out to be equal to its formal oxidation degree (2 or 3). If one turns to the total charge of the *d-system* itself, the latter turns out to be vanishing in agreement with the prescription of Section 1.7.

Effective Hamiltonian for the d-system. Now we consider the expression for H_d^{eff}:

$$(2.110) \quad H_d^{\text{eff}} = H_d + \langle\langle W^c \rangle\rangle_l + \langle\langle W^{rr} \rangle\rangle_l$$

where $\langle\langle \ldots \rangle\rangle_l$ stands for the averaging over the ground state of the l-system. The bare Hamiltonian for d-electrons of TMCs H_d has the form:

$$H_d = U_{dd} \sum_{\mu\sigma} d_{\mu\sigma}^+ d_{\mu\sigma} + \sum_{\mu\nu\sigma} V_{\mu\nu}^{\text{core}} d_{\mu\sigma}^+ d_{\nu\sigma} +$$
$$(2.111)$$
$$+ \frac{1}{2} \sum_{\mu\nu\rho\eta} \sum_{\sigma\tau} (\mu\nu \mid \rho\eta) d_{\rho\tau}^+ d_{\mu\sigma}^+ d_{\nu\sigma} d_{\eta\tau}$$

where $d_{\mu\sigma}^+$ ($d_{\mu\sigma}$) are the creation (annihilation) operators for an electron on the μ-th d-orbital with the spin projection σ; U_{dd} is the core attraction parameter of the

d-electrons; $V_{\mu\nu}^{\text{core}}$ is the matrix element of the operator of interaction of d-electrons with the ligand atom cores; and $(\mu\nu \mid \rho\eta)$ is the two electron integral of the Coulomb interaction.

According to eq. (1.246) it is necessary to average the operators W^c and W^{rr} with the function of the ground state of the l-system Φ^l. For $\langle\langle W^c\rangle\rangle_l$ we get an expression:

$$\langle\langle W^c\rangle\rangle_l = \langle\langle W_1^c\rangle\rangle_l + \langle\langle W_2^c\rangle\rangle_l =$$

$$(2.112) \qquad = \sum_\mu \sum_i g_{\mu i} P_{ii}\hat{n}_\mu + \sum_{\mu,\nu,\sigma} \sum_L V_{\mu\nu}^L P_{LL} \mathsf{d}_{\mu\sigma}^+ \mathsf{d}_{\nu\sigma}$$

The first term in this expression describes the shifts of the d-levels coming from the interactions of d-electrons with electrons on the $4s$- and $4p$-orbitals of the metal atom. The second term represents the interaction of d-electrons with electrons on the valence orbitals of the ligands. The sum of the first and the second terms in eq. (2.111) has the form of the operator of the crystal field induced by the effective charges located on the ligand atoms.

The resonance operator W^r has the form:

$$(2.113) \quad W^r = -\sum_\sigma \sum_{\mu,j} \beta_{\mu j}(\mathsf{d}_{\mu\sigma}^+ \mathsf{l}_{j\sigma} + \mathsf{l}_{j\sigma}^+ \mathsf{d}_{\mu\sigma})$$

where j is the number of MO in the l-system, $\beta_{\mu j}$ is the resonance integral between the μ-th d-orbital and j-th l-MO. Using the general formula eq. (1.244) including the retarded and (ret) advanced (adv) Green's function of the l-system [162, 163]:

$$(2.114) \qquad \begin{aligned} G_{ii}^{ret}(z) &= \langle\Phi^l|\mathsf{l}_i(F_l^{\text{eff}} - z)^{-1}\mathsf{l}_i^+|\Phi^l\rangle \\ G_{ii}^{adv}(z) &= -\langle\Phi^l|\mathsf{l}_i^+(F_l^{\text{eff}} - z)^{-1}\mathsf{l}_i|\Phi^l\rangle \end{aligned}$$

(here F_l^{eff} is the effective one-electron Fock operator for the l-system corresponding to the Hamiltonian H_l^{eff}) and following the HFR approximation accepted for the l-system its Green's function takes the form of eq. (1.209):

$$(2.115) \quad G^{\binom{adv}{ret}}(x, x'; E) = \sum_{i=1}^N \frac{\phi_i^*(x)\phi_i(x')}{E - \varepsilon_i \pm i\delta}$$

After summation over the spin projection σ we get:

$$(2.116) \quad \langle\langle W^{rr}\rangle\rangle_l = \sum_{\mu,\nu,\sigma} W_{\mu\nu}^{cov}\mathsf{d}_{\mu\sigma}^+\mathsf{d}_{\nu\sigma} - 2\sum_{\mu j}\beta_{\mu j}^2 \frac{n_j^2}{\Delta E_{j\mu}}$$

where

$$(2.117) \quad W_{\mu\nu}^{cov} = -\sum_i \beta_{\mu i}\beta_{\nu i}[G_{ii}^{ret}(I_d) + G_{ii}^{adv}(A_d)]$$

and retarded and advanced Green's functions are taken at the I_d, A_d values of their respective arguments (corresponding to the electron extraction from the d-shell and its addition to it, respectively). The poles of the Green's function for the l-system represent the values of the ionization potential I_i and the electron affinity A_i related to the i-th l-MO filled or vacant, respectively. Taking into account that the extracted (added) electron is transferred to the atom (from the atom) of transition metal, not to (from) infinity, the ionization potentials and electron affinities of the l-system are shifted by the energy g_{di} of Coulomb interaction between an electron and a hole placed in the metal d-shell and to the i-th l-MO:

$$(2.118) \quad \begin{aligned} I_i &= -\epsilon_i - g_{di} \\ A_i &= -\epsilon_i + g_{di} \end{aligned}$$

where ϵ_i is the energy of the i-th MO of the l-system.

Finally the effective Hamiltonian for the d-shell H_d^{eff} acquires the form:

$$H_d^{\text{eff}} = C + \sum_{\mu,\nu,\sigma} U_{\mu\nu}^{\text{eff}} d_{\mu\sigma}^{+} d_{\nu\sigma} + \frac{1}{2} \sum_{\mu\nu\rho\eta} \sum_{\sigma\tau} (\mu\nu \mid \rho\eta) d_{\mu\sigma}^{+} d_{\rho\tau}^{+} d_{\eta\tau} d_{\nu\sigma}$$

where C is the constant from eq. (2.116) and effective parameters contain corrections from the Coulomb and resonance interactions of the d-shell with the l-system:

$$(2.119) \quad U_{\mu\nu}^{\text{eff}} = \delta_{\mu\nu} U_{dd} + W_{\mu\nu}^{\text{atom}} + W_{\mu\nu}^{\text{ion}} + W_{\mu\nu}^{\text{cov}}$$

Here

$$(2.120) \quad W_{\mu\nu}^{\text{atom}} = \delta_{\mu\nu} \sum_{i \in s,p} g_{\mu i} P_{ii}$$

The ionic contribution $W_{\mu\nu}^{\text{ion}}$ equals:

$$(2.121) \quad W_{\mu\nu}^{\text{ion}} = \sum_{L} Q_L V_{\mu\nu}^{L}$$

and the resonance (or covalent) contribution $W_{\mu\nu}^{\text{cov}}$ equals

$$(2.122) \quad W_{\mu\nu}^{\text{cov}} = - \sum_{j}^{(MO)} \beta_{\mu j} \beta_{\nu j} \left(\frac{(1 - n_j)^2}{\Delta E_{dj}} - \frac{n_j^2}{\Delta E_{jd}} \right)$$

where $n_j = 0, 1$ is the occupation number for the j-th MO of the l-system; ΔE_{dj} (ΔE_{jd}) is the energy for excitation of an electron from the d-orbital (from the j-th MO) to the j-th MO (to the d-orbital):

$$(2.123) \quad \begin{aligned} \Delta E_{dj} &= I_d + \epsilon_j - g_{dj} \\ \Delta E_{jd} &= -\epsilon_j - A_d - g_{dj} \end{aligned}$$

With this, the construction of the EHCF method has been completed.

2.4.2.3. Semiempirical implementations of the EHCF paradigm

In the context of the EHCF construct described in the previous section, the problem of semiempirical modeling of TMCs' electronic structure is seen in a perspective that is somewhat different from that of the standard HFR MO LCAO-based setting. The EHCF provides a framework which implicitly contains the crucial element of the theory: the nonvanishing cumulant of the two-electron density matrix related to the d-shell. Instead of hardly systematizeable attempts to catch qualitative features of the electronic structure by a more or less sophisticated parametrization for the transition metal atoms, it is now possible to check in a systematic way the value of different parametrization schemes already developed in the "organic" context for the purpose of estimating the quantities necessary to calculate the crystal field according to prescriptions given by the EHCF theory eqs. (2.120) and (2.122). The many-electron states in the d-shell of the metal ion in the complex are described by the FCI with the effective Hamiltonian for the d-subsystem (H_d^{eff}) with the matrix elements which are estimated using any "organic" semiempirical scheme. In such a formulation, the EHCF method was parametrized for calculations of various complexes of metals of the first transition row, with mono- and polyatomic ligands. In [161, 166–168] the parameters for the compounds with donor atoms C, N, O, F, Cl and for doubly and triply charged ions of V, Cr, Mn, Fe, Co, Ni and Cu are described. In fact only the parameters scaling the resonance interaction (one-electron hopping) between the d-shells and the donor atoms were adjusted. These parameters do not depend on the details of the chemical structure of the ligands; rather they are characteristic for each pair of metal-donor atom. The dependence of the exerted effective field on details of geometry and chemical composition of the ligands is to be reproduced in the frame of a standard HFR-based semiempirical procedure. We exemplify this by only one instance of the octahedral complex $MnCl_6^{4-}$. One parameter (β^{MnCl}) has to be fitted to get the data given in Table 2.2. Using this value, the spectrum of the $MnCl_4^{2-}$ complex is reproduced at *its* geometry as shown in Table 2.3.

Further evaluations [164, 165] have demonstrated the applicability of the fitted system of parameters for calculations of the electronic structure and spectra of numerous complexes of divalent cations using merely the CNDO parametrization for the l-system. In [140, 169] the EHCF method is also extended for calculations of the ligands by the INDO, MINDO/3, and SINDO/1 parametrizations. In all calculations the experimental multiplicity (spin) and spatial symmetry of the corresponding ground states were reproduced correctly. The summit of this approach has been reached in [170] by calculations on the *cis*-[Fe(NCS)$_2$(bipy)$_2$] complex. Its molecular geometry is known both for the HS and LS isomers of the said compound. The calculation

Table 2.2. Calculated and experimental d-d transition energies in octahedral $MnCl_6^{4-}$ complex.

	$E^{\mathrm{theor}}\,\mathrm{cm}^{-1}$	$E^{\mathrm{exp}}\,\mathrm{cm}^{-1}$
$^6A_{1g} \rightarrow\,^4 T_{1g}$	18510	18500
$\rightarrow\,^4 T_{2g}$	21520	22000
$\rightarrow\,^4 A_{1g},\,^4 E_g$	23590	23590
$\rightarrow\,^4 T_{2g}(D)$	26460	26750
$\rightarrow\,^4 E_g(D)$	28065	28065
$\rightarrow\,^4 T_{1g}(P)$	32630	36500
$\rightarrow\,^4 A_{2g}(F)$	38140	38400
$\rightarrow\,^4 T_{1g}(F)$	39140	40650

Table 2.3. Calculated and experimental d-d transition energies in octahedral $MnCl_4^{2-}$ complex. Parameter β^{MnCl} is fitted for the $MnCl_6^{4-}$ complex.

	$E^{theor}\,cm^{-1}$	$E^{exp}\,cm^{-1}$
$^6A_1 \to ^4T_1$	21102	21250
$\to ^4T_2$	22446	22235
$\to ^4A_1,^4E$	23020	23020
$\to ^4T_2(D)$	25996	26080
$\to ^4E(D)$	26709	26710
$\to ^4T_1(P)$	30195	27770
$\to ^4A_2(F)$	36444	33300
$\to ^4T_1(F)$	36706	34500
$\to ^4T_2$	37732	36650

Table 2.4. Calculated and experimental d-d transition energies in spin-active [Fe(bipy)$_2$ (NCS)$_2$] complex.

	$E^{theor}\,cm^{-1}$	$E^{exp}\,cm^{-1}$
$^1A_1 \to ^3T_{1g}$	9930	10400
$^1A_1 \to ^1T_{1g}$	17930	18500
$^5T_{2g} \to ^5E_g$	10900	11900
$^5T_{2g} \to ^3T_{1g}$	10700	
$^5T_{2g} \to ^1A_{1g}$	10900	

for both reproduces the respective ground state spins and the spectra of low lying d-d-excitations in a remarkable agreement with experimental data as shown in Table 2.4. Another good example is the treatment of metal porphyrins using the EHCF method. As already mentioned above, for the decades the ab initio methods fail to reproduce the experimental ground state of Fe(II) porphyrine. It is really a complex case since it is an intermediate spin ($S = 1$ – i.e. neither HS nor LS) and spatially degenerate state (^3E). However, applying even very sophisticated methods (including CASPT2 which is considered to be a method of choice for TMCs in the ab initio area) has not yet led to the desired success. According to [171] the HS forms are ground states and the possibility of getting a correct result is rather low, as the gap mounts up to 1 eV in favor of the HS state. Meanwhile the EHCF method in its simplest setting (CNDO type of parametrization employed for the l-system) yields the experimental ground state ^3E without any further adjustment of parameters. This indicates that in fact the problem is not purely numerical inaccuracy of the ab initio methods, but certain structural elements of the theory, which prevent them from obtaining the correct result. Certainly some important configurations are missing in the CASPT2 setting of [171].

2.5. CLASSICAL MODELS OF MOLECULAR STRUCTURE: MOLECULAR MECHANICS

The quantum chemical methods of modeling molecular PES reviewed above do so by directly addressing the electronic structure of the corresponding molecular systems; so the main topic of our discussion was to cover existing methods of doing that. The traditional methods (even simple semiempirical ones) are, however, computationally too demanding to be routinely used for getting PES in the MC or MD contexts. As for those using non-HFR semiempirics, there is not enough experience as yet. This raised the demand for more economical tools for direct modeling of PESs. Such a demand is satisfied by Molecular Mechanics (MM). At first glance, MM seems to ignore the electronic structure of the molecular system under study. Even in the respective community quantum mechanical models are considered excessively complex and superfluous, compared to the problems to be solved. Another important, but not explicitly formulated reason is that the standard QC does not provide a transparent link to intuitively clear concepts of chemistry. According to the current paradigm bonds, valencies, and other characteristics have to appear "by themselves" from the numbers produced by QC program suits. Unfortunately, even a big number of numbers is unable to replace real understanding. In fact MM implicitly takes the electronic structure into account in a different way, in terms different from those of quantum chemistry, rather pertinent to theoretical chemistry: not in terms of MOs or electron densities, but in terms of chemical bonds. Of course, such a picture can be valid only if the bonds themselves are well defined, which again implies certain assumptions concerning molecular electronic structure. We shall turn to this problem later, in connection with MM targeted to coordination compounds. On the other hand, as MM is an indispensable component of hybrid methods we review them briefly now.

2.5.1. Force fields of molecular mechanics

Molecular mechanics [172, 173] is a versatile and currently very popular tool in laboratory and industrial practice. It remains the most practical way of modeling PES for large molecules. Molecular mechanics has as its origin the analysis of the vibrational spectra similar to that described in Chapter 1. The similarity between the behavior of quantum and classical harmonic oscillators stimulated the early works on modeling the vibrational spectra of (organic) molecules in terms of their dynamical matrices, defined as those of the second derivatives of the PES in the vicinity of local minima written in terms of the mass weighted Cartesian shifts of atoms. This construct was (and remains) very successful as it allows one to establish a relation between the vibrational frequencies observed in different types of experiments (IR-absorption and Raman scattering) with the eigenvalues of the dynamical matrices and through the latter with molecular potentials (PES). The MM, however, goes somewhat further and postulates a special general pattern that PES of at least "organic" molecules should obey. This pattern implies the intuitively (but only intuitively – see above) transparent picture of molecular electronic structure expressed in terms of characteristic chemical bonds and groups. In a strict sense MM is only possible if distinguishable and regularly behaving bonds exist in the molecules and these bonds migrate from one molecule to another within a wide class of the latter without changing too much (are transferable or observable in Ruedenberg's terminology). The basic assumption of MM is a hypothesis of a possible direct parametrization of the molecular PES in the form of a sum of transferable bonding and non-bonding potentials as well as some additional cross terms:

$$(2.124) \quad E = E_b + E_{ang} + E_{tors} + E_{imp} + E_{cross} + E_{nb}$$

Each of them is a more or less simple or by contrast sophisticated, but explicit, function of natural nuclear coordinates i.e. bond lengths, valence and torsion angles. Only nonbonding terms are defined as functions of all internuclear separations. The point is, of course, not the possibility of migration between Cartesian coordinates of atoms (nuclei) and some internal molecular coordinates, but the inherent dependence of energy, on *bond lengths,* etc. Among the terms entering the above expression, the energies of bond stretching E_b, energies of valence angle bending E_{ang}, and energies of torsion interactions E_{tors} and E_{imp} (the precise meaning of these terms is explained below) pertain to the bonding contributions. In the case of stretching and bending energies, the additional guess of the Hook-law-like dependence of these contributions on geometry parameters is usually accepted. The geometry variables used in definitions of the bonding energy contributions are the deviations of bond lengths $(l - l_0)$ and of the valence angles $(\alpha - \alpha_0)$ from some ideal values l_0 and α_0, which are parameters of specific implementations of the general MM approach:

$$(2.125) \quad E_b \quad = \frac{1}{2} \sum K_l (l - l_0)^2$$
$$E_{ang} = \frac{1}{2} \sum K_\alpha (\alpha - \alpha_0)^2$$

The elasticity constants K_l and K_α also pertain to the parameter sets. Sometimes (and in modern implementations of MM – quite frequently) the elasticity constants K_l and K_α themselves can be taken to be dependent on geometry, by which anharmonicity effects are taken into account. The summations in the above expression for the bond stretching and angle bending force fields extend respectively to all pairs of atoms connected by a chemical bond or to all triples of consequently bonded atoms with the angle understood as one with the vertex in the middle atom of the triple measured between the "directions" of the bonds connecting the vertex with the end atoms of the triple. The values l_0, α_0, K_l, and K_α are assumed to be specific for the MM atomic types (see below) involved in the formation of the respective bonds. One more aspect relates to the quantities l_0 and α_0. In the literature, these pair- or triple-specific quantities are sometimes referred to as equilibrium ones. This is of course misleading: there is no specific molecule where the l_0 value is the true equilibrium distance or α_0 is a true valence angle, unless this did not happen by accident. By contrast, the equilibrium distances in different molecules appear by optimizing the overall energy expression eq. (2.124): they deviate from the l_0, α_0 values to the extent stipulated by the interference of all other contributions to the energy. The l_0 quantity might be thought to be the equilibrium length of the individual bond between the atoms of given types, which clearly does not exist, as the type definitions themselves are largely based on the bonding environment of the atom at hand. Moreover, the ideal values l_0 and α_0 (and the elasticity constants and other parameters of the current MM methods) are in fact sensitive to the details of the nonbonding and cross interactions accepted in each specific implementation of the method.

Usually the harmonic approximation with some anharmonicity corrections suffice to describe the energy dependence on the interatomic separations of the bonded atoms in the vicinity of the minimum PES. Nevertheless, sometimes, when it attempts either to simultaneously reproduce molecular geometries and the heats formation or simply to cover a wider range of molecular geometries where anharmonicity effects become more pronounced, the bond stretching energy terms are taken in the form:

$$(2.126) \quad E_b = \sum D_0[e^{-\alpha(l-l_0)} - 1]^2$$

known as the Morse potential. One may think that it allows a greater flexibility in parametrization than the simple harmonic function as each bond is characterized by three parameters rather than two. However, it is an illusion, as using the Morse function requires a larger data set to be reproduced so that the D_0 parameters are fixed by the data on "bond energies" and parameter α, the only one free so far, is in fact fixed by the relation:

$$(2.127) \quad K_l = 2D_0\alpha^2$$

The higher anharmonicity constants are all proportional to $D_0\alpha^n$ for the corresponding value of n. The Morse potential is usually thought to be too "rigid" in the sense that the dissociation energy limit D_0 is approached too fast. Another point to be mentioned is that the small interatomic separation limit is not correct for the Morse potential. It takes a finite value for the zero bondlength. Although it may be very large and not accessible in the optimization setting, the incorrect potential behavior at the zero distance may affect the results obtained in the MC and/or MD contexts if no precaution is taken.

Further energy contributions defined in terms of bonds are the so-called torsion and improper torsion contributions:

$$(2.128) \quad E_{tors} = \frac{1}{2} \sum \sum_n V_n(1 + \cos[n(\phi + \psi)])$$

$$E_{imp} = \frac{1}{2} \sum K_{imp}\delta^2 = \frac{1}{2} \sum K'_{imp}d^2$$

In the first expression, the summation is extended to the quadruples of sequentially bonded atoms ABCD and the energy constants V_n are specific for the quadruples of the types of the atoms involved. The torsion angle is that between the planes ABC and BCD. The $n = 1$ term describes a rotation which is periodic by 360°, the $n = 2$ term is periodic by 180°, the $n = 3$ term is periodic by 120° and so on. The V_n constants determine the contribution of atoms A and D to the barrier of rotation around the

B-C bond. Depending on the situation, some of these V_n constants may be zero. In the second expression, the summation extends to quadruples of atoms where three of them are linked by the fourth, forming a vertex. In this case the angle δ is the one between a bond and the plane formed, whereas the distance d is the one between the vertex atom and the plane formed by the other three atoms of the quadruple. Numerical values of the constants (and even their dimensions) K_{imp} and K'_{imp}, of course, differ.

One can easily imagine that the cross terms in eq. (2.124) are numerous and not always well defined. However, they significantly affect the actual values of the bond-defined parameters of the force fields throughout the general parameter fitting procedure and for that reason these terms are used for the purpose of classification of the force fields [174]. Class I force fields are those which do not contain any cross terms. A Class II force field allows for anharmonic terms (e.g. through the use of Morse potentials or of the polynomial of the stretching potential – in practice, terms up to the 6-th power in bond-length variation are used) and explicitly accounts for the coupling between coordinates. Examples of cross terms are

$$
\begin{aligned}
E_{\text{str/bend}} &= k_{ABC}(\alpha_{ABC} - \alpha^0_{ABC})[(l_{AB} - l^0_{AB}) + \\
&\quad + (l_{BC} - l^0_{BC})] \\
E_{\text{str/str}} &= k_{ABC}(l_{AB} - l^0_{AB})(l_{BC} - l^0_{BC}) \\
E_{\text{bend/bend}} &= k_{ABCD}(\alpha_{ABC} - \alpha^0_{ABC})(\alpha_{BCD} - \alpha^0_{BCD}) \\
E_{\text{str/tors}} &= k_{ABCD}(l_{AB} - l^0_{AB})\cos(n\varphi_{ABCD}) \\
E_{\text{bend/tors}} &= k_{ABCD}(\alpha_{ABC} - \alpha^0_{ABC})\cos(n\varphi_{ABCD}) \\
E_{\text{bend/tors/bend}} &= k_{ABCD}(\alpha_{ABC} - \alpha^0_{ABC})(\alpha_{BCD} - \alpha^0_{BCD}) \times \\
&\quad \times \cos(n\varphi_{ABCD})
\end{aligned}
\tag{2.129}
$$

The presence of these higher cross-terms tend to improve the ability of the force field to predict the properties of unusual systems (such as those which are highly strained) and also to enhance its ability to reproduce vibrational spectra. It must be noticed, however, that any of the cross terms listed above have been proven to be truly of the form in which they are written. No attempts have been reported to derive that or any other form of the coupling between different geometry distortions and to estimate the corresponding constants from some independent point of view. The class III force field will also take into account further features such as electronegativity and hyperconjugation. We shall turn to these problems later.

The next group of the energy contributions are the terms collected under the name of interactions of the nonbonded atoms (nonbonding interactions). Historically, the most important among them is the van der Waals interaction of the nonbonded atoms, as the first MM potentials were developed for nonpolar species like alkanes. Normally it is taken in the form:

$$
E_{\text{vdW}} = \sum \epsilon \left[\left(\frac{d_0}{d_{ij}} \right)^{12} - 2 \left(\frac{d_0}{d_{ij}} \right)^6 \right]
\tag{2.130}
$$

called the Lennard-Jones potential, where d_{ij} are the interatomic distances. The summation in the above expression is assumed to extend to all pairs of atoms separated by more than three sequential bonds. The parameters d_0 and ϵ are set on the atomic type basis. For pairs of atoms of different types, they are usually obtained according to the Lorentz–Berthelot combination rules:

$$d_0 = d_0^{AB} = \frac{1}{2}(d_0^{AA} + d_0^{BB})$$

$$\epsilon = \epsilon^{AB} = \sqrt{\epsilon^{AA}\epsilon^{BB}}$$

An alternative form of the van der Waals interaction is represented by the Buckingham (exp-6) potential:

$$(2.131) \quad E_{\text{vdW}} = \sum \epsilon \left[\frac{6}{\zeta - 6} \exp \zeta \left(1 - \frac{d_{ij}}{d_0} \right) - \frac{\zeta}{\zeta - 6} \left(\frac{d_0}{d_{ij}} \right)^6 \right]$$

where ζ is a free parameter. Its use is twofold. Setting $\zeta = 12$ results in the London form of the long-range attraction part of the vdW potential ($\sim d_{ij}^{-6}$); setting $\zeta = 13.772$ makes the second derivative of the potential in its minimum equal to its Lennard-Jones value with no clear argument in favor of any choice. Both Lennard-Jones and Buckingham forms are widely used in MM studies, although sometimes other models of the van der Waals interactions are employed as well.

Next in the list, but of course not of lesser importance among the interactions between nonbonded atoms, is the Coulomb interaction of effective charges residing on atoms in the molecule:

$$(2.132) \quad E_{\text{Coul}} = \sum \frac{q_i q_j}{\varepsilon d_{ij}}$$

The summation extends here to all pairs of atoms (including the bonded ones). In practice the dielectric permittivity ε, sometimes also dependent on the interatomic separation, is used, although this kind of treatment lacks any serious theoretical support. The most sophisticated problem while treating the Coulomb interactions in the above form is where the values of the effective charges have to be taken from. A number of procedures have been proposed in the literature to select effective charges or more generally to parametrize electrostatic properties of molecules in terms of point charges, dipoles, etc. For example, some molecules being highly symmetric do not bear say charge, or dipole, or quadrupole momenta. However, methane, for example, bears some noticeable octupole momentum. Its magnitude can be reproduced by setting charges of $0.14\bar{e}$ on each of the hydrogen atoms and the quadruple of this with the opposite sign on the carbon atom. It must be observed that charges thus derived have no physical significance. In molecules like methane the octupolar moment appears largely due to hybridization (quantum mechanical superposition) between the one-electronic sp-states on the carbon atom. The Mulliken charges or even more the Coulson charges residing on the hydrogens (see below) are microscopic as compared to the above estimate and thus give only a minor contribution to the observed octupolar momentum of methane. In this sense the octupolar momentum of methane is largely a manifestation of the quantum behavior of electrons.

In some cases the electrostatic potential induced by a given molecule is attempted to be reproduced by a set of point charges (or di- and higher multipoles) distributed in the molecule. The practical implementations of this approach, useful for solving some specific problems, are designed to fit the potentials obtained from an ab initio distribution of electrons in a finite number of points "outside" the molecule. Of course one can fit the potential in some points to the model containing the atom-centered charges as used in the expression for the Coulomb energy. It is not, however, clear whether the total Coulomb energy of the defined point charges has anything to do with that of the real (continuous) molecular charge distribution. The same question applies also to the local multipole models of molecular charge distributions derived from fitting molecular electrostatic potential.

An alternative to the above approach is the distributed multipole model, which is a natural generalization of the older concept of effective charges in quantum chemistry. It is based on the observation that the product of two Gaussian functions centered in different points of the real 3-dimensional space (in fact on different nuclei) is itself a Gaussian centered somewhere on a straight line connecting the two original centers. By virtue of this, any product of basis one-electron functions represented by their expansion over the atom-centered Gaussians itself becomes a collection of charge distributions placed in different points in the space. On the other hand, the entire Coulomb energy of a molecular system in the ab initio context can be rewritten in the form of interacting multipoles as the two-electron integrals can be recast in this form (interacting multipoles). If the two basis functions involved in the definition of the above density are centered on the same atom, the corresponding multipoles are centered at the corresponding atoms as well. Otherwise some points in the space, not having a clear physical sense, are involved as the points where the said multipoles are located. The entire picture, however, is consistent. The problem with it is the clear lack of transferability: different basis sets produce different location points for the "space" multipoles as the location points depend on the Gaussian exponents. Also the multipoles derived from analysis of an ab initio electron distribution cannot be believed to go from one molecule to another without change. On the other hand, the picture where the significant part of energy is represented by "classical" force fields (multipole-multipole interactions), although not always atom-centered, looks intellectually very attractive.

All the tricks that are used to avoid calculating the electronic wave function throughout describing the electronic distribution in molecules in the MM context, face the same fundamental problem: the charges (as well as higher multipoles in understanding that a point charge is the electric monopole) are not stable quantities and tend to vary from one molecule to another, even if the same types of atoms are involved. In the general context there is no problem as we are used to thinking that this type of behavior is something one should expect. Varying the electronic distribution when going from one molecule to another is commonly used for explaining many chemical phenomena. Nevertheless, this redistribution is of a quantum nature and addressing it is not welcome in the MM context. This brings the necessity of developing simple schemes of estimating the charges. A simple one dates back to Gasteiger and Marsili [175] which is characterized as a partial equalization of orbital electronegativity. It starts from the Pauling and Mulliken orbital electronegativity, the definition of which is widely used throughout semi-empirical quantum chemistry (see above):

$$(2.133) \quad \chi_A^0 = \frac{1}{2}(IP_A + EA_A)$$

Later Rappé and Goddard [176] suggested equilibrating not the electronegativity, but something like a chemical potential of electrons at each atom, although the method itself is called the "charge equilibration method". Toward this end, the energy of the electrostatic interaction in the molecule is represented as a function of the effective charges in it in the form:

$$(2.134) \quad E_{\text{Coul}}(q_1, ..., q_N) = \sum_A (v_A^0 + \chi_A^0 q_A) + \frac{1}{2}\sum_{AB} q_A q_B J_{AB}$$

with the electronegativities χ_A^0 defined just above, the two-center interaction J_{AB} taken as Coulomb functions of the interatomic separation $e^2 R_{AB}^{-1}$ for $A \neq B$, (by this the above expression becomes geometry dependent) and with J_{AA} equal to $IP_A - EA_A$, which coincides with the estimate of the γ_{AA} parameter in the CNDO approximation eq. (2.36). Later it was realized that at larger distances it makes sense to use the CNDO estimate also for the two-center interaction and to set $J_{AB} = \gamma_{AB}$. The effective charges then appear by minimizing the Coulomb energy with respect to $q_A, A = 1 \div N$ with a charge conservation condition:

$$(2.135) \quad \delta \sum q_A = 0$$

This is equivalent to the system of equations:

$$(2.136) \quad \frac{\partial}{\partial q_A} E_{\text{Coul}}(q_1, ..., q_N) - \mu \frac{\partial}{\partial q_A} \sum q_A = 0$$
$$\sum q_A - Q = 0$$

where the Lagrange multiplier μ takes care for the charge conservation condition. The first (set of) equation(s) yields the set of relations:

$$(2.137) \quad \chi_A^0 + J_{AA} q_A + \sum_{B \neq A} J_{AB} q_B = \mu$$

to be held for each atom A. The left side of each equation can be interpreted as a definition of the chemical potential of the electron at the A atom. In the equilibrium it has a single value for the entire molecular system i.e. for all A's as any intensive quantity should. Solving the above system of linear equations yields the necessary effective charges. Due to the coincidences with the CNDO estimates used throughout, one may hope that the results thus obtained are somehow close to the CNDO estimates of effective charges, known to be reasonable.

2.5.2. Current development and need for extensions

The simple setting characteristic of MM facilitates its wide use and a wealth of implementations, differing in the exact set of parametric forms of the force fields employed for modeling the molecular PES and the particular values of the parameters used. We briefly review the force fields known during the last few decades.

- AMBER (Assisted model building with energy refinement) [177] is the name of a molecular mechanics program. It was parametrized specifically for proteins and nucleic acids. AMBER uses diagonal bond stretching terms harmonic in displacements, the improper torsion terms, and van der Waals nonbonding terms, together with a sophisticated electrostatic treatment employing the charges extracted from fitting the molecular electrostatic potential. No cross terms are included. An option of using a "united atom" e.g. to represent CH_2 moiety by one point mass is provided. Results are very good for the target species, proteins and nucleic acids, but can be somewhat erratic for other systems.

- CHARMM (Chemistry at Harvard macromolecular mechanics) [178] is the name of a molecular mechanics program. It was originally devised for proteins and nucleic acids. It has been by now applied to a range of biomolecules and for studying molecular dynamics, solvation, crystal packing, vibrational analysis, and QM/MM studies. CHARMM uses diagonal bonding terms harmonic in displacements, the improper torsion terms and an electrostatic term.

- CFF (consistent force field) [179] was developed to yield consistent accuracy of results for conformations, vibrational spectra, strain energy, and vibrational enthalpy of proteins. There are several variations of this, such as the Urey-Bradley version (UBCFF), a valence version (see below), and Lynghy CFF. The quantum mechanically parametrized force field (QMFF) was designed to simulate ab initio results. CFF93 is a rescaling of QMFF to reproduce experimental results. These force fields use the diagonal bond stretching terms up to fourth power in displacement and the harmonic improper torsion term. A 6–9 form of the potential is employed for the van der Waals forces together with the electrostatic term to represent nonbonding interaction. A rich variety of cross terms (in fact all those listed above) is foreseen.

- CHEAT (Carbohydrate hydroxyls represented by external atoms) [180] is a force field designed specifically for modeling carbohydrates.

- COSMIC [181] method uses only harmonic potential with respect to displacements in the diagonal force fields. Neither improper torsion nor cross terms are included. The nonbonding interactions are the sum of van der Waals interactions represented by the Morse potential and the charge-based Coulomb energy.

- CVFF [182] is the valence version of CFF. It uses only harmonic expansion with respect to displacements in the diagonal force fields and reduces cross terms selection to some extent.

- DREIDING [183] is an all-purpose organic or bio-organic molecular force field. It has been most widely used for large biomolecular systems. It uses either harmonic or Morse potential for the bond stretching, and the second power polynomials in cosines of the corresponding angles to represent the valence angle bending and out-of-plane bending force fields. The nonbonding interactions are represented by either 6–12 or exp-6 van der Waals and by the charge based electrostatic term. No cross terms are included.

- EAS (Engler, Andose, Schleyer) [184] is quite an old force field designed to model alkanes exclusively. The harmonic potential is used for the bond stretching and cubic anharmonic for the valence angle bending. No out of plane, electrostatic or cross terms are included. The nonbonded interactions are represented by the Buchingham potential.

- ECEPP (Empirical conformational energy program for peptides) [185] is the name of both a computer program and the force field. No flexibility of the valence bonds and angles is assumed at all, so that the corresponding geometry parameters are kept fixed and do not take part in the geometry optimization. The van der Waals term of the 6–12 form and an electrostatic term based on charges are used to represent nonbonding interactions. A 10–12 term describes hydrogen bonds necessary for peptide chemistry.

- EFF (Empirical force field) [186] has been designed just for modeling hydrocarbons. It uses the quartic anharmonic potential for the bond stretching, and the cubic anharmonic for the valence angle bending. No out of plane or electrostatic terms are involved, although the cross terms, except torsion-torsion and bend-torsion ones, are included.

- ESFF has been designed to be universal [187]. The Morse potential is employed for bond stretching, the potential quadratic in the cosine of the valence angles for their bending and the harmonic potential for the out of plane force field. The 6–9 with the charge based electrostatic potential is used for nonbonding interactions. No cross terms are involved.

- GROMOS (Groningen molecular simulation) [188] is the name of both a force field and the program implementing this force field. It is used for studying the dynamics of molecular motion in bulk liquids. It is also used for modeling biomolecules. The harmonic potentials are used for the stretching, bending and out of plane (improper torsion) force fields. Nonbonded interactions are modeled by the 6–12 van der Waals and charge based electrostatic terms. No cross terms are involved.

- MMn with n $= 1–4$ is the series of subsequently developed general-purpose organic force fields [189–191]. All methods of the MMn family use the Buchingham potential for the van der Waals forces. The specific of MMn is occasional employing of the bond dipole based electrostatic energy contribution instead of the charge-based models (see below).

    MM2 [189] uses cubic anharmonic potential to represent the bond stretching, up to sixth power expansion for the valence angle bending, and harmonic field for the out-of-plane deformations. The stretch-bending cross term is included.

    The MM3 method [190] is parametrized for as much as 153 atomic types eventually covering almost all chemical elements in common use. The quartic anharmonic potential is used for the bond stretching, sixth power expansion is used for the valence angle bending, and harmonic field for the out-of-plane deformations. The stretch-bending, bending-bending and stretch-torsion cross terms are included. Electrostatic is optionally charge- or bond-dipole based.

The MM4 method [191] is still under development in terms of extending the number of chemical elements it is parametrized for. The initial published results are encouraging. It uses the sixth power expansion, both for the stretching and for the bending and the improper torsion representation for the out-of-plane deformations. The idea of using the bond-dipole based electrostatic term is abandoned in the MM4 whereas the whole collection of the cross terms is included.

Several important extensions were based on the MM2 platform. Among them the MMP2 designed to incorporate the effects of the conjugate π-systems upon the molecular geometry and torsion barriers must be mentioned. MMX and MMa are variations of MM2.

- MMFF [192] is the molecular force field developed by Merck Inc. It is a general-purpose method, particularly useful for organic molecules. MMFF94 was originally intended for molecular dynamics simulations, but is also engaged in geometry optimization. It uses quartic anharmonic terms for bond stretching, cubic anharmonic term for angle bending and harmonic out of plane potential. The nonbonded interactions are represented by an unusual 7–14 potential together with the charge based electrostatic term. The stretch-bending cross term is added.

- MMGK (Molecular mechanics with Gillespie-Kepert terms) [193] is designed for application to coordination compounds. It is based on CHARMM, but an additional term describing repulsion of some effective interaction centers placed on the coordination bonds is added.

- MOMEC [194] is a force field designed for describing transition metal coordination compounds. It was originally parametrized using harmonic potentials for the stretching, bending and out of plane terms, but the nonbonded interaction is represented by solely exp-6 potential. The metal complex specificity is reflected by the coordination shape maintained by the nonbond interactions between ligands (kind of a point on a sphere model). No cross terms are involved.

- OPLS (optimized potentials for liquid simulation) [195] is designed for modeling bulk liquids. It is also used for modeling the molecular dynamics of biomolecules. It uses harmonic potentials for the stretching and bending. Improper torsion term is used for out of plane forces. Nonbonding contribution is provided by the 6–12 potential and the charge based electrostatics. No cross terms are involved.

- SHAPES [196] is a force field designed for transition metal complexes. It involves the harmonic potential for the stretching, Fourier term for the valence angle bending and improper torsion for the out of plane energies. Nonbonding contributions are represented by the 6–12 and charge based electrostatic terms. No cross terms are involved.

- Tripos [197] is a force field constructed at Tripos Inc. for inclusion in the Alchemy and SYBYL programs. It uses harmonic potentials for the stretching, bending, and out of plane energies. Nonbonding interactions are represented by the 6–12 and charge based electrostatic terms. No cross terms are involved.

- UFF stands for universal force field [176]. This is the next most elaborate force field parametrization in terms of the number of atomic types: 126, eventually covering all Periodic Table known at the time of construction (1992). UFF is most widely used for systems containing inorganic elements. It uses harmonic or Morse potential for the stretching, Fourier representation for the bending and improper torsion for the out of plane terms. An electrostatic term was not originally included in the UFF. The literature accompanying one piece of software recommends using charges obtained with the charge equilibrate method. Independent studies have found the accuracy of results to be significantly better without charges.
- YETI [198] is a force field designed for the accurate representation of nonbonded interactions. It is most often used for modeling interactions between biomolecules and small substrate molecules. The molecular geometry optimization for the component molecules is not previewed so that it has been obtained from some other source, such as AMBER. Then YETI is used to model the docking.

At this point one can conclude that the real picture drawn by the MM methods is much more obscure than the appealing simplicity of its initial formulation. First of all we have to notice that to acquire acceptable quality the MM models require a refined view of atoms in a molecule. In the MM setting it is formalized in the concept of atomic "type". The atomic type is not just the nuclear charge – i.e. atomic number supplied by some set of parameters uniquely characterized by this number, but in addition to that the information on the hybridization of an atom and its specific surrounding is loaded upon the concept of the type. For example, in one of the most widespread MM force fields, the MM2 by Allinger and coworkers [173], 71 atomic types are defined for 28 different nuclei. One of the types is reserved for representing lone pairs, but, for example, 15 different types of carbon atoms are distinguished. In the AMBER force field specifically targeted on the proteins and nucleic acids [177] even finer detalization of carbon atomic types is used. Obviously, it would be desirable if some theoretical reasons are given to somehow restrict or at least systematize this diversity. Also the diversity of the functional forms of the force fields used in different methods seems to need some systematization. For the small displacements the harmonic approximation may be suitable, but for larger displacements a theoretically substantiated form of the potential may be helpful for reducing the number of parameters (see below).

Another problem directly related to the number of atomic types involved is that of actual construction of the parameters' system, provided some guess concerning the form of the force fields is accepted. A simple estimate given in [199] on the example of the MM2 type of parametrization specifying 71 atomic types shows the number of van der Waals parameters to be 142, the number of different stretching parameters to be about 900, and that for the bending parameters, about 27000. Finally the number of the torsion parameters tends to exceed one million. This clearly indicates that the amount of available experimental data of the accuracy required to obtain that many parameters simply does not exist. The task of emulating the raw data using some higher level quantum chemistry technique and then fitting them to the selected functional form of the force field does not seem to be reasonably set either. What would probably help is a theory setting the restriction both on the form of the force field and on the limits of the physically allowable values of parameters or, even better, giving some a priori estimates for them on the basis of some theoretical systematization of the atomic types. The numbers given above can be used to provide a characteristic estimate of the usefulness of this theory: the number of van der Waals parameters is significantly smaller than that for the bond stretching parameters, although the parameters for both the interactions are indexed by pairs of atomic types. This happens due to the possibility of using the combination rules cited above to assign the van der Waals parameters to pairs of atoms of different types. For that reason, only the parameters for each atomic type are needed to evaluate the interaction energy for the pair of atoms of different types. The ultimate criterion for the validity of such a rule is of course the quality of results obtained by using it; however, there are theoretical reasons to think that these rules can be tried. It would be very attractive if a hint allowing one to treat the bending parameters a priori ascribed to triples of atomic types as being specific only for the type of the atom at the apex of the valence angle. By this the number would be reduced drastically – from *ca.* 27000 to *ca.* 70. It is clear that a theory substantiating such a reduction is highly desirable.

Another piece of theoretical reasoning in this realm can start from considering the transferability concept. The term transferability, when applied to the system of MM parameters, refers to the fact that a good parameter system applies to a set of molecules which is large enough and has no necessity to be further adjusted. The MM setting itself does not give any explanation to this important fact: the transferability is taken for granted. The contradictory nature of this hypothesis in its naïve form is obvious: the transferability of the parameters of the molecular potential is restricted by the diversity of the atomic types, for which no restriction is set so far. It is thus not clear whether the parameter sets already defined are transferable enough or a further refined distinction of the narrower atomic types will be necessary.

The low computational costs of the MM methods brought to it a considerable appeal throughout the research community. It is particularly true for those who work largely on applications and are not greatly concerned with the soundness of the theoretical basis, but more with the price/gain ratio. Due to the large degree of uncertainty characteristic of the very basic concepts of MM theory, its further development is rather complicated although desirable. However, too much pragmatism turns out in fact to be not really pragmatic. The reason is of course that very narrow problem setting makes the result to be of low value since it does not apply to anything beyond the scope of the initial setting. This perhaps too general notion is applicable in the case of MM modeling when we address the attempts to extend the traditional MM to more and more complex objects. Among them, the MM treatments of chemical reactivity and those of the coordination compounds must be mentioned. The latter further subdivides into two major classes: coordination compounds of nontransition elements – like alkali and alkaline earth metals or elements like Sn, Sb, and P, and another class represented by the complexes of transition metals with open d-shells occupied by nonbonding electrons.

In the literature [176, 194, 200–204] various MM constructions are presented as effective methods for modeling PES of an arbitrary molecular system. However, in the case of coordination compounds, it is not possible to single out transferable two-center bonds involving the central atom. The number of bonds formed by them (the coordination number) may itself be variable and these variabilities themselves may require the modeling. In [201] an extensive summary of the results of calculations on coordination compounds of a wide variety of metals by the MM methods (as of 1993) is given. During the following decade, much subsequent work, quoted in reviews [194, 205, 206], were performed, in which PES of special classes of metal complexes was parametrized by some MM-like force fields. As it can be seen from a more recent review [203] the conceptual problems manifest themselves in extremely cumbersome and awkward sets of force fields when metal atoms are involved, as compared to traditional 'organic' systems of force fields. For example, it becomes necessary to introduce a double set of optimal valence angles for octahedral (or plane squared) complexes to ensure these characteristic molecular shapes are reproduced in the calculation as are the relative energies of the *cis*- and *trans*-isomers [194, 201]. The number of other parameters also grows rapidly, and it is difficult either to assign a clear physical sense to all of them, or restrict the reasonable interval of their values and thus separate probable ones from the improbable.

The only possibility of introducing some order in this area is to consider the physical mechanisms responsible for bonding. Among the physical processes leading to the bond formation between central atoms and organic ligands bearing donor atoms, we first of all notice a strong electrostatic interaction between the formal charge of the central atom and effective charges in the ligands. These interactions are classical. It may create an impression that the standard MM supplied by Coulomb interaction between effective charges might be sufficient to describe coordination compounds. This idea is however misleading. An important peculiarity is that the charge distributions taking part in these interactions are predetermined by the quantum behavior of electrons in the field induced by nuclei and other electrons. The quantum behavior of the latter spans three types of interactions. The first is the polarization of the ligands by the point (formal) charge of the central ion with the charges in the ligands. The second is the redistribution of the electron density between the ligands and the central ion, which reduces to netto transfer of electronic density from the ligands to the central ion which diminishes its effective charge as compared to its formal charge. This redistribution according to [207] could be taken into account by the method of "equilibration of effective electronegativities" [175]. Meanwhile the equations [175] ignore the presence of off-diagonal elements in the one-electron density matrix in the basis of AOs i.e. of the bond orders which are basically responsible for the bond formation which is the third type of interaction involved. If the quantum chemistry models of previous sections are thoroughly considered, one can find that the off diagonal matrix elements of the one-electron density were responsible for the bonding: the corresponding energy contribution decreases when the inter-atomic separation decreases. By this the models restricted to effective charges ignore the quantum mechanics as applied to electrons, as they ascribe to the latter the Kolmogorov type of statistics (operating with electron *densities* is equivalent to summing of *probabilities* of independent elementary events in order to obtain the probability of the complex event) instead of quantum (summing of probability *amplitudes* of independent elementary events and obtaining the probabilities as the squared sum of the amplitudes). The importance of the off-diagonal elements of the one-electron density matrices taken in the basis of some one-center orbitals is the formal manifestation of the quantum behavior of electrons in the present context. They do not have any classical interpretation and at the same time reflect the basic characteristic of the chemical bonding. Whether in the Hückel method and in other HFR-based semiempirical theories, or in the SLG-based treatments, the energy terms responsible for bonding (attraction) between atoms are proportional to the off-diagonal matrix elements of the density taken between the one-electron states residing on the respective atoms. They appear as a pure consequence of the quantum character of electronic motion in molecular systems, i.e., of the necessity to describe the probabilities as squares of the sums of the probability amplitudes rather than sums of the probabilities. This feature is ignored by the ionic models of the electronic structure of the coordination compounds and also by more refined methods based on the concept of equilibration of effective electronegativities. In the HFR approximation the Coulson bond order (i.e. the off-diagonal matrix element of the one-electron density in the HO basis) of the two-center two-electron bonds can be expressed through the diagonal matrix elements of the one-electron density. This suggests that under certain conditions the off-diagonal elements of the density implicitly present in the bond-stretching terms and thus the bond stretching force fields can be constructed on the basis of the charges only, but this option has not been so far explored to the best of the author's knowledge. In subsequent chapters of this book we shall present possible theoretical reasoning, which allows us to tentatively construct a theory underlying the MM treatment and on its basis to give certain hints concerning the further possible refinements of the MM itself and its extensions towards coordination compounds.

As for hybrid modeling, the problem of the foundations of MM is seen from a somewhat different perspective. A priori there is no limitation for employing that or any other MM scheme as a classical component of a hybrid model. In practice, however, different MM schemes behave differently when tailored to a QM treated part. Indeed, it is not clear how to handle the bond-dipole based electrostatic energy employed in the MM2 and MM3 schemes, if some bond must be broken, as their ends are expected to be treated by different methods. It applies even more to the schemes with charge equilibration. We shall try to describe the problems created by these inconsistencies as related to the current hybrid methods in the next section, with the analysis of the current state of the art, from the point of view of the general theory of electron variables separation.

2.6. HYBRID METHODS OF MODELING COMPLEX MOLECULAR SYSTEMS

After showing in the previous sections that the real place of the hybrid schemes in quantum chemistry is much more important than one may think, as almost every quantum chemistry method developed so far is hybrid explicitly or implicitly, we turn to a description of the existing hybrid methods understood in the narrow sense: namely, those where a part of a system is described by a quantum chemistry method and other parts by MM methods described in previous sections.

Chemical transformation is local. When molecules taking part in it contain hundreds and thousands of atoms, each elementary chemical step touches only their small fragments: one or two bonds are broken or formed by a single act. This kind of behavior, which obviously allows all preparative organic, inorganic and organometallic chemistry to exist, is also the basis for the hybrid QM/MM techniques present in the literature. Since the seminal work by Warshell and Levitt [208], the hybrid QM/MM schemes of calculating large molecular systems acquired an increasing popularity. There is a big variety of hybrid approaches described in the literature [209–218]. Even more, numerous cases of separating electronic variables like π-electron models or even taking into account only valence electrons in semiempirical methods, can be considered as special cases of hybrid schemes, as they also bear the family marks of the QM/MM approach, namely, (i) the separation of the system into parts, and (ii) treating these parts on quantum or classical levels, respectively. In such a broad sense, several other problems in the area of computational chemistry seem to be related to the QM/MM context: these are the problems of embedding in the cluster calculations on solids and their surfaces, with special attention to adsorption and catalysis problems; the problem of description of solute/solvent effects for reactions in condensed media. Also a great variety of different specific schemes referred to as "protocols" implemented in different computational packages are normally considered only from the point of view of their practical feasibility and their fit for a particular applied purpose, rather than in a context of their exact placement among other approximate methods and of the evaluation of the relative precision of that or any other approximation.

In the present section, we employ the theoretical framework developed above to rationalise the state of art in the field of hybrid QM/MM modeling as found in the literature. As we have already mentioned, the idea of treating the chemically transforming part at the quantum mechanical level and the rest at that of molecular mechanics is very naturally based on the whole set of experimental data of synthetic chemistry. It is not reflected in the standard QC techniques. More formally, this idea is expressed in the assumption that the PES of molecular systems of interest (say, of chemically transforming large molecules) can be presented as a sum of quantum chemistry and molecular mechanics contributions. (The origins of this approach date back to the theory of conjugated hydrocarbons presenting the energy as a sum of the π-electron energy and of the σ-frame energy respectively, either calculated by quantum chemistry (Hückel) methods or taken in the harmonic approximation [45]. Obviously, this approach is prototypical of the modern QM/MM techniques).

It is instructive to see what classifications of the QM/MM schemes are developed in the literature. First of all such classifications are based on the types of quantum chemistry and molecular mechanics schemes used. Such a classification is implied by K. Morokuma and his coworkers. As it does not address any internal characteristic features of the obtained hybrid methods, it is rather nomenclature than classification. There are no fundamental restrictions upon the choice of the QM scheme in the hybrid method and in fact almost all of them can be found in the literature: ab initio [219–222, 246], DFT [223–229], and semiempirical [211, 230–232] ones are widely used as in standard quantum chemistry contexts. Remarkably enough, the semiempirical methods still keep their stronghold on the QM/MM context, although in the pure QC studies they have almost completely yielded their place to the ab initio and DFT methods. This is largely a fashion effect rather than the result of a thorough estimation of the real value of the semiempirical methods. In the hybrid context the use of ab initio methods by contrast seems quite strange as in the absence of any general theory of the former, the status of the combination of ab initio QM with purely empirical MM contribution is not clear. The form of wave function usually employed for the QM is the HFR, although methods based on the valence bond approximation are also known [210, 221, 233]. Clearly, using the HFR picture and describing the electronic structure of molecular system in terms of delocalized MOs eventually extended to the entire molecular system is in contradiction with the idea of singling out a "chemically interesting" subsystem or chromophore. Thinking about a QM/MM procedure as of one approximating the "exact" QC approach faces a serious problem: when the molecular system is extended, its MOs become very different from those of the QM. The choice of the MM scheme can also be quite important as it affects the structure of bonding terms near (or, in some schemes, on) the border between the QM and MM parts and the electrostatic polarization of the QM part by the MM-treated environment depends on the charge scheme employed in the MM procedure. Practically, it is more convenient to work with the force field with electrostatics based on the atomic charges rather than with those using the bond dipole scheme as the latter can cause significant errors in the description of polar species, particularly if a polar bond is to be broken when a system is divided into parts. This notion has led the authors of [234] to employ the MM3 force field [190] to replace the bond dipoles with potential-derived atomic point charges. If the force field contains many cross terms coupling bonding force fields (like in the MM3 force field) it may cause additional problems when constructing the junction between the quantally and classically treated parts of the complex system. Some words should be said about ionic force fields using formal ionic charges and employing electrostatic and short-range force fields. These force fields are widely used when treating metal complexes. In the case of these force fields, short-range interactions arising from the MM charges cannot be separated from long-range interactions [218]. This leads to incorrect electrostatic potentials felt by the quantum part so the most popular choice (especially in the description of covalently bound QM/MM systems) is the valence force fields. Different implementations use different force fields: for example, the MM3 force field [190] is used in [234, 235], the CHARMM force field [178] is used in [209, 236], the AMBER force field [177] is used in [237, 238].

From the point of view of general theory described in Section 1.7.2 the relevant classification of QM/MM methods should be based on an assessment of the level to which the key elements of this theory are treated in that or any other specific hybrid scheme. The authors of [234] made a step in this direction and proposed a classification of hybrid schemes based on the interaction between the quantum and classical fragments. Such a classification of the QM/MM schemes is much more informative. It is built around the hypothetical representation of the total energy of the complex system comprising the quantitatively and classically treated subsystems in the form:

$$(2.138) \quad E_{\text{tot}} = E_{\text{QM}} + E_{\text{MM}} + E_{\text{QM/MM}},$$

which is taken for granted. In the above expression E_{QM} stands for the energy of the quantum system, E_{MM} is that of the classical system treated using MM, and $E_{\text{QM/MM}}$ stands for the subsystem interaction known in the context of the QM/MM methods as the quantum-classical junction. The idea of the classification on the basis of eq. (2.138) seems easy, but it implies two important problems: first, eq. (2.138) cannot be ascribed any sense unless the border between the quantum and classical parts is set in a noncontradictory manner. Second, though representing the energy by eq. (2.138) seems very natural it contradicts quantum mechanics: the interactions in quantum mechanics modify the states of both interacting parties, including the classical one. This problem is addressed in detail later.

The origin of eq. (2.138) is well known: the Hamiltonian for the complex system comprising two parts can be (and had been) written in analogous form:

$$(2.139) \quad H_{\text{tot}} = H_{\text{QM}} + H_{\text{MM}} + H_{\text{QM/MM}}$$

long ago [277]. However, in the quantum realm, the Hamiltonian and the energy are not the same thing: the energy is an expectation value of the Hamiltonian over the relevant wave function, describing the state of the system. As it has been shown in Section 1.7.2 reasonable assumption about the form of the wave function of the complex system requires significant work. The psychological aspect of this situation is that the simplest, but the most fundamental problems, are not worked out. The most fundamental aspect not well designed in the standard hybrid techniques is the composition of the system. Computational chemistry in general requires that the composition of the model is uniquely defined. However, even this requirement is sometimes not fully satisfied. This is most obvious in the case of geometry-based procedures of singling out the quantum subsystem. Under this setting the intersubsystem border is understood as a closed surface in the physical three-dimensional space (there is a selection of prescriptions of how to set this border) and the electrons inside it are to be treated by means of quantum mechanics and of those outside it nothing is known at all. This construct tacitly assumes that the electronic variables in this problem are their spatial coordinates. This (coordinate) representation is inconsistent with the quantum chemistry used to describe the quantum system: the latter universally uses for variables describing electrons not the three-dimensional coordinates in the physical space, but the occupation numbers of the one-electron states, the expansion coefficients of the one-electron states over some basis, CI amplitudes, etc. The orbitals used always extend beyond the geometrically set border; also, the electrons occupying similar, although nonspecified orbitals outside the "quantum" subsystem defined on the geometry, grounds penetrate the latter. In the solute-solvent context [239,240] when there is no chemical bond between the quantum and classical parts of the system, all this might be acceptable, although nobody has ever given any estimate of the amount of error introduced by these procedures. The situation becomes worse in the contexts of modeling enzymatic or heterogenous reactions. Authors of [234] notice that if the border is set in the coordinate space and from the geometry reasons, it may be necessary to break a chemical bond between a classical subsystem X and the quantum part Y; the composition of Y (and of X too) is not uniquely defined as there are at least three possibilities: the cation Y^+, the anion Y^-, and the radical $Y^\cdot$. As previously, the number of electrons in the geometrically separated quantum system is not well defined i.e. it is not a good quantum number (strongly fluctuating), and in the last case (of the radical) also the spin projection (and thus the spin itself) of the quantum subsystem is not defined. From the general theoretical point of view everything is quite obvious: the geometry based separation of the complex system into parts contradicts the basic principles of quantum mechanics. At the molecular scale of lengths it makes no sense to talk about any "borders" in the coordinate space, as they cannot separate anything – the de Broglie wave lengths of electrons involved are comparable with the size of say the cavity in the continual insulator model of the solvent-solute interaction. When it comes to breaking the bonds on the geometry grounds, the situation becomes even more severe as the geometry based constructs must be judged by comparing the de Broglie lengths not with *inter*molecular but with the *intra*molecular scale of lengths, which results in an even less favorable estimate. In plain words one can say that the geometry based separation in parts guarantees that the number of electrons in the quantum subsystem is poorly defined. At the same time, no quantum chemical method so far developed can deal with a noninteger number of electrons. In a general theoretical setting it is not impossible: in the superconductivity theory one manipulates with the wave functions of undefined number of electrons and in the grand canonical ensemble setting the number of particles is not fixed either. The only problem is that the QC methods employed to describe the quantum parts in the hybrid methods deal neither with the Bardeen-Cooper-Shrieffer functions of superconductors nor with the grand canonical ensemble, but by contrast require the number of electrons to be integer and fixed. This point is so obvious that no method description has ever mentioned that. But the geometry based singling out of the quantum subsystem does not guarantee that it is satisfied – in fact, just the opposite.

Anticipation of serious trouble coming from indefiniteness of the chemical composition (which obviously includes the number of electrons or equivalently the total charges) of the subsystems led to the suggestion that seemed at the first glance to solve all problems at once by filling the cut bonds with atoms (usually hydrogens) or more complex groups artificially added to the quantum system. This allows one to keep the number of electrons in the quantum subsystem fixed at some reasonable value. However the cost of this remedy is rather high: one definiteness is paid by another indefiniteness, namely by that of the number and type of nuclei in the quantum system. Within such a setting, the QC calculation is performed for a subsystem whose chemical composition does not coincide with that of the quantum subsystem it is assumed to represent. This inevitably leads to serious artefacts and potentially to errors.

With the above reservations we turn back to the interaction based classification of the hybrid QM/MM methods which allows us to distinguish the mechanical embedding, polarization embedding etc. We shall consider them subsequently.

2.6.1. Mechanical embedding: IMOMM, IMOMO, ONIOM, etc

Examples of this type of modeling are the IMOMM [213] and IMOMO [241] schemes developed by Morokuma with coworkers when both QM and MM systems are not polarized by each other and their interaction is represented by classical force fields only. In fact, mechanical embedding does not assume any explicit interaction between the subsystems treated by classical and quantum chemical methods. The simplest model for implementation is the so-called "subtractive scheme" implemented for example in [241, 242]. In the framework of these schemes some part of the system is assigned to be treated at a quantum level on geometry grounds. If the QM fragment acquires dangling bonds under this singling out, they are saturated by some groups (hydrogens, methyl groups etc.). This forms the so-called *model system*. Then the entire system is calculated by an MM method of choice (or, more generally, by whatever lower-level method, may be even by a QM one of say smaller basis set or lower correlation account or by a semiempirical one with a less developed system of molecular integrals) and the hybrid total energy is then obtained by adding the QM (higher-level) calculated energy of the model subsystem and subtracting the MM calculated energy of the model subsystem. The expression for the total energy in the simplest case of two (quantum and classical) subsystems reads:

$$(2.140) \quad E_{\text{tot}} = E_{\text{low}}(\text{real}) + E_{\text{high}}(\text{model}) - E_{\text{low}}(\text{model}),$$

where "high" and "low" refer to the levels of approximation while "real" and "model", basically to the size of the calculated system. The above expression obscures the need for explicit formulation of the properties of the boundary between subsystems. The simplest implementation of this scheme is provided by the IMOMM methodology [213]. The analogous IMOMO scheme is the procedure of combining two QM (MO-based) methods. These approaches combining two subsystems are called two-layered [243]. The subtractive approach combining more than two regions (i.e., including some intermediate buffer subsystems) is called ONIOM [243]. The expression for the total energy in the ONIOM scheme is written analogously to eq. (2.140):

$$E_{\text{tot}} = E_{\text{low}}(\text{real}) + E_{\text{high}}(\text{S} - \text{model}) - E_{\text{med}}(\text{S} - \text{model}) +$$
$$+ E_{\text{med}}(\text{I} - \text{model}) - E_{\text{low}}(\text{I} - \text{model}),$$

where "med" corresponds to some medium level of approximation, while the notations S-model and I-model correspond to the small and intermediate sizes of the model systems, respectively.

Expression eq. (2.140) stipulates that the interactions between quantitatively and classically treated subsystems are somehow included in the difference of the energy of the model system calculated by the higher- and lower-level methods. Obviously it is not the true interaction of the subsystems treated at different levels, but just an interpolation. In this context it makes no sense to discuss physical contributions to the intersubsystem interactions such as the electrostatic polarization of the QM region by the MM-treated environment (and vice versa) or anything else of that sort. Clearly, the description of a reaction center or a chromophore requiring truly quantum description by MM can be quite problematic. The errors in this approach vary irregularly while the geometry changes: the MM schemes can describe perfectly the system near the equilibrium and totally fail near the transition state (saddle point). Also, as it is discussed in its proper place (Section 2.5) the MM itself is not developed enough to cover all classes of molecules, which produces additional uncertainty in the energy estimates. Saturation of broken bonds by hydrogens (or other groups) in an uncontrollable manner affects the results of electronic structure calculations. To level out these effects, the model system is suggested to be chosen to be large enough, which makes the procedure both expensive and obscure. In fact, the validation of the subtractive scheme is based on a more or less accidental compensation of errors. While using the ONIOM scheme it is explicitly prescribed by its authors [243] to estimate the errors in energy incurred by transition from a model molecule to a more realistic one and to use that pair of high-/low-level methods for which these errors are close enough.

The accuracy of the "subtractive" hybrid schemes i.e., mechanical embedding, can be evaluated by an analysis of the characteristic numerical examples. The problems caused by the ad hoc way of construction of this method are clearly seen when it fails. In most cases, the failures can be seen on rather simple molecular objects which are used as tests for those or other hybrid schemes. For example, an application of the IMOMM method to analyse the conformational properties of *cis*-butane, performed in [235], shows that marking two terminal carbon atoms of the molecule as QM leads to valence angles of $129.9°$, while the same quantity obtained by pure QM calculation yields $117°$ and the pure MM calculation gives the value $116.1°$ for these angles. We see that in this case the transition to the hybrid QM/MM procedure destroys even the results of pure MM calculations, which are by themselves quite acceptable. Problems also arise when multilayered schemes are used. For example, the energy of the reaction of the oxidative addition of H_2 to $Pt(P(t\text{-}Bu)_3)_2$ calculated within the B3LYP:HF:MM3 scheme is by 7.9 kcal/mol smaller than that calculated by the B3LYP:HF:HF method [243]. It turns out that the choice of the description for the third layer (the most inert and the most distant from the reaction center of the system) rather than of the second one, turns out to be crucial for the description of the energy of this reaction, which is itself estimated to be $\sim$4 kcal/mol [243]. It demonstrates the complete failure of this scheme of junction construction.

2.6.2. Polarization embedding

Further elaboration of the hybrid models stipulated by the necessity to model chemical processes in polar solvents or in the protein environment of enzymes, or in oxide-based matrices of zeolites, requires the polarization of the QM subsystem by the charges residing on the MM atoms of the classically treated solvent, or protein, or oxide matrix. This polarization is described by renormalizing the one-electron part of the effective Hamiltonian for the QM subsystem:

$$(2.141) \quad h_{\mu\nu}^{\text{pol}} = h_{\mu\nu} - \sum_{M} q^{M} V_{\mu\nu}^{M}$$

where summation over M is extended to the atoms in the MM (classical) subsystem, q^{M} is its effective charge and $V_{\mu\nu}^{M}$ is the matrix element of the Coulomb potential induced by the unit charge located at the atom M between the one-electron states μ and ν in the quantum system. Obviously the terms of this type appear in the general expression of eq. (1.258) for the effective Hamiltonian for the R-system. This type of modeling is quite common in the literature [209, 244]. A general objection against this type of treatment is that it violates the principle that *actio* should be equal to *reactio* [245] since no effect on the part of the QM system upon the MM system is assumed in eq. (2.141). Employing effective MM charges in the hybrid QM/MM schemes of the polarization embedding in terms of the classification [234] raises several problems. First of all, it is the source from where the MM charges are expected to come. Generally, as mentioned in the MM Section 2.5, the effective charges may appear from procedures of fitting molecular potential i.e. PES to some form including the Coulomb interaction of the effective charges. For example in the atom-atomic schemes describing crystals of unsaturated hydrocarbons (such as benzene) the effective charges residing on the carbon (hydrogen) atoms amount to $\pm 0.153.\bar{e}$. This value is exaggerated and cannot be reproduced by any semiempirical QC methods, which are generally known to produce the charges correctly to the extent that the experimental dipole moments are decently reproduced. A correct setting should probably include the QC precalculated effective charges to be located on the MM atoms with the hope of reproducing the result of the QC calculation.

2.6.3. Link atoms, capping atoms, etc

While taking into account the polarization of the quantum subsystem by the charges residing in its classical part, the (mostly the hydrogen) atoms, recklessly used to saturate the bonds broken when the model system is cut from the whole, acquire much more importance. The atom saturating the bond broken when a border is set on geometry grounds is called the link atom. Since the early times of quantum chemistry and even now, it is very common to make the molecular problem tractable by neglecting polyatomic and presumably chemically nonactive substituents, replacing them with hydrogen atoms. The QM/MM methodology takes the bulky substituents explicitly into account. The most straightforward way to treat covalently bound QM and MM parts is the link atom method. This approach is implemented in the industrial and semi-industrial packages [246–249]. The important drawback of the link atom method (and especially of the "dummy groups" method) is introducing additional nuclear degrees of freedom for which no reasonable equation of motion (or equivalently no energy minimum condition) can be derived. Such problems and the positioning of link atoms have been addressed by numerous authors. For example, the authors of [213] have proposed a special procedure for geometry optimization with rigid restrictions imposed on the position of the link atom. More complex is the so-called scaled position link atom method (SPLAM) [245]. It requires corrections to bonding, dipole, and van-der-Waals terms to be introduced. In some cases it works significantly better than the simple link atom method, but the status of results is still unclear. In any case numerical estimates performed in [245] cannot be considered as being very successful. The typical failure of this version of the QM/MM scheme is that it gives the results which are quite close to the pure MM ones. For example, considering the water dimer has shown that the BLYP QM method predicts the OO distance and the HO...H angle (2.98 Å and 123°, respectively) to be very close to the experiment (2.98 Å and 122°), while the MM scheme gives significantly worse results (2.77 Å and 162°). The SPLAM hybrid scheme gives values that almost coincide with those of the MM (2.78 Å and 163°). In this case the use of the hybrid QM/MM approach seems to be senseless. It proves in fact that some important contributions are missed in the approach of [245].

The problem of geometry optimization in the link atom schemes is sometimes addressed by modifying the expression for the total energy of the molecular system. It is shown in [234] that the difference between QM and MM interaction energies for link atoms enters into the total energy and strongly depends on the link atom position since the dependence of energy on geometry given by the QM theory and the MM force fields are very different. Practically it leads to the collapse of the fictitious link atom with the boundary atom (that in the MM treated region whose bond with a QM atom had been broken). The characteristic result is that the equilibrium position of the link atom is poorly defined and cannot be rationalized. One of the possible prescriptions for avoiding the collapse of the link atom proposed in [234] is to use for geometry optimization some potential energy function not coinciding with the total energy of the molecular system:

$$(2.142) \quad E_{\text{pot}} = E_{\text{tot}} - E_{\text{link}}$$

Such an approach also seems to be quite artificial. An example of employing it is given in [250] (cited in [245]) where it is shown that strong deviations of the link atom equilibrium position from the line connecting atoms forming a covalent bond are possible and lead to serious problems. Also, the vibrational spectra calculated with the optimization of the link atom position are much worse than those derived from the MM force field itself. When it comes to estimating chemically interesting quantities in the QM/MM approach, the calculated proton affinity for small gas phase aluminosilicate clusters turns out to be very sensitive to the length of the bond between the boundary QM atom and the hydrogen link atom [251].

The link atom scheme and all the concepts which arise in this context of hybrid modeling, turn out to be extremely contradictory as too many different aspects of the intersubsystem junction are loaded upon them. The link atom arises as an auxiliary tool to describe an atom on the poorly defined boundary between the subsystems. In this setting a noncontradictive object of the general theory – the frontier atom – one which bears hybrid orbitals ascribed to different parts of the complex system, is split between the link atom and the boundary MM atom. The orbitals centered on the link atom are included in the basis of the orbitals of the quantum subsystem. They are used to mimic the behavior of the orbitals of the frontier atom. However, as it is easy to understand, they perform this job rather poorly. The AO basis of the link (hydrogen) atom (even if it chosen to be rich enough) has nothing to do with that of a heavy MM atom. (We notice that within the frame of ab initio methods there is no reasonable mechanism which could help to level out this difference). In any hybrid scheme there are no tools to reasonably model the direction of the orbitals of the MM atom and the proposed medications are worse than the diseases they are supposed to cure. For example, a very artificial construct is employed: to mimic, say, the C-C bond, the link hydrogen atom must be placed not in the point where the C atom ascribed to the MM subsystem is placed but in some other point. In this case when the interactions between classical and quantum subsystems are treated explicitly, the coordinates of the link atom cannot be used to estimate the MM energy of its interactions with other MM atoms. By this, the coordinates of the MM atom represented by a link atom are still necessary. Then the link atom has coordinates of its own, which are excessive as the coordinates of the MM atom of the broken bond are not dropped from the entire set of the nuclear coordinates. This has resulted in many papers where authors attempt to solve an unsolvable problem: how to fix (and whether it is worth fixing) the position of a hydrogen link atom so that everything comes out more or less decently.

The combination of taking into account the polarization of the quantum subsystem by the effective charges residing on the MM atoms with the link atom construct results in further inconsistencies. They manifest themselves in a vividly discussed dilemma: whether to include the Coulomb interaction between the charge on the MM atom and that on the link atom representing it in the QM subsystem or to drop it completely. In this case, using eq. (2.141) for the action of external charges on the quantum subsystem becomes particularly problematic. In addition to the inconsistency of the orbital basis of the link atom with that of the MM atom which it represents in the QM part of the complex system, the effective charge of the link atom may differ from the effective charge of the represented MM atom. The most severe problem is that the link atom falls into the electrostatic field of the MM atom bearing the charge q_X (and *vice versa*), whereas the distance between them is much smaller than the length of the broken bond. This introduces a lot of confusion.

Originally [209], neither electrostatic or van-der-Waals interactions between the link atoms and the MM subsystem were taken into account. As we have demonstrated above it is definitely wrong. Later, different prescriptions to omit some real physical interactions and by this to compensate unphysical ones were proposed: the authors of [252] neglect the Coulomb interactions between the QM subsystem and the MM group closest to it, the authors of [253] force the charge on the boundary MM atom to be zero. Such omissions of interaction terms cannot be justified and must be considered special tricks for masking the errors caused by other inconsistencies in junction construction. Another prescription [254] is shifting the values of the charges with subsequent compensation of this perturbation by introducing fictitious dipoles. Typically, the possibility of manipulating the interactions of the artificial link atom is considered an advantage of the special flexibility of this approach [255]. However, the price of this flexibility is too high as it leads to complete uncertainty in the results obtained and thus marks down any possible predictive capacity of the QM/MM approach. These contradictions can be resolved on the basis of the general theory. From its point of view the problem of interaction between the MM atom and the corresponding link atom does not exist, but is replaced by that of the Coulomb interaction between electron densities located on the frontier atom, but ascribed to different subsystems. Then the answer is obvious:

$$(2.143) \quad \sum_{rr' \in A} \sum_{mm' \in A} (rr' \parallel mm') \langle\langle \mathrm{m}^+ \mathrm{m}' \rangle\rangle_M \langle\langle \mathrm{r}^+ \mathrm{r}' \rangle\rangle_R.$$

and one has to take it into consideration rater than drop it. The problems arise as a consequence of attempts to paint it as as a two-center interaction. In fact it is not a two-center, but a one-center, one and is not equal to any reasonable interaction between the boundary and link atoms. Many whimsical methods are proposed in the literature [256] to redistribute charges in the classical subsystem so that conserving their overall balance levels out the nonphysically strong interaction between the link atom and that of the MM atom. These tools are implemented for example within the "industrial" package QUASI [249].

Turning back to the intersubsystem interactions considered as a basis of the classification of the hybrid QM/MM schemes, we first address those intersubsystem junctions which are represented by classical bonding terms. In this case an important question arises: which terms should be included and which should not? The most widespread method is to include classical bonding force fields only when at least one MM atom is involved [209, 237]. The main general problem with such a setting is double counting of interactions since some of them are taken into account also by a quantum mechanical procedure. To overcome this inconsistency the authors of [252] proposed to calculate only those improper dihedral fields of the MM subsystem from which both outer atoms come. In general the methodologies based on deleting the terms in the expression for the total energy of the system related to the link atoms are very difficult to systematize. Practical implementation of these methods is very difficult as in the framework of this scheme serious artefacts appear quite expectedly.

The saturation of dangling bonds by hydrogen is not the only possible way proposed in the literature. The main reason for using other types of saturating groups/atoms is the intention to improve the description of the polarity and other properties of the broken bond. The saturation groups known in the literature are pseudohalogens [257, 258] and "dummy groups" [259].

Both the mechanical embedding and the polarization embedding type models within the classification of [234] suffer by missing the effects of the quantum subsystem on the classical one. This type of energy contribution may be crucially important. It is precisely that which is implicitly taken into account by the Born-Onsager solvent-solute model of continual insulator and its descendants. In these models the solute polarizes the solvent and the polarization implicitly understood is related to the nuclear orientational motions of the permanent dipoles of the molecules constituting the solvent. More subtle effects are related to the electronic polarization of the classical system. These effects are ignored in most of the modeling packages currently available. The practical scheme including classical treatment of the MM system polarization is provided by [260]. This type of approaches is, however, rarely used though the MM polarization was taken into account even in the early scheme [208] where it had been done by using atomic polarizabilities. Here of course the electronic polarization of the otherwise classical subsystem is meant. Even earlier a very nice classical model of electronic polarizability was constructed [261] to be used in calculations of the impurities in ionic crystals, to describe the response of the surrounding ions upon the variation of the electronic distribution in the impurity and later employed in [262] to construct a quantum-classical junction for describing strongly ionic crystals.

The polarization of the classical subsystem by the quantum one turns out to be of crucial importance when it goes about coordination compounds of metal ions with high formal charges. Despite some electron back donation, the effective charge on the metal remains rather high and thus the organic environment is expected to be significantly polarized. The general theory is that this situation prescribes employing the adequate polarizability technique for the classical subsystem. Its use will be demonstrated in a due course.

The classification of the hybrid QM/MM methods proposed in [234] is not complete. It does not include some self-consistent schemes like [215] and does not consider the possibility of charge transfer between subsystems [263]. The authors of [264] have imposed a requirement of intersubsystem self-consistency on the construction of a junction. It means that the charge transfer between subsystems should be taken into account. Practical implementation of this requirement was performed by using special iterative procedures of double (intrafragment and interfragment) self-consistent (DSC) calculations. It leads to explicitly taking into account the electron transfer between the subsystems (and also of the polarization of the QM subsystem). This methodology, however, cannot be justified as the self-consistent field procedures are separately applied to systems with strongly fluctuating numbers of electrons that also lead to poor definition of the fragments themselves (and of their quantum numbers). According to the results of the previous section, the electron transfers should be considered virtual ones and taken into account in the perturbative fashion. This point is confirmed numerically as the application of the DSC scheme to the iron picket-fence porphyrine has led to improbably large intersubsystem charge transfer of 3.6 electrons [264].

2.6.4. Local SCF and analogous methods

As we see, using the geometry based division into subsystems and link atom scheme creates many problems. The main lesson to be learned from it is that the intersubsystem boundary must be set by a method adequate for the quantum realm. This cannot be done in the direct – coordinate – space, but only in the Hilbert space, by selecting the subspaces of the entire orbital space and defining the quantum subsystem in terms of these subspaces and developing the effective Hamiltonians acting in these subspaces. If these subspaces are correctly chosen, the problem of defining the subsystem's composition in terms of electron numbers can be reasonably solved. At the same time, the reasons for introducing the link atoms are satisfied by this construct: the orbitals necessary to describe the fixed number of electrons in the quantum subsystem appear in the model. By this, the boundary between subsystems becomes a "grey" zone between quantum and classical subsystems. However, no excessive nuclear coordinates appear as the orbitals to be used are assumed to be centered at the *frontier* atoms of the system i.e. those which by definition bear orbitals belonging to the quantum subsystem. The closest analogy is provided by the π-electron approximation, where from this point of view, each carbon is a frontier atom. In the general case, the planar symmetry of course does not take place and some other considerations must be used to define these states: a spectacular example is provided by a tentative "π-electronic" model of fullerenes or nanotubes which are reasonably treated by the corresponding "π"-orbitals directed approximately along the radii of the corresponding sphere or ellipsoid or cylinder, but require some more elaborated procedure for their development rather simple reference to symmetry.

Models of this type are present in the literature. The simplest ones are based on the use of local orbitals. It is the local self-consistent field (LSCF) approach [216, 231, 265, 266]. In it the chemical bonds between QM and MM regions are represented by strictly local bond orbitals (SLBOs). The BOs can be obtained by the a posteriori localization procedures known in the literature. The localized orbitals thus obtained have some degree of delocalization, i.e. they have non-zero contributions of the AOs centered on the atoms not incident to a given bond (or a lone pair) ascribed to this particular BO. These contributions are the so-called "tails" of the localized orbitals and neglecting them yields the strictly local BOs (SLBOs) which are used in the LSCF scheme. The QM part of the system is described by a set of delocalized MOs while the boundary is modeled by the frozen SLBOs.

An important assumption made in the LSCF construction is that the SLBOs are transferable within a wide class of molecules. The frozen character of the boundary SLBO causes the sensitivity of the results obtained within the LSCF scheme to the size of the QM region. The electronic structure of the QM region is described by the HFR (or synonymously self consistent field i.e. SCF) procedure developed for the modified Fock operator, which includes Coulomb and exchange interaction with the SLBOs and Coulomb interaction with the MM charges of eq. (2.141). Technically the problem is identified as one of defining the charge to be set at the MM atom X represented in the quantum system by some link atom. Following the solution in [216], the effective charge of an MM atom bearing a bond to be treated at the quantum level must be set to $q_X + 1$ (q_X is the effective charge on X in the MM setting) on the ground that this atom supplies to the quantum subsystem one electron which must be compensated by the corresponding core charge. In its original implementation, the LSCF method was developed with an additional condition of fixing the positions of the atoms of the environment and thus it was not suitable for geometry optimization. This restriction has been lifted in [267], but this requires additional adjustment of the force field parameters. The authors of [267] have noticed that in the framework of the original LSCF scheme the ion-nuclei interactions are underestimated and the variation of the overlap between boundary basis functions due to variations of the boundary bond length is not taken into account. To correct these defects they introduced the boundary bond potential of the form:

$$(2.144) \quad E_{X-Y} = (A + Br + Cr^2)e^{Dr} + \frac{E}{r}$$

where the exponential term is introduced to mimic the overlap dependence on the interatomic separation and the last contribution describes an interaction of effective charges. Parameters A–E are numerically optimized. In fact, the first contribution should describe the intrabond resonance energy and one has to consider the potential eq. (2.144) mainly as a correction for the bonding (the overlap dependent contribution). At the same time the values of the parameters A–E obtained by the authors [267] seem not to correspond to their declared physical meaning. The same conclusion can be drawn from the energy profile for the process of the boundary bond elongation – for large values of bond length the difference between the LSCF and SCF (HFR) curves increases drastically since the nonphysically large Coulomb contribution becomes prevailing. The optimized bond lengths in the SCF and the LSCF/MM approaches can differ significantly and also remarkably and unpredictably depend on the details of the environment the bond is assigned to (for example, C(QM)-N(MM) bond is by 0.024 Å longer than the bond N(QM)-C(MM) with the same hybridization). The problems with the correction of bond description by the potential eq. (2.144) are caused by the number and type of factors it is designed to reproduce – the precise form of boundary orbitals is a function of geometry (see eqs. (3.141) and (3.136) in Chapter 3), elements of the intrabond density matrices (especially, of the two-electron ones) also depend on the geometry (see below). Also the expression for the bond energy cannot be arbitrarily postulated but must be defined from the analysis of a particular QM expression. The angular dependence of the bond potential is totally neglected in the potential eq. (2.144), while it appears naturally from the derivation of intersubsystem junction due to angular dependence of resonance interactions. One can see that the LSCF approach has an important drawback in construction: the parameters for the SLBOs should be determined from model molecules for each new system. Construction of the extensive database of the SLBO parameters is considered a strategy within this approach by its authors.

A careful numerical examination and comparison of the link atom and LSCF techniques were performed in [255] where the CHARMm force field [178] and the AM1 method [68] as respective MM and QM procedures have been used. In the case of the link atom procedure, two options were used: QQ – the link atom does not interact with the MM subsystem and HQ – link atom interacts with all MM atoms. The main conclusion of this consideration is that the LSCF and the link atom schemes are nevertheless of similar quality. The error in the proton affinity induced by these schemes is several kcal/mol. It is noteworthy that all the schemes work rather poorly as tools, even for that simple problem of describing the conformational properties of n-butane. The large charge on the MM atoms in the proximity of the QM subsystem (especially on the boundary atom) cause further significant errors in the proton affinities for all methods (especially in the case of the LSCF approach where the error can be of tens of kcal/mol). This is not surprising as the stability and transferability of intrabond electron densities is broken here. It proves that the simple electrostatic model is not appropriate for these schemes and that a detailed analysis of the junction form is necessary in the general case. The numerical analysis shows that the error induced by the HQ model is smaller than that induced by the QQ model. As the HQ model explicitly includes unphysical interactions with the artificial link atom, it means that these interactions partly compensate for errors in the junction construction to some uncertain extent. Practically, the link atom interacts with the MM atoms even in the case of the QQ model but this happens indirectly *via* the interactions of the QM subsystem with the MM atoms. In the case of the QQ model, the non-compensated charge on the QM subsystem (without link atom) interacts with the MM atomic charges. It causes significant polarization of the QM subsystem, which is confirmed by numerical estimates of charges in it.

The principles similar to those of the LSCF are used for junction construction in [134] based on the fragment SCF method. Another model thoroughly elaborated to consider the effect of motions of environment atoms on the ab initio level is that of [268]. In the framework of this model the procedure of freezing the SLBO was refined and the area of delocalization of SLBOs extended from two boundary atoms to two boundary groups. At the same time an attempt to obtain numerical results of good quality has forced these authors to introduce some very artificial procedures throughout the construction of the junction. Among them we mention placing very large fictitious positive point charges on the bond. This model also requires extensive parametrization: in order to reproduce the energetics of alanine dipeptide and tetrapeptide, 27 parameters describing the interface between subsystems were adjusted. Essential parametrization is quite an important problem of this model. It would be fair to say that reproducing conformational energies for a polar system like polypeptide is a difficult test for any computational methodology. The authors of [268] formulate the essential requirements to the bond for which molecule can be cut: (i) this bond should not have significant multiple bonding character, and (ii) this bond should be far away from the region where significant electron re-distribution occurs. The first requirement seems to be quite reasonable if a bond is described by a single localized orbital, while the second one reflects a lack of some contributions to the energy, which become important if a smaller QM subsystem is chosen. The estimate of these contributions is not possible due to intersubsystem junction in this method.

It must be realized that the terminology in the QM/MM area is rather shaky. Many methods bear names not directly reflecting what is really done. For example, the method based on the effective fragment potential (EFP) construction described in [269] is in some respects close to the LSCF methodology. In this case the boundary is modeled by a buffer region consisting of several localized molecular orbitals which are obtained by a QM calculation performed for the entire system or for a subset of the latter. The orbitals obtained in the calculation for the entire system are set frozen in the EFP calculation. The orbitals of the QM part are forced to be orthogonal to those of the buffer region and the effect of the more distant environment is represented by an EFP. The important idea of this approach is to make the distance (in the physical space) between the QM and EFP regions large enough to make it possible to present the corresponding intersubsystem interactions as nonbonding ones. At the same time the freezing of the buffer one-electron states can be a source of significant errors as the changes in polarization contributions coming from the buffer region are neglected throughout the geometry optimization. The numerical example given by the authors [269] is quite characteristic. They have calculated the proton affinities of lysine and the H-bonded and non-H-bonded tripeptide Gly-Lys-Gly by the QM/buffer/EFP method. If the buffer region is chosen to be formed by γ- and δ-CH_2 groups of the lysine chain (i.e., quite far from the reaction center) the QM/buffer/EFP calculation gives the value of the proton affinity to be 2.2 kcal/mol higher than the reference QM one for all these molecules. It unequivocally testifies that these 2.2 kcal/mol constitute the error of junction construction in this case, which seems to be quite large. Moreover, the difference between QM/buffer and QM/buffer/EFP results in the proton affinities of lysine and non-H-bonded tripeptide Gly-Lys-Gly of only 0.2 kcal/mol, i.e. the effect of the environment described by the EFP is by an order of magnitude smaller than the error produced by the junction.

2.6.5. Methods based on effective potentials

A further important development in junction construction is provided by the methods invoking the concept of effective or model potentials [270, 281]. Taking into account the modern derivation of the effective potentials based on the GF representation of the wave function due to Seijo [36] it is natural to expect that similar treatment can be performed for the cores of more general form than the atomic cores described in Section 2.2. Further development of the idea of using pseudopotential theory in relation with the problem of hybrid modeling of the complex systems brings one to the construct of the effective group potential (EGP) introduced in [271, 272] and further developed and used in [273–275]. The general construct of the EGP evolves along the same line as that of the effective core potential (ECP – see details above). This leads to the form following logically from the general theory where it is demonstrated that the effect of the electrons not considered explicitly reduces to one-electron operator affecting those considered explicitly used throughout the calculations:

$$(2.145) \quad \hat{F}_{\text{eff}} = \hat{h} + \hat{W}_{\text{EGP}} + \hat{\Sigma},$$
$$\hat{W}_{\text{EGP}} = \sum_{rr'} \alpha_{rr'} r^+ r'.$$

The general theory provides an explicit form for the matrix elements $\alpha_{rr'}$ (eq. (1.244)). In the EGP setting, they are considered as parameters and obtained by fitting (following the norm of the difference between the exact and the model Fock operators criterion) of the model Fock operator acting in the restricted subspace of orbitals and the exact Fock operator. This procedure allows one to reproduce the anisotropy characteristic for the inert environment: the EGP of the ammonia molecule represented in the quantum subsystem only by its lone pair has the overall symmetry of the C_{3v} group, describing by this the effective interaction of electrons in the quantum subsystem with those occupying the one-electron states of the N-H bonds, which are not considered explicitly. An analogous picture is observed for the case of Cp^- anion of which only five π-orbitals are taken explicitly and the rest is considered as the core. The difference with the traditional π-electron theories here is that the π-orbitals of the Cp^- units are used to represent the whole cyclopentadienyl anions in a situation where there is no planar symmetry at all: namely to describe a coordination compound $CpIn(CO)_4$, which demonstrates the secondary importance of the purely symmetry considerations for the electronic variables separation. It is also fair to say that the idea of describing Cp complexes of metals using only π-electrons of the ligands belongs to Shustorovich and Dyatkina [276].

If a part of the inert environment is taken by the lattice of the ionic or covalent crystal, use of the pseudopotential theory is proposed in papers [36, 278–283]. The source of the latter are the electrons and nuclei of the crystal, except the area taken by the explicitly considered quantum system. This approach allows staying the ab initio context to determine the form of the orbitals in the QM subsystem. In the pseudobond approach [284, 285] the free valence of a QM atom is saturated with a special atom located exactly at the position of the neighbor MM atom. The basis set and number of electrons of this pseudoatom are set to be equal to those of the fluorine atom. The electronic structure of the broken C-C bonds are mimicked by a special adjustment of the effective core potential of this pseudo fluorine atom. Another approach based on the use of effective potentials is proposed in [286]. A series of one-electron quantum capping potentials replacing the link atom is developed by modifying conventional effective core potentials: the spherical shielding and Pauli exchange repulsion terms replace the dropped valence electrons. The capping potentials are adjusted to reproduce all-electron geometries and charge distributions. The analysis of this scheme shows that the error induced by a capping potential is significant (especially for angular distortions) but generally smaller than that in the simple link atom scheme.

The EGP method and its analogs seem very promising for the hybrid method's construction, employing ab initio QC procedures as the QM counterpart. Nevertheless, due to the details of the pseudopotential-based theories construction it is not possible to prove their transferability in the case of the variation of the geometry of the classically treated part of the complex system. In fact, transferability does not have place, which most clearly manifests itself at the intersubsystem frontier. The accepted procedure of defining the pseudopotentials assumes that they must be recalculated whenever the classical environment changes its geometry. It actually happens already in the ECF setting where the pseudopotential is reduced to that of the filled atomic cores and varies when the nuclear positions vary. When it comes to a more elaborate core group (like NH bonds in ammonia molecules) changing positions of H-atoms changes the electronic structure of these bonds and thus affects the form of the filled orbitals of the "spectator" groups. So recalculating the electronic structure of the environment is required whenever it may be necessary, for example, at each step of the optimization procedure, which is not practically possible.

2.6.6. Orbital carrier spaces for quantum subsystems

Both the LSCF and pseudopotential-based methods face a similar problem throughout their practical implementation. It includes the dependence of the orbitals ascribed to the quantum system on the geometry of the classical system. The most transparent example comes from the EGP treatment of the NH_3 molecule. Within this setting the NH_3 molecule is described by an effective two-electron atom N# bearing an HO and exerting the pseudopotential of the C_{3v} symmetry reproducing the presence of the N-H bonds. As already mentioned, variation of geometry (for example of the pyramidalization angle of the NH_3 molecule) changes the parameters of this pseudopotential at least for that obvious reason that the space of orbitals treated explicitly must be orthogonal to the occupied states not considered explicitly. Since the latter depends on geometry, the former does the same. Even more important is the fact that the matrix elements of effective Fockian, pertinent to the only explicitly considered lone pair of the NH_3 molecule, are expectedly dependent on the pyramidalization angle. In the EGP setting these dependencies would require a complete recalculation of the effective Fock operator at each move in the classical subsystem and even worse – this recalculation has to include a quantum (HFR) treatment of the classical part, making the whole enterprise basically senseless. A similar situation occurs in the LSCF theory where SLBOs are geometry-dependent. Within the LSCF theory the idea has been put forward to use extensive data bases for SLBOs in different environments, but it does not solve the problem of the geometry dependence of SLBOs, which clearly cannot be covered by any database. To solve this and to escape the need for constructing large databases of the SLBOs required by the LSCF approach, Gao et al. proposed the generalized hybrid orbital (GHO) method [287,288]. This approach is intended to interpolate the shapes and orientations of the HOs residing on the frontier atoms. The first important step in the GHO method, which will also be used in our derivation, is dividing the hybrid orbitals into active and auxiliary ones – the former are added to the QM subsystem. In the original LSCF approach, four orbitals of the boundary atom are included in the self-consistent procedure. In the GHO approach, only one active orbital per frontier atom is included in the set of orbitals spanning the space where the QM procedure evolves. Therefore, with the exception that the HO used in the QM procedure is not mandatory a pure p-orbital, boundary atom in the GHO scheme is very close to the atom in the π-electron approximation. It allows one to fit some adequate semiempirical parameters for the boundary atoms. The possibility of choosing reasonable and transferable semiempirical parameters for HOs of such a defined boundary atom must rely upon a very subtle procedure of determination of the HOs. Conversely, the GHO scheme uses some very crude assumptions [288] about the form and the direction of HOs, based on pure geometry grounds, which are in fact equivalent to (i) fixing the C_{3v} symmetry of the local MM environment and (ii) assuming that all the HO directions coincide with the directions of bonds. Practically, these conditions are satisfied only for methane molecules. Moreover, even for these assumptions the s-/p-ratio for the active orbital is determined by incorrect formula working only for equivalent active and auxiliary orbitals or for purely p active orbital. The real structure of the HOs as a function of the geometry will be described below and we shall see that neither of the assumptions accepted in [288] is actually fulfilled. Neither a possible asymmetry of the geometry of the boundary atom environment nor the chemical nature of neighboring groups are taken into account by the GHO method. In practical implementation of [287, 288] all the HOs are determined by directions of the bonds between the boundary atom and its three MM neighbors. Since the s-/p-ratio and direction of the active HO in the GHO method are the functions of only MM atom positions, the form of this HO may be far from being optimal, to say nothing of its behavior with geometry variations. The non-variational form of the HOs brought the authors of [287] to the necessity of making significant and hardly justifiable renormalization of the Hamiltonian and the MM force field parameters to reproduce correct bond lengths, directions of auxiliary orbitals along the corresponding bonds, and the effective charges on the atoms. For example, in the case of carbon atom the resonance parameter β_s of the MNDO method had to be changed by more than 10 eV; the MM C-C ideal bond length parameter r_0 had to be changed by 0.05 Å.

2.6.7. Basic problems of hybrid methods

From the above survey it is clear that despite the attractivity of the idea of hybrid treatment of the complex systems, not much is known about it from the theoretical point of view. Moreover, the most problematic schemes of separating the system into quantum and classical parts – link atom and mechanical embedding – are implemented in two most widespread program suits of QC: GAMESS [247] and GAUSSIAN [248]. The general picture of how this progress (if any) evolves in the QM/MM area resembles the treatment of a difficult patient by symptomatic medications: each of them helps to lift some visible signs of trouble, but results in strong side effects to be cured by further remedies, while the overall process does not manifest any signs of improvement. The general methodology of constructing hybrid modeling schemes evolves in a completely different way. The problem is solved by including the QM/MM methods in the general context of dividing a complex system into parts. When a hybrid scheme is going to be constructed, the authors are expected to solve, in the same or another manner, several key problems. First, the variables describing the complex system must be defined; second, the parts into which the latter is to be divided must be identified; third, the interactions between the parts must be adequately constructed. These problems are referred to as the problems of frontier and junction construction. They relate both to the variables describing the electronic structure of the system as well as the nuclear coordinates describing the spatial structure of the complex system. A wide range of intuitive solutions, not having a solid basis, are proposed in the literature in response to these questions. However, there is no consistency among them, or with the general theory, and it leads to many serious problems. Among them mention must be made of uncontrollable numerical effects as well as incurable contradictions. In practice, one normally notices some technical inconsistencies, which can be overcome by setting absurd values to the junction parameters or arbitrarily adding or omitting terms in the energy of the complex system, etc. These inconsistencies are cited in abundance and reviewed in this chapter. Also, the number of solutions proposed in the literature is impressive. It is very difficult to impose any system upon them and the proposed ones seem quite awkward. It is not the fault of the authors of these systems: it is hardly possible to set any order to the collection of solutions if "adding unicorn's milk completely changes the properties of the brew".

A hint comes from a distant area, as the problem must be solved not by separating areas in the the direct three-dimensional coordinate space, which brings up many contradictions, but by singling out subspaces in the Hilbert space. This approach is much more natural, as QC uses, not the coordinate representation, but the representation of the occupation numbers and these latter are the adequate set of variables so that in their terms the separation of the system into parts can be performed sequentially. Under this condition one may assume that the QM descriptions of different accuracy are applied to different numbers of electrons residing in different orthogonal subsets of the one-electron states. The classical example of such a method of singling out a subsystem from a more complex system is of course the π-electron approximation, known from the very early years of quantum chemistry. One can see that such a method of dividing the system into parts satisfies the very important requirement that the number of particles in the subsystem is a good quantum number and on the other hand it demonstrates that ascribing orthogonal orbitals even centered at the same atom to different subsystems does not pose any problems: neither conceptual nor technical ones. It is shown above that even quantum chemistry itself can be considered from this perspective.

The general theory of electronic structure of complex systems and their PES are based on the tacit assumption that the basis orbitals are well defined orthonormal functions, which can be conveniently divided into two (or more if necessary) classes. The reality is much more tough and results in serious conceptual problems in all the existing packages offering hybrid modelization techniques in their respective menus. These have been addressed in the previous section. Now we address the meaning of the results obtained so far. In fact, up to this point, we obtained the description suitable for *any* hybrid QM/QM method. Within this context, the distribution of orbitals between the subsystems treated by different methods is only a formal exercise. We can follow any method and if the condition of the smallness of the fluctuations of electron numbers in the selected subspaces is satisfied, the chosen distribution of the orbitals and electrons between the subsystems must be considered an acceptable one. The intersubsystem frontier in this setting is reasonably defined as a set of atoms that bear the (hybrid) orbitals ascribed to different subsystems. As with any "mathematical" definition, it has an area of applicability limited by the condition that the terms entering the definition keep their sense. It is clearly valid for the semiempirical domain where it is possible (in a noncontradictory manner) to form HOs which are attached to a given atom. The postulated orthogonality of AOs is of course of great help here. In the ab initio context, it would be a much more complex task due to nonorthogonality of the original AO basis set. This problem becomes severe with large basis sets with numerous diffuse functions. In this case, additional work is needed. At the same time, as it also happens with mathematical definitions, some cases we normally do not expect to also fall under them. These are of course all atoms in the π-approximation, which turn all to be the frontier ones.

The orbitals (or better to say the exact form of the carrier spaces for the electron groups) in the QM/QM context can be found by adjusting the form of the hybrids centered on the frontier atoms on the basis of some more or less arbitrary criteria. Some solutions of this sort are known in the literature. Theoretically, sound treatment of this problem must be based on the variational principle so that the HOs on the frontier atoms are determined together with other electronic variables from the energy minimum condition. This type of approach is implemented for example in the VB2000 program suit [289]. The situation with the hybrid methods in the strict sense, namely with QM/MM, is more complex, as no source of flexibility is assumed in such a setting. Nevertheless, following the instructive example of the π-approximation, we assume that some electronic structure variables can be implied for the classically treated part of the complex system. The classical description is here only an approximation for the energy dependence on nuclear coordinates, which is ultimately determined by quantum laws of electron motion in the field induced by nuclei. Then, generalizing the auxiliary HOs by Gao [217], we can think about the whole set of one-electron states spanning the carrier space for electrons residing in that part of the system that is classically described. The frontier atoms in this case are those that bear both the active and auxiliary HOs in Gao's terminology or those belonging to R- and M-systems in our terms. The M-system itself is then assumed to be treated using the classical force fields. Of course, we could close our eyes to the problem of the origin of these force fields, as it is done in the context of the MM theory. In the context of the hybrid QM/MM theories, however, we need a much more elaborated picture, which would be equivalent to (or at least a good model of) the variational procedure of determination of the orbitals centered on the frontier atoms mentioned above. On the other hand, it is desirable that the MM force fields are somehow consistent with the used QM picture. One may think of the QM/MM treatment as an approximation of the QM treatment of the specified level for the entire system. The latter must be done by a certain QM method. Then the QM/MM description of the same system can be considered as one which approximates the "exact" treatment of the whole by a hybrid procedure. This process can be continued up to the limit when the QM system disappears and the whole system is described by an MM procedure. Approaches of that sort are currently implemented in those force fields which are parametrized against results obtained by some quantum chemistry method. It is not, however, precisely what we want or need: if a standard QM method is used, there is no transparent way to keep, throughout this transition, the necessary elements of the electronic structure. Therefore we propose a way to the hybrid methods in a narrow sense, which, at first glance may seem to be too indirect. We propose to develop a transition procedure, connecting an appropriate (and approximate) but still quantum mechanical description of the M-system of the complex system to a classical and eventually the force field type description of the latter, but allowing us to single out the effects of the M-system's geometry and composition on the one-electron states of the frontier atoms, which could potentially be ascribed to the quantum subsystem. This is done by setting the problem somewhat wider: as that of the sequential derivation of the MM – a classical or rather a mechanistic – model of PES from an appropriate quantum mechanics description. If we establish a sequence of the moves (transformations and approximate estimates) leading from an appropriate QM theory to a mechanistic (ultimately a force field MM) model of PES, we can exploit this asset as follows: we can apply this sequence of moves not to the molecular system in its entirety, but to some part of the system. For the part where these moves are applied, we obtain the required mechanistic description and this part then becomes the classical part of the complex system, whereas the rest remains for the quantum description.

Analyzing the semiempirical QC methods in relation to their suitability for developing the hybrid QM/MM methods reveals certain problems. Using the HFR form of the electron trial wave function together with the ZDO type of parametrization results in the decomposition of the total energy of a molecular system into a sum of mono- and diatomic increments:

$$(2.146) \quad E_{total} = \sum_{A} E_A + \sum_{A<B} E_{AB}$$

where A and B denote the atoms in the molecule under consideration. The increments (in the CNDO type of parametrization) have the form [39]:

$$E_A = \sum_{\mu \in A} P_{\mu\mu} U_{\mu\mu} + \tfrac{1}{2} \sum_{\mu\nu \in A} \left(P_{\mu\mu} P_{\nu\nu} - \tfrac{1}{2} P_{\mu\nu}^2 \right) \gamma_{AA},$$

$$(2.147) \quad E_{AB} = - \sum_{\mu \in A} \sum_{\nu \in B} \left(2 P_{\mu\nu} \beta_{\mu\nu}^{AB} + \tfrac{1}{2} P_{\mu\nu}^2 \gamma_{AB} \right) +$$

$$+ \left(Z_A Z_B R_{AB}^{-1} - P_{AA} V_{AB} - P_{BB} V_{BA} + P_{AA} P_{BB} \gamma_{AB} \right)$$

where $U_{\mu\mu}$ is the energy of the orbital μ in the field of the atomic core A, $-\beta_{\mu\nu}^{AB}$ is an off-diagonal element of the one-electron Hamiltonian, γ_{AB} — the two-center two-electron Coulomb integral, V_{AB} describes the interaction of the valence electron of the atom A with the core of the atom B,

$$(2.148) \quad P_{AA} = \sum_{\mu \in A} P_{\mu\mu}$$

is the atomic electron population and $P_{\mu\nu}$ are the matrix elements of the one-electron density matrix which in the AO basis have the form:

$$(2.149) \quad P_{\mu\nu} = 2 \sum_{\lambda \in occ} c_{\mu\lambda} c_{\nu\lambda},$$

as expressed in terms of the MO LCAO coefficients $c_{\mu\lambda}$. The above (HFR) form of the total energy does not provide any common ground with the MM picture. The pairs AB in eq. (2.146) do not correspond to the bonds, but run over all possible pairs of atoms. It is not possible to demonstrate a priori the transferability of any of the involved quantities from one molecule to another. Moreover, the variable describing the electronic structure in approaches of this type – MO LCAO coefficients $c_{\mu\lambda}$ – are not transferable. Further, even if the density matrix elements can be shown to be transferable by massive numerical experiments, their transferability cannot be proven by more general theoretical means. The atom-triples specific contributions to the bending energies or atom-quadruples specific contributions to the torsion energies cannot be transparently extracted from eqs. (2.146) and (2.147).

We can, however, specify criteria for the quantum chemical (quantum mechanical) method to be compatible with the MM picture. The standard MM description assumes the use of the local concepts such as chemical bonds and lone pairs. To be on common ground with MM we have to use a quantum chemistry method which expresses molecular electronic structure and electronic energy in similar local terms and reproduces molecular properties with sufficient accuracy. The variables characterizing the wave function of this method have to be transferable in a broad sense of the term "transferability", i.e., the form of any bond-related functions (e.g. the bond energy dependence on interatomic separation) must also be transferable. The electronic structure variables (ESVs) of the MO LCAO theories (expansion coefficients of MOs) are not transferable as is the whole MO LCAO picture of molecular electronic structure. The contradiction between the sufficiently local character of chemical structure and its delocalized description in the MO LCAO based theories is well known in the literature for almost half a century (see e.g. [290]). Many criteria are proposed in the literature, which allow one to localize the canonical MOs. It may appear that these localized orbitals can be considered as a transferable basis for constructing QM/MM junctions. However, one notices that if the HFR derived MOs are localized by a procedure, the overall many electron function – the corresponding ground state Slater determinant – does not change and thus the expectation values of the observables which reflect the degree of delocalization of the electronic structure (for example the electron number fluctuations [291]) remain the same, irrespective of the type of MOs used — canonical or localized ones. So, what we need is a semiempirical QC method presenting the trial wave function of electrons in terms of local objects: i.e. bonds and lone pairs. This is described in Section 2.4.

We would like to add some comments on the conceptual prerequisites for formal constructions (deduction, i.e. derivation) of the mechanistic models of PES. We notice first that the standard MM models [173] are based on the concepts relevant to the problems of molecular IR spectroscopy [292]. Mainly in its framework, it is possible to think about a molecule as about a set of point masses moving under the action of forces dependent on their relative coordinates according to harmonic law. This approximation for the molecular PES has led to the corresponding concept of the "molecular mechanics atom". Its further development was guided by the wish to load the information concerning the "force fields" – rules defining the form of the PES – upon it. In keeping with its origin, this concept has the corresponding area of application – rather wide, but still (as for any concept) limited. The problems, as we could make out in the present chapter, appear when one tries to go beyond the applicability area of this concept, when someone tries to load the information necessary for describing the electronic structure elements on it, when the QM/MM methods are being constructed. The concept of atom in MM is just not structured enough. Putting this in a completely programmer's manner, one may say that this "object" does not have the "data fields" (table columns) necessary for recording this information. However, the problem taken by the community as a technical complication is in fact deeply conceptual. It must elaborate the *concepts* relevant to the problem of constructing hybrid QM/MM methods. Following V.A. Fock [293, 294], new concepts arise in the theory when passing from more general and exact theories to approximate methods. Results presented in Section 2.4 show that an appropriate candidate for the role of a more general and exact theory to be used as a starting point for constructing the hybrid methods is the SLG-based semiempirical QC. The reason is its local character, which allows it to bring back into the theory such important chemical concepts as "bond", "lone pair", and "hybridization", which have disappeared from the HFR based QC as its construction elements. Among these concepts, that of the "bond" is fundamental also for the MM, but the other two look very awkward in the frame of the MM, particularly as it applies to lone pairs frequently represented by a separate atomic type. The concept necessary for our purposes (e.g. "atom of deductive molecular mechanics") will naturally arise in the course of the following simplification of the picture of the electronic structure as compared to that which appears in the semiempirical SLG-based QC of organic molecules.

The general scheme of the derivation of mechanistic models of PES from a QC description of molecular electronic structure reduces to the following moves. In the SLG based semiempirical QC there are the variables of two classes (electronic structure variables – ESV): geminal amplitudes and variables describing the HOs. ESVs of these classes must be (approximately) estimated, although the ESVs depend on molecular composition and geometry.

In Chapter 3 we introduce the formal construction and testing of an "intermediate" procedure bridging QM and MM procedures. This will be a mechanistic treatment, derived from the quantum description of the molecular system. Then this technique will be used to define the one-electron states of the frontier atoms – the key elements of the intersubsystem border/junction: the shapes of the one-electron states at the frontier atoms, their electronic densities and the response of either subsystem to the variables characterizing each subsystem.

References

1. IUPAC *Compendium of Chemical Terminology* (2 ed.), 1997.
2. T. Helgaker, P. Jørgensen and J. Olsen. *Molecular Electronic-Structure Theory*, Wiley, Chichester, 2000.
3. S. Wilson. *Electron Correlation in Molecules*, Clarendon, Oxford, 1984.
4. P.R. Surján. *Second Quantized Approach to Quantum Chemistry*, Springer, Berlin, 1989.
5. I. Mayer. *Simple Theorems, Proofs, and Derivations in Quantum Chemistry*, Kluwer/Plenum, Dordrecht/New York, 2003.
6. D.R. Hartree. *Proc. Cambridge Phylos. Soc.*, **45**, 230, 1948.
7. C.C.J. Roothaan and P.S. Bagus. *Method. Comp. Phys.*, **2**, 47, 1963.
8. D.F. Feller and K. Ruedenberg. *Theor. Chem. Acta*, **52**, 231, 1979.
9. M.W. Schmidt and K. Ruedenberg. *J. Chem. Phys.*, **71**, 3951, 1979.
10. L.A. Curtiss, K. Raghavachari, G.W. Trucks and J.A. Pople. *J. Chem. Phys.*, **94**, 7221, 1991.
11. L.A. Curtiss, K. Raghavachari, G.W. Trucks, V. Rassolov and J.A. Pople. *J. Chem. Phys.*, **109**, 7764, 1998.
12. J.M.L. Martin. *J. Chem. Phys.*, **97**, 5012, 1992.
13. J.M.L. Martin. *J. Chem. Phys.*, **100**, 8186, 1994.
14. D. Feller and K.A. Peterson. *J. Chem. Phys.*, **108**, 154, 1998.
15. D. Feller. *J. Chem. Phys.*, **96**, 6104, 1992.
16. D.E. Woon and T.H. Dunning Jr. *J. Chem. Phys.*, **99**, 1914, 1993.
17. K.A. Peterson, R.A. Kendall and T.H. Dunning Jr. *J. Chem. Phys.*, **99**, 9790, 1993.
18. D.E. Woon and T.H. Dunning Jr. *J. Chem. Phys.*, **101**, 8877, 1994.
19. C.M. Schwartz. *Phys. Rev.*, **126**, 1015, 1962.
20. C.M. Schwartz. *Methods Comp. Phys.*, **2**, 47, 1963.
21. D.P. Carrol, H. Silverstone and R.M. Metzger. *J. Chem. Phys.*, **71**, 4142, 1979.
22. J.M.L. Martin. *Chem. Phys. Lett.*, **259**, 679, 1996.
23. A.K. Wilson and T.H. Dunning Jr. *J. Chem. Phys.*, **106**, 8718, 1997.
24. W. Klopper. *J. Chem. Phys.*, **102**, 6168, 1995.
25. W. Kutzelnigg and J.D. Morgan. *J. Chem. Phys.*, **96**, 4484, 1992.
26. J.D. Morgan and W. Kutzelnigg. *J. Phys. Chem.*, **97**, 2425, 1993.
27. T. Helgaker, W. Klopper, H. Koch and J. Noga. *J. Chem. Phys.*, **106**, 9639, 1997.
28. B.O. Roos and K.P. Lawley. eds. *Ab Initio Methods in Quantum Chemistry. Part II*, vol. 69 of *Adv. Chem. Phys.*, pp. 399, Wiley, Chichester, 1987.
29. G.D. Purvis and R.J. Bartlett. *J. Chem. Phys.*, **76**, 1910, 1982.
30. R.A. Chiles and C.E. Dykstra. *Chem. Phys. Lett.*. **80**, 69, 1981.
31. E.A. Hylleraas. *Z. Phys.*, **54**, 347, 1929.
32. H. Hellmann. *J. Chem. Phys.*, **3**, 61, 1935.
33. H. Hellmann. *Acta Fizicochem.*, **USSR 1**, 913, 1935.
34. H. Hellmann. *Acta Fizicochem.*, **USSR 4**, 225, 1936.
35. H. Hellmann. *Einführung in die Quantenchemie*, F. Deuticke, Leipzig und Wien, 1937.
36. L. Seijo and Z. Barandiaran. In *Computational Chemistry: Reviews of Current Trends*, (ed. J. Leszczynski) vol. 4, World Scientific, Singapore, 1999.
37. J.N. Bardsley. *Chem. Phys. Lett.*, **7**, 517, 1970.
38. L. Szasz. *Pseudopotential Theory of Atoms and Molecules*, Wiley, New York, 1985.
39. J.A. Pople and D.L. Beveridge. *Approximate Molecular Orbital Theory*, McGraw-Hill, New York, 1970.
40. E. Hückel. *Z. Phys.*, **70**, 204, 1931.
41. R. Pariser and R. Parr. *J. Chem. Phys.*, **21**, 466, 767, 1953.
42. J.A. Pople. *Trans. Farad. Soc.*, **49**, 1375, 1953.
43. L.C. Young. *Lectures on Calculus of Variations and Optimal Control Theory*, W.B. Saunders, Philadelphia, 1969.
44. P.G. Lycos and R.G. Parr. *J. Chem. Phys.*, **24**, 1166, 1956; *ibid.*, **25**, 1301, 1956.

45. J.E. Lennard-Jones. *Proc. Roy. Soc. A.*, **158**, 280, 1937.
46. C.H. Martin, R.L. Graham and K.F. Freed. *J. Chem. Phys.*, **99**, 7833, 1993.
47. C.H. Martin and K.F. Freed. *J. Chem. Phys.*, **101**, 4011, 1994.
48. C.H. Martin and K.F. Freed. *J. Chem. Phys.*, **101**, 5929, 1994.
49. C.H. Martin and K.F. Freed. *J. Phys. Chem.*, **99**, 2701, 1995.
50. K.F. Freed. In *Semiempirical Methods of Electronic Structure Calculations. Part A. Techniques*, (ed. G.A. Segal), Plenum, New York, 1977.
51. Clark T. *Computational Chemistry*, VCH Verlag, Berlin, 1991.
52. M. Wolfsberg and L. Helmholz. *J. Chem. Phys.*, **20**, 837, 1952.
53. R. Hoffmann. *J. Chem. Phys.*, **39**, 1397, 1963.
54. I.B. Bersuker. *Electronic Structure and Properties of Coordination Compounds*, Khimiya, Moscow [in Russian], 1976.
55. H. Bash, A. Viste and H.B. Gray. *J. Chem. Phys.*, **44**, 10, 1966.
56. I. Fisher-Hjalmars. In *Modern Quantum Chemistry. Istanbul Lectures*, vol. 1, ed. by O. Sinanoğlu, Academic, New York, 1965.
57. K.R. Roby. *Chem. Phys. Lett.*, **11**, 6, 1971; K.R. Roby, *Chem. Phys. Lett.*, **12**, 579, 1972.
58. F. Weinhold and J.E. Carpenter. *J. Mol. Struct.*, **165**, 189, 1988.
59. D.B. Cook, P.C. Hollis, and R. McWeeny. *Mol. Phys.*, **13**, 553, 1967.
60. W. Koch. *Z. Naturforsch. A. Phys. Sci.*, **48**, 819, 1993.
61. E.U. Condon and G.H. Shortley. *Theory of Atomic Spectra*, Cambridge University Press, New York, 1935.
62. J.F. Mulligan. *J. Chem. Phys.*, **19**, 347, 1951.
63. R.G. Parr and G.R. Taylor. *J. Chem. Phys.*, **19**, 497, 1951.
64. M.J.S. Dewar and W. Thiel. *J. Am. Chem. Soc.*, **99**, 4899, 1977; M.J.S. Dewar and W. Thiel, *ibid.*, **99**, 4907, 1977.
65. M.J.S. Dewar and N.L. Hojvat (Sabelli). *J. Chem. Phys.*, **34**, 1232, 1961.
66. G. Klopman. *J. Amer. Chem. Soc.*, **86**, 4550, 1964.
67. R.C. Bingham, M.J.S. Dewar and D.H. Lo. *J. Am. Chem. Soc.*, **97**, 1285, 1975.
68. M.J.S. Dewar, E.G. Zoebisch, E. Healy and J.J.P. Stewart. *J. Am. Chem. Soc.*, **107**, 3902, 1985.
69. J.J.P. Stewart. *J. Comp. Chem.*, **10**, 209, 1989.
70. G.I. Csonka and J. Ángyán. *J. Mol. Struct. (THEOCHEM)*, **393**, 31, 1997.
71. M.P. Repasky, J. Chandrasekhar and W.L. Jorgensen. *J. Comput. Chem.*, **23**, 1601, 2002.
72. M.J. Filatov, O.V. Gritsenko, and G.M. Zhidomirov. *Theor. Chim. Acta*, **72**, 211, 1987.
73. D. Nanda and K. Jug. *Theor. Chem. Acta*, **57**, 95, 1980.
74. M.J.S. Dewar, C. Jie and J. Yu. *Tetrahedron*, **49**, 5003, 1993.
75. http://www.cachesoftware.com/mopac/Mopac2002manual
76. M. Kolb and W. Thiel. *J. Comput. Chem.*, **14**, 775, 1993.
77. W. Weber and W. Thiel. *Theor. Chem. Acc.*, **103**, 495, 2000.
78. D.W. Clack, N.S. Hush and S.R. Yandle. *J. Chem. Phys.*, **57**, 3503, 1972.
79. D.W. Clack. *Mol. Phys.*, **27**, 1513, 1974.
80. D.W. Clack and W. Smith. *J. Chem. Soc. Dalton Trans.*, 2015, 1974.
81. A.D. Bacon and M.C. Zerner. *Theor. Chim. Acta*, **53**, 21, 1979.
82. M.C. Böhm and R. Gleiter. *Theor. Chem. Acta*, **59**, 127, 1981.
83. M.C. Böhm and R. Gleiter. *Theor. Chem. Acta*, **59**, 153, 1981.
84. T.R. Cundari. *J. Chem. Soc, Dalton Trans.*, 2771, 1998.
85. T.R. Cundari and J. Deng. *J. Chem. Inf. Comp. Sci.*, **39**, 376, 1999.
86. M. Mohr, J.P. McNamara, H. Wang, S.A. Rajeev, J. Ge, C.A. Morgado and I.H. Hillier. *Faraday Discuss.*, **124**, 413, 2003.
87. T.R. Cundari, J. Deng and W. Fu. *Int. J. Quant. Chem.*, **77**, 421, 2000.
88. W. Thiel and A.A. Voityuk. *J. Phys. Chem.*, **100**, 616, 1996.
89. K.R. Adam, I.M. Atkinson and L.F. Lindoy. *J. Mol. Struct.*, **384**, 183, 1996.
90. V.A. Basiuk, E.V. Basiuk and J. Gómez-Lara. *J. Mol. Struct. (THEOCHEM)*, **536**, 17, 2001.
91. L. Zülicke. *Quantenchemie. Band 1*, Deutscher Verlag der Wissenschaften, Berlin, 1973.

92. R. McWeeny. *Methods of Molecular Quantum Mechanics* (2 ed.) Academic, London, 1992.

93. I.G. Kaplan. *Int. J. Quant. Chem.*, **89**, 268, 2002.

94. I.G. Kaplan. *J. Mol. Struct.*, **838**, 39, 2007.

95. I.G. Kaplan. *Symmetry of Many-Electron Systems*, Academic, New York, 1975.

96. C.J. Ballhausen. *Indroduction to Ligand Field Theory*, McGraw Hill, New York, 1962.

97. I.B. Bersuker. *Electronic Structure and Properties of Coordination Compounds* [in Russian] Khimiya, Moscow, 1976.

98. D.T. Sviridov and Y.F. Smirnov. *Theory of Optical Spectra of Transition-Metal Ions* [in Russian] Nauka, Moscow, 1977.

99. J.C. Slater. *Adv. Quant. Chem.*, **6**, 1, 1972.

100. L.D. Landau and E.M. Lifshits. *Quantum Mechanics. Prescription of Theoretical Physics*, vol. 3, Pergamon, London, 1977.

101. A. Messiah. *Quantum Mechanics*, vol. 2, 1958.

102. P. Hohenberg and W. Kohn. *Phys. Rev.*, **136**, B864, 1964.

103. W. Kohn and L.J. Sham. *Phys. Rev.*, **140(4A)**, A1133, 1965.

104. R.K. Nesbet. *Int. J. Quant. Chem.*, **85**, 405, 2001.

105. R.K. Nesbet. *Int. J. Quant. Chem.*, **100**, 937, 2004.

106. T. Ziegler. *Chem. Rev.*, **91**, 651, 1991.

107. J. Li, P. Beroza, L. Noodleman and D.A. Case. *Quantum mechanical modeling of active sites in metalloproteins.* In L. Banci and P. Comba, (ed), *Molecular Modeling and Dynamics of Bioinorganic Systems, Proceedings of NATO ASI Workshop*, Kluwer, Dordrecht, pp. 279, 1997.

108. H. Chermette. *Coord. Chem. Rev.*, **178–180**, 699, 1998.

109. C.E. Schäffer and C. Anthon. *Coord. Chem. Rev.*, **226**, 17, 2002.

110. P.E.M. Siegbahn. *Faraday Discuss.*, **124**, 289, 2003.

111. M. Freindorf and P.M. Kozlowski.. *J. Phys. Chem. A.*, **105**, 7267, 2001.

112. L. Rulíšek and Z. Havlas. *Int. J. Quant. Chem.*, **91**, 504, 2003.

113. T. Ziegler, A. Rauk and E.J. Baerends. *Theor. Chem. Acta*, **43**, 261, 1977.

114. C.C.J. Roothaan. *Rev. Mod. Phys.*, **32**, 179, 1960.

115. B.N. Plakhutin and G.M. Zhidomirov. *Zh. Strukt. Khim.*, **27**(2), 3, 1986.

116. O. Gunnarsson and B.I. Lundqvist. *Phys. Rev. B.*, **13**, 4274, 1976.

117. U. von Barth. *Phys. Rev. A.*, **20**, 1693, 1979.

118. R. Bosque and F. Maseras. *J. Comp. Chem.*, **21**(7), 562, 2000.

119. M.C. Zerner. Intermediate neglect of differential overlap calculations on the electronic spectra of transition metal complexes, In N. Russo and D.R. Salahub, (eds), *Metal-Ligand Interactions: Structure and Reactivity, NATO ASI Workshop*, Kluwer, Dordrecht, pp. 493, 1996.

120. P. Winget, C. Selçuki, A.H.C. Horn, B. Martin and T. Clark. *Theor. Chem. Acc.*, **110**, 254, 2003.

121. S. Diner, J.-P. Malrieu and P. Claverie. *Theor. Chim. Acta*, **13**, 1, 1969.

122. J.-P. Malrieu. The PCILO method, in *Semiempirical Methods of Electronic Structure Calculation. Part A: Techniques*, ed. by G.A. Segal, Plenum, New York, 1977.

123. C. Edmiston and K. Ruedenberg. *Rev. Mod. Phys.*, **35**, 457, 1963.

124. J.M. Foster and S.F. Boys. *Rev. Mod. Phys.*, **32**, 300, 1960.

125. W. von Niessen. *J. Chem. Phys.*, **56**, 4290, 1970.

126. P.R. Surján. *The Two-Electron Bond as a Molecular Building Block.* In Z.B. Maksić (ed), *Theoretical Models of Chemical Bonding, Part 2*, Springer, Heidelberg (Topics Curr. Chem.) pp. 205, 1989.

127. V.M. Tatevskii. *Classical Theory of Molecular Structure and Quantum Mechanics* [in Russian] Khimiya, Moscow, 1973.

128. N.F. Stepanov, M.E. Erlykina and G.G. Filipov. *Methods of Linear Algebra in Physical Chemistry* [in Russian] MSU Publ., Moscow, 1976.

129. I. Røeggen. *Int. J. Quant. Chem.*, **20**, 817, 1981; *ibid.* **22**, 149, 1982; **31**, 951, 1987; *J. Chem. Phys.*, **89**, 441, 1988.

130. A.M. Tokmachev and A.L. Tchougréeff. *Russ. J. Phys. Chem.*, **73**, 259, 1999.

131. A.M. Tokmachev, A.L. Tchougréeff and I.A. Misurkin. *Russ. J. Phys. Chem.*, **74**, S205, 2000.

132. A.M. Tokmachev and A.L. Tchougréeff. *J. Comp. Chem.*, **22**, 752, 2001.

133. A.M. Tokmachev and A.L. Tchougréeff. *J. Phys. Chem. A.*, **107**, 358, 2003.

134. G. Náray-Szabó. *Comput. Chem.*, **24**, 287, 2000.

135. V.A. Fock, M.G. Veselov and M.I. Petrashen. *J. Exp. Theor. Phys.*, **10**, 723, 1940 [in Russian].

136. S. Weinbaum. *J. Chem. Phys.*, **1**, 593, 1933.

137. A.L. Tchougréeff and A.M. Tokmachev. *J. Comp. Meth. Sci. Eng.*, **2**, 309, 2002.

138. J.-P. Daudey, J.-P. Malrieu and O. Rojas. *Localization and Delocalization in Quantum Chemistry*, Vol. 1. *Atoms and Molecules in the Ground State*, O. Chalvet, R. Daudel, S. Diner and J.-P. Malrieu, eds., Reidel, Dordrecht, 1975.

139. D. Cremer. *J. Am. Chem. Soc.*, **99**, 1307, 1977.

140. A.M. Tokmachev. *PhD Thesis* [in Russian], Karpov Institute, Moscow, 2003.

141. V.N. Tutubalin. *Probablity Theory and Stochastic Processes* [in Russian] MSU, Moscow, 1992.

142. A.M. Tokmachev and A.L. Tchougréeff. *J. Phys. Chem., A.* **107**, 358, 2003.

143. W. Kohn. *Phys. Rev. Lett.*, **76**, 3168, 1996.

144. M. Schütz, M. Hetzer and H.-I. Werner. *J. Chem. Phys.*, **111**, 5691, 1999.

145. J. Rubio, A. Povill, J.P. Malrieu and P. Reinhardt. *J. Chem. Phys.*, **107**, 10044, 1997.

146. M.C. Strain, G.E. Scuseria and M.J. Frisch. *Science*, **271**, 51, 1996.

147. A. Derecskei-Kovacs, D.E. Woon and D.S. Marynick. *Int. J. Quan. Chem.*, **61**, 67, 1997.

148. J.M. Cullen. *Int. J. Quant. Chem.*, **56**, 97, 1995.

149. P.E. Schipper. *J. Phys. Chem.*, **90**(18), 4259, 1986.

150. H.A. Bethe. *Ann. Physik*, **3**, 133, 1929.

151. C.K. Jørgensen. *Absorption Spectra and Chemical Bonding in Complexes*, Pergamon, Oxford, 1962.

152. F.A. Cotton and G. Wilkinson. *Advanced Inorganic Chemistry*, Wiley-VCH, NY, 1999.

153. C.E. Schäffer. *Pure Appl. Chem.*, **24**, 361, 1971.

154. C.E. Schäffer and C.K. Jørgensen. *Mol. Phys.*, **9**, 401, 1965.

155. C.E. Schäffer. *Struct. Bond.*, **5**, 68, 1968.

156. C.E. Schäffer. *Pure Appl. Chem.*, **24**, 361, 1970.

157. C.E. Schäffer. *Struct. Bond.*, **14**, 69, 1973.

158. C.E. Schäffer. *Theor. Chim. Acta.*, **34**, 237, 1974.

159. A.V.P. Lever. *Inorganic Electronic Spectroscopy*, Elsevier, Amsterdam, 1984.

160. M. Gerloch and R.G Wooley. *Prog. Inorg. Chem.*, **31**, 371, 1983.

161. A.V. Soudackov, A.L. Tchougréeff and I.A. Misurkin. *Theor. Chim. Acta*, **83**, 389, 1992.

162. A.A. Abrikosov, L.P. Gor'kov and I.E. Dzyaloshinskii. *Methods of Quantum Field Theory in Statistical Physics.* (2 ed.) [in Russian], Dobrosvet, Moscow, 1998.

163. D.J. Thouless. *The Quantum Mechanics of Many-Body System*, Academic, New York, 1972.

164. A.V. Soudackov, A.L. Tchougréeff and I.A. Misurkin. *Zh. Fiz. Khim.*, **68**(7), 1256, 1994.

165. A.V. Soudackov, A.L. Tchougréeff and I.A. Misurkin. *Zh. Fiz. Khim.*, **68**(7), 1264, 1994.

166. A.V. Soudackov, A.L. Tchougréeff and I.A. Misurkin. *Int. J. Quant. Chem.*, **57**, 663, 1996.

167. A.V. Soudackov, A.L. Tchougréeff and I.A. Misurkin. *Int. J. Quant. Chem.*, **58**, 161, 1996.

168. A.V. Sinitsky, M.B. Darkhovskii, A.L. Tchougréeff and I.A. Misurkin. *Int. J. Quant. Chem.*, **88**, 370, 2002.

169. A.M. Tokmachev and A.L. Tchougréeff. *Khim. Fiz.*, **18**(1), 80, 1999.

170. A.L. Tchougréeff, A.V. Soudackov, I.A. Misurkin, H. Bolvin and O. Kahn. *Chem. Phys.*, **193**, 19, 1995.

171. K. Pierloot. *Mol. Phys.*, **101**, 2083, 2003.

172. V.G. Dashevskii. *Conformations of Organic Molecules* [in Russian], Khimiya, Moscow, 1974.

173. U. Burkert and N.L. Allinger. *Molecular Mechanics*, ACS, Washington, 1982.

174. M.J. Hwang, T.P. Stocktisch and A.T. Hagler. *J. Amer. Chem. Soc.*, **116**, 2515, 1994.

175. W.J. Mortier, K. Van Genechten and J. Gasteiger. *J. Am. Chem. Soc.*, **107**, 829, 1985; J. Gasteiger and H. Saller. *Angew. Chem. Int. Ed. Engl.*, **24**, 687, 1985; L.G. Hammarstrom, T. Liljefors and J. Gasteiger. *J. Comp. Chem.*, **9**, 424, 1988.

176. A.K. Rappé, C.J. Casewit, K.S. Colwell, W.A. Goddard III and W.M. Skiff. *J. Am. Chem. Soc.*, **114**, 10024, 1992.

177. W.D. Cornell, P. Ciepak, C.I. Bayly, I.R. Gould, K. Merz, D.M. Ferguson, D.C. Spellmeyer, T. Fox, J.W. Caldwell and P. Kollman. *J. Am. Chem. Soc.*, **117**, 5179, 1995.

178. B.R. Brooks, R.E. Bruccoleri, B.D. Olafson, D.J. States, S. Swaminathan and M. Karplus. *J. Comp. Chem.*, **4**, 187, 1983.

179. O. Ermer and S. Lifson. *J. Am. Chem. Soc.*, **95**, 4121, 1973; M.J. Hwang, J.P. Stocktisch and A.T. Hagler, *J. Am. Chem. Soc.*, **116**, 2515, 1994.

180. M.L.C.E. Kouwijzer and P.D.J. Grootenhuis. *J. Phys. Chem.*, **99**, 13426, 1995.

181. S.D. Morley, R.J. Abraham, I.S. Haworth, D.E. Jackson, M.R. Saunders and J.G. Vinter. *J. Comput. Aided Mol. Des.*, **5**, 475, 1991.

182. S. Lifson, A.T. Hagler and P. Dauber. *J. Am. Chem. Soc.*, **101**, 5111, 5122, 5131, 1979.

183. S.L. Mayo, B.D. Olafson and W.A. Goddard III. *J. Phys. Chem.*, **94**, 8897, 1990.

184. E.M. Engler, J.D. Andose and P.V.R. Schleyer. *J. Am. Chem. Soc.*, **95**, 8005, 1973.

185. G. Nemethy, K.D. Gibsen, K.A. Palmer, C.N. Yoon, G. Paterlini, A. Zagari, S. Rmllsey and H.A. Sheraga. *J. Phys. Chem.*, **96**, 6472, 1992.

186. J.L.M. Dillen. *J. Comput. Chem.*, **16**, 595, 610, 1995.

187. S. Barlow, A.L. Rohl, S. Shi, C.M. Freenlan and D. O'Hara. *J. Am. Chem. Soc.*, **118**, 7578, 1996.

188. W.F. van Gunsteren, S.R. Billeter, A.A. Eising, P.H. Hünenberger, P. Krüger, A.E. Mark, W.R.P. Scott and I.G. Tironi. Biomolecular Simulation: The GROMOS96 Manual and User Guide; vdf Hochschulverlag AG an der ETH Zürich and BIOMOS b.v.: Zrich, Groningen, 1996.

189. N.L. Allinger. *J. Am. Chem. Soc.*, **99**, 8127, 1977.

190. N.L. Allinger, Y.H. Yuh and J.-H. Lii. *J. Am. Chem. Soc.*, **111**, 8551, 1989; J.H. Lii and N.L. Allinger. *J. Am. Chem. Soc.*, **111**, 8566, 8576, 1989; N.L. Allinger, X. Zhou and J. Bergsma. *J. mol. Struct. Theochem.*, **312**, 69, 1994.

191. N.L. Allinger, K. Chen and J.-H. Lii. *J. Comput. Chem.*, **17**, 642, 1996; N. Nevins, K. Chen and N.L. Allinger. *J. Comput. Chem.*, **17**, 669, 1996; N. Nevins, J.-H. Lii and N.L. Allinger. *J. Cornput. Chem.*, **17**, 695, 1996; N.L. Allinger, K. Chen, J.A. Katzenellenbogen, S.R. Wilson and G.M. Anstead. *J. Cornput. Chem.*, **17**, 747, 1996.

192. T.A. Halgren. *J. Comput. Chem.*, **17**, 490, 1996.

193. I.V. Pletnev and V.L. Melnikov. *Russ. Chem. Bull.*, **46**, 1278, 1997.

194. P. Comba and T. Hambley. *Molecular Modeling of Inorganic Compounds*, VCH, NY, 1995.

195. W.L. Jorgensen and J. Tirado-Rives. *J. Am. Chem. Soc.*, **110**, 1657, 1988.

196. V.S. Allured, C.M. Kelly and C.R. Landis. *J. Am. Chem. Soc.*, **113**, 1, 1991.

197. M. Clark, R.D. Cranler III and N. van Opdenbosch. *J. Comput. Chem.*, **10**, 982, 1989; J.R. Maple, M.-J. Hwang, T.P. Stocktisch, U. Dinur, M. Waldnlan, C.S. Ewig and A.T. Hagler. *J. Comput. Chem.*, **15**, 162, 1994.

198. A. Vedani. *J. Comp. Chem.*, **9**, 269, 1988.

199. A.R. Leach. *Molecular Modelling. Principles and Applications* (2 ed.), Pearson Education, Harlow et al., 2001.

200. A.K. Rappé, K.S. Colwell and C.J. Casewit. *Inorg. Chem.*, **32**, 3438, 1993.

201. B.P. Hay. *Coord. Chem. Rev.*, **126**, 177, 1993.

202. B.P. Hay, L. Yang, J.-H. Lii and N.L. Allinger. *J. Mol. Struct. (THEOCHEM)*, **428**, 203, 1998.

203. P. Comba and R. Remenyi. *Coord. Chem. Rev.*, 238, **239**, 9, 2003.

204. H. Erras-Hanauer, T. Clark and R. Van Eldik. *Coord. Chem. Rev.*, 238, 233, **239**, 2003.

205. P. Comba. *Coord. Chem. Rev.*, **182**, 343, 1999.

206. M. Zimmer. *Coord. Chem. Rev.*, **212**, 133, 2003.

207. J. Kozelka. In *Metal Ions in Biological Systems*, vol. 33 (ed. A. Sigel and H. Sigel) Marcel Dekker, New York, 1996.

208. A. Warshel and M. Levitt. *J. Mol. Biol.*, **103**, 227, 1976.

209. M.J. Field, P.A. Bash and M. Karplus. *J. Comp. Chem.*, **11**, 700, 1990.

210. F. Bernardi, M. Olivucci and M.A. Robb. *J. Am. Chem. Soc.*, **114**, 1606, 1992.

211. D. Bakowies and W. Thiel. *J. Comp. Chem.*, **17**, 87, 1996.

212. I.H. Hillier. *J. Mol. Struct. (THEOCHEM)*, **463**, 45, 1999.

213. F. Maseras and K. Morokuma. *J. Comp. Chem.*, **16**, 1170, 1995.

214. J. Sauer and M. Sierka. *J. Comp. Chem.*, **21**, 1470, 2000.

215. M.A. Thompson. *J. Phys. Chem.*, **100**, 14492, 1996.

216. X. Assfeld and J.L. Rivail. *Chem. Phys. Lett.*, **263**, 100, 1996.

217. J. Gao. Methods and applications of combined quantum mechanical and molecular mechanical potentials, In *Reviews in Computational Chemistry*, K.B. Lipkowitz and D.B. Boyd. (eds.), **9**, VCH, New York, 1996.

218. P. Sherwood. Hybrid quantum mechanics/molecular mechanics approaches, In J. Grotendorst, *Modern Methods and Algorithms of Quantum Chemistry*, Proceedings (2 ed.), NIC Series, 3, John von Neumann Institute for Computing, Jülich, 285, 2000.

219. M. Freindorf and J. Gao. *J. Comp. Chem.*, **17**, 386, 1996.

220. F.J. Luque, M. Bachs, C. Alemán and M. Orozco. *J. Comp. Chem.*, **17**, 806, 1996.

221. Y. Mo and J. Gao. *J. Comp. Chem.*, **21**, 1458, 2000.

222. T. Yoshida, N. Koga and K. Morokuma. *Organometallics*, **15**, 766, 1996.

223. I. Tuñón, I.T.C. Martins-Costa, C. Millot, M.F. Ruiz-López and J.L. Rivail. *J. Comp. Chem.*, **17**, 19, 1996.

224. D. Wei and D.R. Salahub. *Chem. Phys. Lett.*, **224**, 291, 1994.

225. M. Sierka and J. Sauer. *Faraday Discuss.*, **106**, 41, 1997.

226. R.V. Stanton, D.S. Hartsough and K.M. Merz. *J. Phys. Chem.*, **97**, 11868, 1993.

227. P.D. Lyne, M. Hodoscek and M. Karplus. *J. Phys. Chem. A.*, **103**, 3462, 1999.

228. X.P. Long, J.B. Nicholas, M.F. Guest and R.L. Ornstein. *J. Mol. Struct.*, **412**, 121, 1997.

229. S.A. French, A.A. Sokol, S.T. Bromley, C.R.A. Catlow, S.C. Rogers, F. King and P. Sherwood. *Angew. Chem.*, **113**, 4569, 2001.

230. P.L. Cummins and J.E. Gready. *J. Comp. Chem.*, **18**, 1496, 1997.

231. V. Théry, D. Rinaldi, J.-L. Rivail, B. Maigret and G.G. Ferenczy. *J. Comp. Chem.*, **15**, 269, 1994.

232. G.G. Ferenczy and J.G. Ángyán. *J. Comp. Chem.*, **22**, 1679, 2001.

233. J. Åqvist and A. Warshel. *Chem. Rev.*, **93**, 2523, 1993.

234. D. Bakowies and W. Thiel. *J. Phys. Chem.*, **100**, 10580, 1996.

235. T. Matsubara, S. Sieber and K. Morokuma. *Int. J. Quant. Chem.*, **60**, 1101, 1996.

236. A.J. Turner, V. Moliner and I.H. Williams. *Phys. Chem. Chem. Phys.*, **1**, 1323, 1999.

237. U.C. Singh and P.A. Kollman. *J. Comp. Chem.*, **7**, 718, 1986.

238. B. Waszkowycz, I.H. Hillier, N. Gensmantel and D.W. Payling. *J. Chem. Soc. Perkin Trans.*, **2**, 2025, 1991.

239. J. Tomasi and M. Perisco. *Chem. Rev.*, **94**, 2027, 1994.

240. O. Tapia and G. Johannin. *J. Chem. Phys.*, **75**, 362, 1981.

241. S. Humbel, S. Sieber and K. Morokuma. *J. Chem. Phys.*, **105**, 1959, 1996.

242. R.D.J. Froese, S. Humbel, M. Svensson and K. Morokuma. *J. Phys. Chem. A.*, **101**, 227, 1997.

243. M. Svensson, S. Humbel, R.D.J. Froese, T. Matsubara, S. Sieber and K. Morokuma. *J. Phys. Chem.*, **100**, 19357, 1996.

244. M.A. Thompson, E.D. Glendening and D. Feller. *J. Phys. Chem.*, **98**, 10465, 1994.

245. M. Eichinger, P. Tavan, J. Hutter and M. Parrinello. *J. Chem. Phys.*, **110**, 10452, 1999.

246. U. Eichler, Ch. M. Kölmel and J. Sauer. *J. Comp. Chem.*, **18**, 463, 1997.

247. M.W. Schmidt, K.K. Baldridge, J.A. Boatz, S.T. Elbert, M.S. Gordon, J.H. Jensen, S. Koseki, N. Matsunaga, K.A. Nguyen, S.J. Su, T.L. Windus, M. Dupuis and J.A. Montgomery. *J. Comput. Chem.*, **14**, 1347, 1993.

248. M.J. Frisch, G.W. Trucks, H.B. Schlegel, G.E. Scuseria, M.A. Robb, J.R. Cheeseman, V.G. Zakrzewski, J.A. Montgomery, Jr., R.E. Stratmann, J.C. Burant, S. Dapprich, J.M. Millam, A.D. Daniels, K.N. Kudin, M.C. Strain, O. Farkas, J. Tomasi, V. Barone, M. Cossi, R. Cammi, B. Mennucci, C. Pomelli, C. Adamo, S. Clifford, J. Ochterski, G.A. Petersson, P.Y. Ayala, Q. Cui, K. Morokuma, D.K. Malick, A.D. Rabuck, K. Raghavachari, J.B. Foresman, J. Cioslowski, J.V. Ortiz,

A.G. Baboul, B.B. Stefanov, G. Liu, A. Liashenko, P. Piskorz, I. Komaromi, R. Gomperts, R.L. Martin, D.J. Fox, T. Keith, M.A. Al-Laham, C.Y. Peng, A. Nanayakkara, C. Gonzalez, M. Challacombe, P.M.W. Gill, B. Johnson, W. Chen, M.W. Wong, J.L. Andres, C. Gonzalez, M. Head-Gordon, E.S. Replogle and J.A. Pople. *GAUSSIAN98, Revision A.7*, Gaussian, Pittsburgh, PA, 1998.

249. P. Sherwood, A.H. de Vries, M.F. Guest, G. Schreckenbach, C.R.A. Catlow, S.A. French, A.A. Sokol, S.T. Bromley, W. Thiel, A.J. Turner, S. Billeter, F. Terstegen, S. Thiel, J. Kendrick, S.C. Rogers, J. Casci, M. Watson, F. King, E. Karlsen, M. Sjovoll, A. Fahmi, A. Schäfer and C. Lennartz. *J. Mol. Struct. (THEOCHEM)*, **632**, 1, 2003.

250. D. Bakowies. *PhD Thesis*, Universität Zürich, Zürich, 1994.

251. G.J. Kramer and R.A. van Santen. *J. Am. Chem. Soc.*, **115**, 2887, 1993.

252. K.P. Eurenius, D.C. Chatfield, B.R. Brooks and M. Hodoscek. *Int. J. Quant. Chem.*, **60**, 1189, 1996.

253. M.J. Harrison, N.A. Burton and I.H. Hillier. *J. Am. Chem. Soc.*, **119**, 12285, 1997.

254. A.H. de Vries, P. Sherwood, S.J. Colins, A.M. Rigby, M. Rigutto and G.J. Kramer. *J. Phys. Chem. B*, **103**, 6133, 1999.

255. N. Reuter, A. Dejaegere, B. Maigret and M. Karplus. *J. Phys. Chem. A.*, **104**, 1720, 2000.

256. I. Antes and W. Thiel. *J. Phys. Chem. A.*, **103**, 9290, 1999.

257. Hyperchem, Inc., *Hyperchem Users Manual; Computational Chemistry*, Hypercube, On, Canada, 1994.

258. P. Cummins and J. Gready. *J. Phys. Chem. B.*, **104**, 4503, 2000.

259. S. Ranganathan and J. Gready. *J. Phys. Chem. B.*, **101**, 5614, 1997.

260. M.A. Thompson and G.K. Schenter. *J. Phys. Chem.*, **99**, 6374, 1995.

261. K.B. Tolpygo. *Usp. Fiz. Nauk.*, **74**, 269, 1961.

262. L.N. Kantorovich. *J. Phys. C. Solid State Phys.*, **21**, 5041, 1988; L.N. Kantorovich. *J. Phys. C. Solid State Phys.*, **21**, 5057, 1988; L.N. Kantorovich and B.P. Zapol. *J. Chem. Phys.*, **96**, 8420, 1992; L.N. Kantorovich and B.P. Zapol. *J. Chem. Phys.*, **96**, 8427, 1992; L.N. Kantorovich. *Int. J. Quant. Chem.*, **78**, 306, 2000.

263. A.M. Tokmachev, A.L. Tchougréeff and I.A. Misurkin. *J. Mol. Struct. (THEOCHEM)*, **506**, 17, 2000.

264. I.B. Bersuker, M.K. Leong, J.E. Boggs and R.S. Pearlman. *Int. J. Quant. Chem.*, **63**, 1051, 1997.

265. G. Monard, M. Loos, V. Théry, K. Baka and J.L. Rivail. *Int. J. Quant. Chem.*, **58**, 153, 1996.

266. S. Antonczak, G. Monard, M.F. Ruiz-López and J.L. Rivail. *J. Am. Chem. Soc.*, **120**, 8825, 1998.

267. N. Ferré, X. Assfeld and J.-L. Rivail. *J. Comp. Chem.*, **23**, 610, 2002.

268. D.M. Philipp and R.A. Friesner. *J. Comp. Chem.*, **20**, 1468, 1999.

269. M.S. Gordon, M.A. Freitag, P. Bandyopadhyay, J.H. Jensen, V. Kairys and W.J. Stevens. *J. Phys. Chem. A.*, **105**, 293, 2001.

270. R. Poteau, I. Ortega, F. Alary, A. Ramirez Solis, J.-C. Barthelat and J.P. Daudey. *J. Phys. Chem. A.*, **105**, 198, 2001.

271. P. Durand and J.-C. Barthelat. *Theor. Chim. Acta*, **38**, 283, 1975.

272. P. Durand and J.-P. Malrieu. *Ab Initio Methods in Quantum Chemistry*. Part I, K.P. Lawley, I. Prigogine, S.A. Rice eds., Wiley-Interscience, New York, 1987.

273. F. Bessac, F. Alary, Y. Carissan, J.-L. Heully, J.-P. Daudey and R. Poteau. *J. Mol. Struct. (THEOCHEM)*, **632**, 43, 2003.

274. F. Alary, R. Poteau, J.-L. Heully, J.-C. Barthelat and J.-P. Daudey. *Theor. Chim. Acc.*, **104**, 174, 2000.

275. R. Poteau, I. Ortega, F. Alary, A. Ramirez Solis, J.-C. Barthelat and J.-P. Daudey. Effective Group Potentials. 1. Method, *J. Phys. Chem. A.*, **104**, 198, 2001.

276. E.M. Shustorovich and M.E. Dyatkina. *Zh. Strukt. Khimii.*, **1**, 109, 1960 [in Russian]; *J. Struct. Chem., (USSR)* **1**, 98, 1960 [in English].

277. J. Åqvist and A. Warshel. *Chem. Rev.*, **93**, 2523, 1997.

278. I.V. Abarenkov and I.M. Antonova. *In Problems of Quantum Theory of Atoms, Molecules and Solids*, ed. by I.V. Abarenkov StPbSU Publ., 1996.

279. I.V. Abarenkov and I.V. Kryukov. *In Problems of Quantum Theory of Atoms, Molecules and Solids*, ed. by I.V. Abarenkov, StPbSU Publ., 1996.

280. I.V. Abarenkov and I.I. Tupitsyn. *J. Chem. Phys.*, **115**, 1650, 2001.

281. I.V. Abarenkov and I.M. Antonova. *Int. J. Quant. Chem.*, **96**, 263, 2004.

282. I.V. Abarenkov and K.V. Smelkov. *Int. J. Quant. Chem.*, **96**, 255, 2004.

283. J.A. Mejias and J.F. Sanz. *J. Chem. Phys.*, **102**, 850, 1995.

284. Y. Zhang, T.S. Lee and W. Yang. *J. Chem. Phys.*, **110**, 46, 1999.

285. Y. Zhang, H. Liu and W. Yang. *J. Chem. Phys.*, **112**, 3483, 2000.

286. J.A. DiLabio, M.M. Hurley and P.A. Christiansen. *J. Chem. Phys.*, **116**, 9578, 2002.

287. J. Gao, P. Amara, C. Alhambra and M.J. Field. *J. Phys. Chem. A.*, **102**, 4714, 1998.

288. P. Amara, M.J. Field, C. Alhambra and J. Gao. *Theoret. Chem. Acc.*, **104**, 336, 2000.

289. Jiabo Li and Roy McWeeny. *Int. J. Quant. Chem.*, **89**, 208, 2002.

290. C.A. Coulson. *Valence*, Oxford University Press, Oxford, 1961.

291. J.-P. Malrieu. *The Concept of the Chemical Bond. Theoretical Models of Chemical Bonding, Part 2*, ed. by Z.B. Maksić, Springer, Heidelberg, 1989.

292. E.B. Wilson, J.C. Decius and P.C. Cross. *Molecular Vibrations. The Theory of Infrared and Raman Vibrational Spectra*, McGraw-Hill, New York, 1955.

293. V.A. Fock. *Usp. Fiz. Nauk.*, **16**, 1070, 1936 [in Russian].

294. V.A. Fock. Fundamental importance of approximate methods in theortical physics. In *Philosophical Problems of Physics*, Leningrad, 1974 [in Russian].

295. J.J.P. Stewart. *J. Mol. Model.*, **13**, 1173, 2007.

3

DEDUCTIVE MOLECULAR MECHANICS: BRIDGING QUANTUM AND CLASSICAL MODELS OF MOLECULAR STRUCTURE

Abstract   In the previous chapter we used the general scheme of electron variable separation to analyze current hybrid methods and suggest improvements to them. The situation in which we find ourselves is that the variable separation technique proposed in Chapter 1 can and should be used to sequentially construct hybrid methods by applying the GF form of the trial wave function for the complex molecular system. The prerequisite for such an enterprise is that the orbitals of the system can be divided into complementary orthogonal carrier subspaces for the quantally and classically treated subsystems of the complex system. This prerequisite is, however, not taken for granted unless for some reason the required subspaces can be defined on symmetry grounds (as in the case of π-systems in the Hückel and other similar methods), and that is what we shall provide in this chapter. The way it is done here may seem too indirect. It is, however, necessary to follow this route. The key relation to be established is that between the geometry of the classically treated part of the complex system and the orbitals spanning the carrier space for its quantally treated part. Clearly the orbitals located on the frontier atoms are most sensitive to the geometry variations occurring in the classically treated subsystem right next to the frontier. However, to get this dependence we need a general theory relating forms of the orbitals to the geometries of the molecules. The required theory has to be constructed in terms of local quantities, i.e. hybrid orbitals rather than molecular orbitals, which is what we provide in this chapter.

3.1. MOTIVATION. MOLECULAR MECHANICS AND ADDITIVE SCHEMES. STEREOCHEMISTRY AND VSEPR THEORY

Deductive molecular mechanics (DMM), described in detail in this chapter, is developed with the purpose of serving as a tool for deriving mechanistic models of molecular energy (classical force fields) starting from a suitable quantum mechanics (QM) description of molecular structure. The current situation within such a setting is that despite its long history and a wealth of successful applications, the MM approach remains, not a substantiated theory, but a vaguely defined empirical tool. It is clear that the molecular PES can be expanded up to the second order in nuclear displacements near the equilibrium geometry and this fact is often considered a general argument in favor of the possibility of using the MM-type expansions for the total energy [1, 2]. This type of reasoning, however, is very questionable as the transferability of the elements of the dynamic (second derivatives) matrix between different molecules in a wide enough class of the latter has not been proven. The transferability of the MM force fields is in fact the real content of this approach. Nevertheless, the reasons to treat that or any other force field as a transferable between two specific molecules or classes of molecules are either purely pragmatic or are decided on school-wise grounds [3]. In this chapter we take a step towards quantitative analysis of the transferability of the MM force fields.

Our derivation[1] is based on the assumption that the trial wave function underlying the MM description is one of the antisymmetrized products of strictly local geminals (SLG) described in Section 2.4 and in [9–12]. The key feature of the underlying QM method is its locality, recovering the concepts of chemical bonds and lone pairs on the basis of a non-Hartree-Fock electronic trial wave function. The employed form of the trial wave function allows us to obtain a natural representation of molecular energy in terms of these local objects. The electronic structure variables (ESVs) in this approximation become essential components of a logical framework for the transition from the QM to an MM description as they allow construction of the potential energy surfaces (PES) by proper consideration of the response of the equilibrium value of ESVs to the variations of molecular geometry and composition. In this chapter the ESVs defining density matrix elements and basis one-electron states (hybrid orbitals – HOs) in the SLG approximation are thoroughly analyzed. The transferability of the density matrix elements with respect to the parameters of molecular Hamiltonian and to the geometry variations and the linear response relations for the HOs are proven to take place under very nonrestrictive preconditions. Special attention is paid to numerical estimates of the ESVs' features, giving an "experimental" support to approximate expressions of molecular energy.

---

[1]Mathematics in this chapter is rather different from that in chapter 1 and incidentally not common for the usual chemistry curriculum. It is largely based on the Lie groups and particularly on the $SO(4)$ group. The importance of the $SO(4)$ group for problems of atomic physics was discovered by V.A. Fock [4] who demonstrated that the Schrödinger equation for the hydrogen atom possesses such a symmetry which explains the so-called "accidental degeneracy" of its electronic spectrum – the coincidence of the energies in the electronic *shell* with given principal quantum number n, irrespective of the value of the azimuthal quantum number l distinguishing the *subshells*. In the heavier atoms the above symmetry does not take place so the rôle of the $SO(4)$ group is completely different: it is not a symmetry group for the system any more, but the dynamical group: one spanning the whole set of accessible hybridization states of the system, so that the energy of the system becomes a function of the element of this group. Then the problem of parametrizing – introducing a convenient set of coordinates on this group – arises. It is done generally using the theory of Lie groups. The sources on Lie groups are numerous and we mention only some of them [5, 6] – classical tracts on the groups of interest: the first of two more "physical"; the second – mathematical chef- d'oeuvre. Very clear explanation is given in [7]. As an introductory text, [8] is fine.

An important general motivation for the search for transferability of certain elements of molecular electronic structure is a well-known possibility to express the numerous experimental characteristics of molecules as additive functions of the increments of the respective characteristics attributable to the parts of these molecules. The most striking is the precision of this approach called the "additive scheme". The sets of parameters describing heats of formation, dipole moments, polarizabilities etc. in terms of atomic and/or bond increments had been developed a long time ago, yet in the 40s and 50s of the last century. Clearly the classical MM theory [1, 2] as we know it and as briefly described in Section 2.5 is a result of refining these concepts in the direction of including more and more subtle effects of geometry dependence.

A realm not directly related to the additivity concept – stereochemistry – is also a source of data supporting our way of constructing the mechanistic model of PES. Stereochemistry can be regarded as a qualitative tool to rationalize the spatial patterns the atoms and groups follow when attached to each other. For a century, two fundamental concepts shaped this area: that of the tetrahedral carbon atom introduced by van t'Hoff and Le Bel [13,14] and that of the pyramidal nitrogen atom. From a general theoretical point of view, the preferable molecular shapes are ultimately controlled by the dependence of energy on valence angles. Despite its long development history, a common viewpoint of the origin of this angular dependence of molecular energy has not been developed yet. On the one hand, even very simple quantum chemical methods reproduce the observed features of molecular geometry with remarkable precision [15]. In very general terms, it is clear that the forms of the coordination polyhedra are controlled by the relation between the bonding (two-center) interactions which favor the population of excited and ionized states of an atom under consideration and the excitation and ionization energies themselves, which tend to keep an atom in its ground (unhybridized) state [16]. Nevertheless, there exists a certain gap between a purely theoretical, qualitative understanding of the ultimate source of the observed stereochemical features and how these features are simulated in the current MM force fields.The reason is that no sequential derivation has been proposed to bridge the two banks of the river: general theoretical understanding on the one hand and specific force fields of MM or empirical rules in stereochemistry on the other. Moreover, the Gillespie scheme [17–19], designed largely for systematization of qualitative stereochemical data, ascribes the bending energy to interactions between electron pairs residing in the valence shell of the atom of interest. This theoretical construct is known under the name of valence shell electronic pair repulsion (VSEPR). According to it, the angular dependence of energy appears due to Coulomb repulsion between electron pairs. In the literature, several attempts to reconcile this qualitative and intellectually very attractive picture with the results of the quantum chemical calculations can be found. These attempts, reviewed in [20], turned out to be discouraging, however. It has been found that the energy terms responsible for the molecular shape formation cannot be identified with the interpair Coulomb interactions. This finding applies both to sp^3 carbon and sp^3 nitrogen stereochemistry. In any case quantum chemistry calculations do not provide any explanation for *why* molecules have that or another stereochemistry: they simply fix the same fact by means of yet another — now numerical — experiment. On the other hand, one cannot ignore the enormous heuristic strength of the Nyholm-Gillespie *rules* ascribing that or any other interaction strength to lone pairs and variously polar bonds.

Previous attempts to sequentially construct additive systematics for molecular energies (which *a fortiori* include MM) reviewed in [20] had the following common points: the transferability hypothesis, one-determinant approximation for the underlying QM wave function, and a posteriori localization of the orbitals. These features collectively prevented authors reviewed in [20] from constructing a sequential route from the QM description of the molecular electronic structure to any additive systematics. The reason is that the real derivation of any additive systematics must include both a proof of transferability and a procedure of defining the relevant local states (whether transferable or not). The derivation of MM from QM consists of several steps:

- groups of electrons responsible for the observed effects on molecular energy and geometry to be reproduced in the target MM description are to be identified;
- approximate methods sufficient for the description of selected electron groups and reproducing the target effects are identified and formalized in the structure of the corresponding trial wave function;
- electronic structure variables (ESVs) describing the identified groups in the above sufficient approximation are to be selected;
- in terms of this set of ESVs, an intermediate mechanistic description (deductive molecular mechanics – see below – DMM) of the PES is to be constructed;
- intermediate ESVs can be excluded for example in the framework of the linear response theory (if this latter applies) or by any other relevant method in order to get the classical (MM) model of PES.

In the subsequent sections we perform the outlined program (in different versions) with respect to simple organic molecules.

3.2. CHARACTERISTIC FEATURES OF MOLECULAR ELECTRONIC STRUCTURE IN SLG APPROXIMATION

Now we are ready to start the derivation of the intermediate scheme bridging quantum and classical descriptions of molecular PES. The basic idea underlying the whole derivation is that the *experimental fact* that the numerous MM models of molecular PES and the VSEPR model of stereochemistry are that successful, as reported in the literature, must have a *theoretical explanation* [21]. The only way to obtain such an explanation is to perform a derivation departing from a certain form of the trial wave function of electrons in a molecule. QM methods employing the trial wave function of the self consistent field (or equivalently Hartree-Fock-Roothaan) approximation can hardly be used to base such a derivation upon, as these methods result in an inherently delocalized and therefore nontransferable description of the molecular electronic structure in terms of canonical MOs. Subsequent a posteriori localization procedures prescribed in the literature as tools allowing one to obtain localized one-electron states to be used as building blocks of local descriptions in fact create more problems rather than provide solutions. First, the localization procedures are numerous and the fact that they give close (but not identical) results simply makes the choice of a unique one more difficult, as there is no clear selection criterion. Second, irrespective of the localization procedure used the a posteriori localized one-electron states always have some residual amplitudes on other atoms of the molecules known as "tails". Neither the subsequent "tail cutting" nor leaving them "as is" can hardly be formalized or reconciled with the general requirement for transferability of the whole picture. All these observations force us to undertake a search for an alternative to the traditional QM methods, to be used as a starting point for deriving a mechanistic description. The main criterion for such a method is that it must describe the electronic structure in terms relevant for the target MM picture i.e. in those of bonds and lone pairs. The method satisfying these criteria is described in Section 2.4. It uses the geminal form of the electronic wave function [22] and strictly local HOs [23] as the one-electron basis set to construct it. Now we are going to analyze the characteristics of molecular electronic structures, which appear from the SLG based methods, to fulfill the first step of our program.

3.2.1. Transferability of density matrix elements in the SLG picture

Numerical experiments performed using the SLG based semiempirical methods show that the bond and lone pair geminals (see Section 2.4) are fairly transferable from one molecule to another and that the electronic energy functional in this approximation may be naturally rewritten using really local quantities, such that their local nature is guaranteed by construction. The total molecular energy can then be recast to the form of eq. (2.88) which represents the molecular PES as a sum of local increments. These increments depend on the ESVs of two classes (i) those defining the hybridization of atomic basis sets and (ii) the one- and two-electron density matrix elements characteristic for each bond in the molecule. In this section we concentrate on the proof of transferability of the electron density matrix elements as they appear in the SLG approximation. The density matrix elements are in turn expressed through the geminal amplitudes, coming from a diagonalization of the effective bond Hamiltonians. Thus any analysis of the properties of the density ESVs starts from a description of this latter.

3.2.1.1. Effective bond Hamiltonians

Within the original SLG approach [11, 12] and Section 2.4 the geminals are characterized by the amplitudes (see eq. (2.60)) u_m, v_m, and $\sqrt{2}w_m = z_m$, which simplifies the normalization condition eq. (2.62) for the amplitudes to: $u_m^2 + v_m^2 + z_m^2 = 1$. The effective Hamiltonians for each bond geminal are expressed in terms of the molecular integrals in the HO basis. Obviously for a geminal expanded over three singlet

two-electron basis configurations the optimal values of the configuration amplitudes
are the solutions of the eigenvector problem (see also [24]):

$$(3.1) \qquad \begin{pmatrix} \mathfrak{R}_m & \mathfrak{D}_m & 0 \\ \mathfrak{D}_m & \mathfrak{C}_m & \mathfrak{D}_m \\ 0 & \mathfrak{D}_m & \mathfrak{L}_m \end{pmatrix} \begin{pmatrix} u_m \\ z_m \\ v_m \end{pmatrix} = \epsilon^m \begin{pmatrix} u_m \\ z_m \\ v_m \end{pmatrix}$$

corresponding to its lowest eigenvalue.

The matrix elements of the effective bond Hamiltonians are defined as (with the
MINDO/3 parameterization for the Hamiltonian taken for the sake of definiteness):

$$
\begin{aligned}
\mathfrak{R}_m &= 2U_m^r + (r_m r_m | r_m r_m)^{R_m} - 4\gamma_{R_m L_m} P_m^{ll} + \\
&\quad + 2 \sum_{B \neq R_m} \gamma_{R_m B} Q_B + 2 \sum_{\substack{t_{m_1} \in R_m \\ m_1 \neq m}} g_{r_m t_{m_1}}^{R_m} P_{m_1}^{tt} \\
(3.2) \qquad \mathfrak{L}_m &= 2U_m^l + (l_m l_m | l_m l_m)^{L_m} - 4\gamma_{R_m L_m} P_m^{rr} + \\
&\quad + 2 \sum_{B \neq L_m} \gamma_{L_m B} Q_B + 2 \sum_{\substack{t_{m_1} \in L_m \\ m_1 \neq m}} g_{l_m t_{m_1}}^{L_m} P_{m_1}^{tt} \\
\mathfrak{C}_m &= \tfrac{1}{2}(\mathfrak{R}_m + \mathfrak{L}_m) - \Delta\gamma_m; \quad \mathfrak{D}_m = -\sqrt{2}\beta_{r_m l_m}^{R_m L_m}
\end{aligned}
$$

where

$$(3.3) \qquad \Delta\gamma_m = g_m - \gamma_{R_m L_m}, \quad g_m = \frac{1}{2} \sum_{t \in \{r,l\}} (t_m t_m | t_m t_m)^{T_m}$$

The calculations of [11, 12] performed on organic compounds of different classes
(alkanes, alcohols, amines etc.) have demonstrated the remarkable stability of all the
geminal related ESVs. The values of the polarity $P_m^{rr} - P_m^{ll}$ do not exceed 0.07 by
absolute value for the compounds containing carbon, nitrogen, and hydrogen atoms
(for the situation with oxygen and fluorine see below). Also the ionicity (the overall
weight of the ionic configurations $u_m^2 + v_m^2$) for a rich variety of bonds has a stable value of about 0.4. The bond orders $2P_m^{rl}$ all acquire values between 0.92 and
1.0. These features, though not completely unexpected, as the transferability of the
parameters of the single bonds in organic compounds is well known experimentally,
require a theoretical explanation. This is given below.

Pseudospin representation and the perturbative estimates of the bond-geminal ESVs.
To provide the required explanation, we notice that the effective Hamiltonians for
the bond geminals can be represented as a sum of the unperturbed part which, when
diagonalized yields invariant, i.e. *exactly transferable*, values of the ESVs, and of a
perturbation responsible for the specificity of electronic structure for different chemical compositions and environments of the bond.

Pseudospin operator of the bond geminal.　　Let us introduce a pseudospin operator $\widehat{\tau}_m$ corresponding to the pseudospin value $\tau_m = 1$. The matrices of its components in the basis of the configurations defining the geminal are given by:

$$
\hat{\tau}_{zm} = \begin{pmatrix} 1 & 0 & 0 \\ 0 & 0 & 0 \\ 0 & 0 & -1 \end{pmatrix}, \quad \hat{\tau}_{+m} = \begin{pmatrix} 0 & \sqrt{2} & 0 \\ 0 & 0 & \sqrt{2} \\ 0 & 0 & 0 \end{pmatrix}
$$

(3.4)

$$
\hat{\tau}_{-m} = (\hat{\tau}_{+m})^{\dagger}
$$

The configurations corresponding to $\langle \hat{\tau}_{zm} \rangle = \pm 1$ are the ionic ones with both electrons located on the same end of the chemical bond (right or left, respectively). In terms of the pseudospin operator, the elements of the density matrices in eqs. (2.78), (2.81) can be presented as follows:

(3.5)
$$
P_m^{tt} = \tfrac{1}{2}(1 + t\langle \hat{\tau}_{zm} \rangle), \; P_m^{rl} = \tfrac{1}{2}\langle \hat{\tau}_{+m} \rangle, \; P_m^{lr} = \tfrac{1}{2}\langle \hat{\tau}_{-m} \rangle
$$
$$
\Gamma_m^{tt} = \tfrac{1}{2}(\langle \hat{\tau}_{zm}^2 \rangle + t\langle \hat{\tau}_{zm} \rangle), \; \Gamma_m^{rl} = \Gamma_m^{lr} = \tfrac{1}{2}(1 - \langle \hat{\tau}_{zm}^2 \rangle)
$$

The effective bond Hamiltonians also can be rewritten in terms of the pseudospin operators. Indeed, the effective Hamiltonian eq. (3.1) for each of the bond geminals can be presented in the form:

$$
\hat{H}_m^{\text{eff}} = \hat{H}_{0m}^{\text{eff}} + \Delta_{Im}\hat{\tau}_{zm}^2 + \Delta_{Pm}\hat{\tau}_{zm}
$$

(3.6)
$$
\hat{H}_{0m}^{\text{eff}} = \begin{pmatrix} \mathfrak{C}_m & \mathfrak{D}_m & 0 \\ \mathfrak{D}_m & \mathfrak{C}_m & \mathfrak{D}_m \\ 0 & \mathfrak{D}_m & \mathfrak{C}_m \end{pmatrix}
$$
$$
\Delta_{Im} = \frac{1}{2}(\mathfrak{R}_m + \mathfrak{L}_m) - \mathfrak{C}_m = \Delta\gamma_m, \quad \Delta_{Pm} = \frac{1}{2}(\mathfrak{R}_m - \mathfrak{L}_m)
$$

with two perturbation terms. The perturbation proportional to $\hat{\tau}_{zm}^2$ controls the relative contribution of the ionic and covalent configurations to the bond geminal, and the perturbation proportional to $\hat{\tau}_{zm}$ describes the asymmetry (polarity) of the bond. The geminal amplitudes obtained by diagonalizing the unperturbed bond Hamiltonians $\hat{H}_{0m}^{\text{eff}}$ and the density or $\hat{\tau}$-type ESVs thus obtained are perfectly invariant:

(3.7)
$$
u_{0m} = v_{0m} = w_{0m} = \tfrac{1}{2}, \; \left(z_{0m} = \tfrac{1}{\sqrt{2}} \right), \; P_{0m}^{tt'} = \tfrac{1}{2}, \; \Gamma_{0m}^{tt'} = \tfrac{1}{4}
$$
$$
\langle \hat{\tau}_{zm} \rangle_0 = 0, \; \langle \hat{\tau}_{zm}^2 \rangle_0 = \tfrac{1}{2}, \; \langle \hat{\tau}_{+m} \rangle_0 = \langle \hat{\tau}_{-m} \rangle_0 = 1
$$

They do not depend either on the kinds of atoms or on molecular geometry. The unperturbed effective Hamiltonians $\hat{H}_{0m}^{\text{eff}}$ themselves, of course, depend on all these parameters, so that the energies of bonds even in this simple picture are composition- and geometry dependent, due to the corresponding dependence of the matrix elements of the Hamiltonian, but not the ESVs under consideration. The structure of the problem squeezes the whole multidimensional manifold of matrix elements (and even more dimensional manifold of the parameters defining the matrix elements) into two independent quantities Δ_{Im} and Δ_{Pm}. One can see that the invariant values of the ESVs eq. (3.7) are rather close to the exact SLG values which appear from numerical experiments. These are almost independent of the particular parameterization used.

This is also not surprising by itself, but the existence of the values of ESVs dependent on absolutely nothing of course gives the explanation to this fact. Now we shall obtain estimates of the precision to which the found transferability holds.

Perturbative estimate of ESVs with respect to noncorrelated bare Hamiltonian. The specificity of each bond and molecule in the approach based on the SLG expressions for the wave function is taken into account perturbatively by using the linear response approximation [25]. We need perturbative estimates of the expectation values of the pseudospin operators which, in their turn, give values of the density matrix elements according to eq. (3.5). According to the general theory (Section 1.3.3.2) the linear response $\delta\langle\hat{A}\rangle$ of an expectation value of the operator $\hat{A}$ to the time independent perturbation $\lambda\hat{B}$ of the Hamiltonian (λ is the parameter characterizing the intensity of the perturbation) has the form:

$$(3.8) \qquad \delta\langle\hat{A}\rangle = \lambda\langle\langle\hat{A};\hat{B}\rangle\rangle = \lambda\left\langle\left[\mathrm{Ad}_{\hat{H}^{(0)}}^{-1}\hat{B},\hat{A}\right]\right\rangle$$

where the zero frequency response function for the ground state is given by the relation:

$$(3.9) \qquad \langle\langle\hat{A};\hat{B}\rangle\rangle = 2\sum_{i\neq 0}\langle 0|\hat{A}|i\rangle(\epsilon_0 - \epsilon_i)^{-1}\langle i|\hat{B}|0\rangle$$

Inserting the amplitudes u_{0m}, v_{0m}, and z_{0m} of the geminals for the ground states of the bare effective Hamiltonian and of two corresponding excited states of the bond and two excitation energies of the effective Hamiltonian $\hat{H}_{0m}^{\mathrm{eff}}$ (which are equal to $\sqrt{2}|\mathfrak{D}_m|$ and $2\sqrt{2}|\mathfrak{D}_m|$) to the general expression for the response functions of eq. (1.70) one can immediately check that in the case when the operators $\hat{A}$ and $\hat{B}$ are the components of the pseudospin operator or its squared z-component, only the diagonal response functions are nonvanishing:

$$(3.10) \qquad \begin{aligned} \langle\langle\hat{\tau}_{zm};\hat{\tau}_{zm}\rangle\rangle &\neq 0 \\ \langle\langle\hat{\tau}_{zm}^2;\hat{\tau}_{zm}^2\rangle\rangle &\neq 0 \end{aligned}$$

so that the first-order responses of the expectation values of the pseudospin operator components to the perturbations proportional to $\hat{\tau}_{zm}$ and $\hat{\tau}_{zm}^2$ can be written as:

$$\langle\hat{\tau}_{zm}\rangle = \delta\langle\hat{\tau}_{zm}\rangle = \frac{\mathfrak{L}_m - \mathfrak{R}_m}{4\beta_{r_m l_m}^{R_m L_m}}$$

$$(3.11) \qquad \delta\langle\hat{\tau}_{zm}^2\rangle = -\frac{\Delta\gamma_m}{8\beta_{r_m l_m}^{R_m L_m}}$$

$$\delta\langle\hat{\tau}_{+m} + \hat{\tau}_{-m}\rangle = 0$$

The last row is most important here. It demonstrates that in the linear response approximation the off-diagonal matrix element of the one-electron density matrix (the Coulson bond order) does not change, (i.e. is invariant even for the different atoms forming the bond and even more to the geometry changes). This result suggests the stability of the bond orders with some precision. However, this result should not be overemphasized. In the present form it is a consequence of the HFR type of the two-electron wave function implied by the present treatment. As one can check, the decomposition of the effective bond Hamiltonian eq. (3.6) is equivalent to making the HFR approximation for its ground state. The result is that the unperturbed part defined by eq. (3.6) yields the symmetric ground state with the total weight of the ionic configurations equal to that of the covalent one. This coincides with the result of the HFR approximate treatment of the symmetric bond. For this reason the formulae for the bond-order variation and the two-electron density matrix elements (and the HFR approximation itself) are not valid at larger interatomic separations (the denominators in eq. (3.9) proportional to $|\mathfrak{D}_m|$ become too small). The exact solution of the SLG problem has correct asymptotic behavior for single bonds even at infinite interatomic separations. This attractive feature of the SLG model is lost in the perturbative treatment based on the Hamiltonian separation of eq. (3.6) as it is in the HFR approximation as well. This is the reason why we reconsider this problem in a correlated setting.

Perturbation of the density matrix elements for correlated ground state. To overcome the failure of the perturbative treatment of the ESVs describing one- and two-electron densities described above, let us reconsider a symmetric two-electron two-center bond. The new treatment corresponds to a different decomposition of the effective bond Hamiltonian eq. (3.6). We assume that the contribution to the effective bond Hamiltonian, which is proportional to $\hat{\tau}_{zm}^2$, is included in the unperturbed (zero order) Hamiltonian. The problem then reduces to diagonalizing a 2×2 matrix. It can be easily solved and the ESVs (elements of density matrices), as they appear from this solution, are:

$$(3.12) \quad \Gamma_m^{tt'} = \frac{1}{4}\left(1 - tt'\frac{1}{\Gamma(\zeta_m)}\right), \ P_m^{tt} = \frac{1}{2}, \ P_m^{rl} = \frac{\zeta_m}{2\Gamma(\zeta_m)}$$

where

$$(3.13) \quad \zeta_m = 4\beta_{r_m l_m}^{R_m L_m}/\Delta\gamma_m, \ \Gamma(\zeta_m) = \sqrt{1 + \zeta_m^2}$$

Smaller interatomic separations characteristic for the real bonds correspond to the limit $\zeta_m \gg 1$ and the ESVs in this limit have the following asymptotic behavior:

$$(3.14) \quad \Gamma_{0m}^{tt'} = \frac{1}{4}\left(1 - tt'\frac{1}{\zeta_m}\right), \ P_m^{rl} = \frac{1}{2}\left(1 - \frac{1}{2\zeta_m^2}\right)$$

Using the separation of the effective Hamiltonian into the unperturbed part and the perturbation, the total ionic contribution to the geminal is calculated exactly (variationally). Only the bond polarity needs to be estimated perturbatively in the linear response approximation, but now the correlated ground state of the symmetric effective bond Hamiltonian is taken for evaluating the response function. In this context, it is convenient to use a dimensionless bond asymmetry parameter:

$$(3.15) \quad \mu_m = \frac{\mathfrak{L}_m - \mathfrak{R}_m}{\Delta\gamma_m\Gamma(\zeta_m)}$$

instead of the original perturbation parameter Δ_{Pm}. Inserting the necessary ground and excited states parameters to the definition of the response function $\langle\langle\hat{\tau}_{zm};\hat{\tau}_{zm}\rangle\rangle$ one obtains:

$$(3.16) \quad \langle\hat{\tau}_{zm}\rangle = \mu_m \frac{\Gamma(\zeta_m) - 1}{\Gamma(\zeta_m) + 1}$$

for the polarity of the bond between the atoms with the fixed hybridizations. It vanishes for infinite interatomic separation as it should for the exact wave function. The bond ionicity i.e. the *sum* of the ionic contribution to the wave function and the Coulson bond order are not affected in the linear response approximation.

The bond polarity is given by eq. (3.16) even if the second-order correction is considered (i.e., the contribution to the bond polarity proportional to μ_m^2 is absent). The second-order corrected expectation values of the pseudospin operators defining the bond ionicity and bond order by contrast are not vanishing and have the following forms:

$$(3.17) \quad \begin{aligned} \langle\hat{\tau}_{zm}^2\rangle &= \langle\hat{\tau}_{zm}^2\rangle_c \left[1 + \mu_m^2 \frac{2\Gamma(\zeta_m) + 1}{2(\Gamma(\zeta_m) + 1)}\right] \\[2ex] \langle\hat{\tau}_{+m} + \hat{\tau}_{-m}\rangle &= \langle\hat{\tau}_{+m} + \hat{\tau}_{-m}\rangle_c \left[1 + \mu_m^2 \frac{2\Gamma(\zeta_m) + 1 - \Gamma^2(\zeta_m)}{2(\Gamma(\zeta_m) + 1)^2}\right] \end{aligned}$$

where the quantities with the subscript c correspond to the estimates of eq. (3.12) i.e. with respect to the correlated ground state of the symmetrized bond Hamiltonian.

Lone pairs. An archetypal form of the two-electron group, different from the two-center bond studied above, and incidentally much more simple in the given formulation, is a lone pair. As it is mentioned in Section 2.4 the lone pair in the SLG context is described by a degenerate geminal containing the contribution of only one ionic configuration. For the sake of definiteness we set it to be the right-end ionic configuration of the corresponding degenerate bond (the amplitude u_m becomes equal to unity, see eq. (2.61)). The ESVs related to the lone pair can be readily evaluated:

$$(3.18) \quad \begin{aligned} \langle\hat{\tau}_{zm}\rangle &= \delta\langle\hat{\tau}_{zm}\rangle = 1, \ \delta\langle\hat{\tau}_{zm}^2\rangle = \frac{1}{2} \\[2ex] P_m^{rl} &= P_m^{lr} = 0, \ \Gamma_m^{rr} = 1, \ \Gamma_m^{ll} = \Gamma_m^{rl} = \Gamma_m^{lr} = 0 \end{aligned}$$

These quantities are perfectly invariant and transferable from one molecule to another and basically characterize (within the accepted approximation, of course) the qualitative difference between the atoms of different chemical elements by the number of lone pairs they bear.

Numerical experiments concerning the density ESVs' transferability. The above analytical results have been supplied by numerical estimates done to get a feeling of the real sense of the "first" and "second" order approximations. Numerical results on the ESVs $\langle\hat{\tau}_{zm}\rangle$, $\langle\hat{\tau}_{zm}^2\rangle$, and $\langle\hat{\tau}_{+m}\rangle$ obtained by the SLG method eq. (3.1) using the MINDO/3 parameterization and by the approximate formulae of eqs. (3.9), (3.12),

(3.14), (3.16), and (3.17), for some characteristic bonds in small molecules were obtained and analyzed [26]. These results show that in the case of bonds with small polarity, all the formulae perform very well. Twenty two bonds in twelve molecules have been considered. In the set including H_2, H_2O, CH_4, HF, and CH_3F molecules as representative members, the values of the presumably small parameter characterizing the relative weight of the covalent and ionic contributions in these bonds ζ^{-1} all fall in the range between 0.127 (for H_2) and 0.355 (for the C-F bond in CH_3F). For all studied C-H bonds (primary and secondary ones as well as geminal – in the chemical sense of this word – ones to the electronegative atom) the values of ζ^{-1} span the range between 0.180 and 0.187. These values of ζ^{-1} cover a very narrow range of the ionicities/covalencies of the bonds under study. For example, for all studied cases the overall weights of the ionic contributions span the range between 0.361 and 0.464. The average ionicity is then 0.403 and the standard deviation over the considered data set is 0.023. This corresponds to the precision of 6%. The bond polarity parameter μ does not exceed the value of 0.540, which is reached in the HF molecule.

The most precise approximations are given by eqs. (3.16) and (3.17) yielding results which perfectly coincide with the exact (SLG-MINDO/3) ones even for very polar O-H and F-H bonds. This may be qualified as using estimates of the second order in μ, provided the bond polarity (or equivalently $\langle \hat{\tau}_{zm} \rangle$) are linear in μ if the orders up to the second are considered. Estimates obtained in the limit $\zeta_m \gg 1$ by the formulae eq. (3.12) give reasonable results for the ESVs of the bonds in not too polar molecules at their equilibrium geometries. The bond- and atom-specific corrections of the first and second order in ζ_m^{-1} and μ_m acquire the form:

$$\Gamma_m^{tt'} = \frac{1}{4} + \delta\Gamma_m^{tt'}, \; P_m^{tt'} = \frac{1}{2} + \delta P_m^{tt}$$

$$(3.19) \quad \delta P_m^{tt} = \frac{t_m \mu_m}{2}; \; \delta P_m^{rl} = -\frac{\mu_m^2}{4} - \frac{1}{4\zeta_m^2}$$

$$\delta\Gamma_m^{tt} = \frac{t_m \mu_m}{2} - \frac{1}{4\zeta_m}; \; \delta\Gamma_m^{rl} = -\frac{\mu_m^2}{2} + \frac{1}{4\zeta_m}$$

The bond polarity parameter μ_m affects remarkably (in the first order) only the diagonal density matrix elements; the off-diagonal ones acquire the corrections of the second order in μ_m.

The stability of the values of bond order is even more striking. In the described data set the values of the bond orders span the range from 0.929 to 0.992 with an average of 0.974 and standard deviation of 0.017. This corresponds to the precision of 1.7%. Of course the high stability is explained by the validity of the above limit, which in its turn is due to the fact that the difference between one- and two-center electron-electron repulsion integrals ($\Delta\gamma_m$) at interatomic separations characteristic of chemical bonding is much smaller than the resonance interaction at the same distance. The most important reason for the stability (i.e. of transferability) of the bond orders is that they deviate from the ideally transferable value in the second order in two small parameters ζ^{-1} and μ.

Further analysis allows us to single out two types of contributions to the parameter μ_m. It can be broken down into a sum of a component which is, however, dependent on the hybridization of the entering orbitals corresponding to the bond itself and the rest describing the environment of the bond:

$$(3.20) \qquad \mu_m = \mu_{0m} + \mu_{1m}$$

The intrinsic bond-related part is:

$$\mu_{0m} = \frac{1}{\Delta\gamma_m \Gamma(\zeta_m)} [2(U_m^l - U_m^r) +$$

$$(3.21) \qquad + (l_m l_m | l_m l_m)^{L_m} - (r_m r_m | r_m r_m)^{R_m} +$$

$$+ \sum_{\substack{t_{m_1} \in L_m \\ m_1 \neq m}} g_{l_m t_{m_1}}^{L_m} n_{m_1} - \sum_{\substack{t_{m_1} \in R_m \\ m_1 \neq m}} g_{r_m t_{m_1}}^{R_m} n_{m_1}]$$

where n_m is equal to 1 for a chemical bond incident to the atom at hand and to 2 for a lone pair at this atom. This contribution is characteristic of the pair of atoms $R_m L_m$ with given ratios of the s- and p -weights in the HOs $|r_m\rangle$ and $|l_m\rangle$ ascribed to the bond at hand and clearly depending on the chemical nature of these atoms through the specific values of atomic parameters and the numbers of lone pairs they bear.

The contribution to the bond asymmetry coming from the environment of the bond is:

$$\mu_{1m} = \frac{1}{\Delta\gamma_m \Gamma(\zeta_m)} [2 \sum_{B \neq L_m} \gamma_{L_m B} Q_B - 2 \sum_{B \neq R_m} \gamma_{R_m B} Q_B +$$

$$(3.22) \qquad + 4\gamma_{R_m L_m} \langle \widehat{\tau}_{zm} \rangle + 2 \sum_{\substack{t_{m_1} \in L_m \\ m_1 \neq m}} g_{l_m t_{m_1}}^{L_m} t_{m_1} \langle \widehat{\tau}_{zm_1} \rangle -$$

$$- 2 \sum_{\substack{t_{m_1} \in R_m \\ m_1 \neq m}} g_{r_m t_{m_1}}^{R_m} t_{m_1} \langle \widehat{\tau}_{zm_1} \rangle]$$

In the molecules with only weakly polar bonds, one can expect that the external part μ_{1m} is small. In the test set of molecules mentioned above, the values of μ_{1m} do not exceed 0.02, which is not more than 10% of the total for the asymmetry parameter μ_m. In the molecules containing many polar bonds, the effect of randomly distributed effective atomic charges almost vanishes, leading to small values of external contributions to the bond polarity parameters μ_{1m}. The only situation when one can expect the environment to affect the characteristics of the otherwise transferable bond, is when the bond under consideration appears in close vicinity with a few strongly charged atoms arranged in such a way that their fields sum up to a nonzero overall field directed along the bond. That is, clearly, one of the situations which elaborated MM parameterizations mark as special ones, requiring specific values of parameters.

The break up of μ_m into external and internal contributions according to eq. (3.20) yields a transferable, environment-independent approximating function for the ESVs by substituting into eqs. (3.16) and (3.17) the internal contribution of the parameter μ_{0m} instead of its total value μ_m. The numerical results show that this approach leads to the approximate ESVs perfectly coinciding with those obtained by the SLG method itself that demonstrates the applicability of such a scheme. Comparing estimates obtained with the use of the exact values of parameters μ and their intrabond estimates μ_0 allows us to single out the effects of the environment. It turns out that in the test set, the estimates using μ and μ_0 coincide up to the third decimal digit. For example, the primary C-H bonds in the ethane and propane molecules result in coinciding values of the ESVs estimated by using μ_0. Therefore, the small difference between the ESVs of these bonds obtained by the SLG method is caused by the slightly different environment, i.e. by the μ_1 values which are equal to 0.003 and 0.006, respectively. The small magnitude of the deviations between the precise results obtained by the SLG-MINDO/3 method and approximately estimated ESVs can be rationalized by the smallness of their effect on the total energy of the molecule. For example, even in the case of the polar water molecule, the approximation of the bond ESVs by their values estimated using the μ_0 value leads to an increase in the total energy by only 0.014 kcal/mol as compared to the exact SLG calculation. This clearly indicates that in most cases the effect of μ_1 can be neglected. This is of course a conclusion derived on the basis of numerical experiment. As such it also requires a theoretical explanation and thus additional work to be done.

3.2.2. Mathematical description of hybridization

In the previous section we performed the first part of our program of bridging the gap between the quantum and classical descriptions of molecular electronic structure and molecular PES. This is reduced so far to singling out a certain class of ESVs, namely the geminal amplitudes, or more precisely the elements of the one- and two-electron density matrices in the basis of HOs. The equilibrium values of these ESVs turned out to be quite stable i.e. transferable according to the numerical experiments and to the above theoretical consideration, which demonstrated the reasons for such behavior. Here we address another component of the whole picture, another group of the ESVs – those describing the basis of HOs in which the transferability of the density matrix elements holds. Numerical experiments performed in the SLG-MINDO/3 approximation demonstrate that the HOs are much more sensitive to all variations of the molecular Hamiltonian, both those induced by chemical composition and those induced by variations of molecular geometry.

In the framework of the SLG scheme, the basis orbitals are defined by orthogonal transformations of AOs for each atom with an sp-valence shell. The energy eq. (2.88) is the function of the parameters defining these transformations. The 4×4 $SO(4)$ matrices h^A of transformation from the AO to the HO basis are set on each heavy atom A. In general any $n \times n$ orthogonal matrix can be presented as a product of $n(n-1)/2$ Jacobi matrices of the form

$$
(3.23) \quad J_{ij}(\omega_{ij}) =
\begin{pmatrix}
1 & & & & & & \\
& \ddots & & & & & \\
& & \cos\omega_{ij} & & \sin\omega_{ij} & & \\
& & & 1 & & & \\
& & & & \ddots & & \\
& & -\sin\omega_{ij} & & \cos\omega_{ij} & & \\
& & & & & & \ddots
\end{pmatrix}
$$

Each of these matrices describes a rotation (by an angle $\omega_{ij}, i < j; j = 1 \div n$) in a two-dimensional subspace (plane) of the n-dimensional space. In a specific case of the $SO(3)$ group – that of rotations of the physical $\mathbb{R}^3$ space – the matrices of rotations around coordinate axes $x, y,$ and z are:

$$
(3.24) \quad
\begin{pmatrix}
1 & 0 & 0 \\
0 & \cos\omega_{yz} & \sin\omega_{yz} \\
0 & -\sin\omega_{yz} & \cos\omega_{yz}
\end{pmatrix}
;
\begin{pmatrix}
\cos\omega_{xz} & 0 & -\sin\omega_{xz} \\
0 & 1 & 0 \\
\sin\omega_{xz} & 0 & \cos\omega_{xz}
\end{pmatrix}
;
$$

$$
\begin{pmatrix}
\cos\omega_{xy} & \sin\omega_{xy} & 0 \\
-\sin\omega_{xy} & \cos\omega_{xy} & 0 \\
0 & 0 & 1
\end{pmatrix}
$$

It is possible that the choice of three rotation axes is done in a different manner. In fact the most widely used one differs from the above and represents an arbitrary rotation as a product (sequential performing) of rotations around the axis z, than around y and then around the new axis z. The corresponding angles are called the Euler angles, but we do not use this type of parameterization of rotations in this book. In the case of 4-dimensional orbital space, the number of necessary angular variables equals six. Three of them (pseudorotation angles $\vec{\omega}_b = (\omega_{sx}, \omega_{sy}, \omega_{sz})$ with subscripts indicating pairs of basis AOs mixed by the corresponding 2×2 Jacobi rotations) define the structure of the HOs (s-/p-mixing) while the other three (quasirotation angles $\vec{\omega}_l = (\omega_{yz}, -\omega_{xz}, \omega_{yz})$) define the $SO(3)$ matrix performing rotation of the set of four HOs as a whole (the prefix quasi refers here to the fact that no physical body rotates under the action of these matrices, only the system of HO's). The $SO(4)$ group (i.e. the group of 4×4 orthogonal matrices with unit determinant) is a so-called "dynamic" group of the manifold of the HOs of a given atom in a sense that it produces the whole possible variety of hybridizations at each heavy atom [8] while acting on the AOs set residing on the latter. The matrix $h(A)$ generating the set of HOs centered on the atom A is thus a matrix product:

$$
(3.25) \quad h(A) = h(\vec{\omega}^A) = R(\vec{\omega}_l^A) H(\vec{\omega}_b^A)
$$

where the matrix multipliers responsible for the orientation (R) of the whole set of the HOs at a given atom and for the hybridization (H) i.e. for the relative weights of the s- and p-orbitals in the HOs, are themselves the products of the corresponding Jacobi matrices:

$$(3.26) \quad \begin{aligned} R(\vec{\omega}_l) &= J_{yz}(\omega_{yz})J_{xz}(\omega_{xz})J_{xy}(\omega_{xy}) \\ H(\vec{\omega}_b) &= J_{sz}(\omega_{sz})J_{sy}(\omega_{sy})J_{sx}(\omega_{sx}) \end{aligned}$$

Generally, the transformation of orbitals caused by a pseudorotation forms a set of HOs which is known as the hybridization pattern (like sp^3, sp^2 etc.) which is more or less stable, while the set of quasirotation angles is totally non-transferable and depends on the orientation of the molecule in space.

The mathematical description of hybridization is based on employing the algebraic group structure of the hybrids' manifold. Because of it, any small variation of HOs in a vicinity of a given set of HOs given by a 4×4 orthogonal matrix h can be expressed using another $SO(4)$ matrix H close to the unity matrix:

$$H = I + \delta^{(1)}H + \delta^{(2)}H$$

$$(3.27) \quad h' = Hh \approx h + \delta^{(1)}h + \delta^{(2)}h$$

$$\delta^{(1)}h = \delta^{(1)}Hh, \quad \delta^{(2)}h = \delta^{(2)}Hh$$

where the hybrids' set h' is close to the initial set h. In order to go further we need the general form of matrix H and its expansion up to the second order in the vicinity of the unity matrix. To obtain this, we address the parameterization of the $SO(4)$ manifold in the vicinity of the unity matrix using the following construction.

From the theory of continuous (Lie) groups it is known that their properties are completely defined by the commutation relations between some matrices known in this context as infinitesimal operators of the Lie group. By definition the infinitesimal operator is a partial derivative of the matrix representing an element of the Lie group with respect to a variable defining this element calculated for the zero values of all variables of this type, which corresponds to taking the partial derivatives in the point corresponding to the unity matrix which is also the unity element of the Lie group under consideration. Obviously there are as many independent infinitesimal operators as variables needed to define the elements of the group. On the other hand the choice of the set of infinitesimal operators is by no means unique as selecting another set of coordinates representing the group elements in the vicinity of the unity also defines another set of infinitesimal operators – the partial derivatives with respect to the new coordinates. For the $SO(4)$ group of interest, the infinitesimal operators are defined with respect to the above six angular variables forming the simplest coordinate map on the $SO(4)$ manifold:

$$(3.28) \quad \begin{aligned} B_\gamma &= \left. \frac{\partial h(\vec{\omega})}{\partial \omega_{s\gamma}} \right|_{\vec{\omega}=0} = \left. \frac{\partial H(\vec{\omega}_b)}{\partial \omega_{s\gamma}} \right|_{\vec{\omega}_b=0} \\ &\text{and} \\ \epsilon_{\alpha\beta\gamma}L_\gamma &= \left. \frac{\partial h(\vec{\omega})}{\partial \omega_{\alpha\beta}} \right|_{\vec{\omega}=0} = \left. \frac{\partial R(\vec{\omega}_l)}{\partial \omega_{\alpha\beta}} \right|_{\vec{\omega}_l=0} \end{aligned}$$

Obviously the zero values of the angular variables define the unity matrix of the $SO(4)$ group; $\epsilon_{\alpha\beta\gamma}$ is the complete antisymmetric – Levi-Cività – tensor. The commutation properties of the infinitesimal operators defined in this manner are inconvenient. On the one hand the subset L_γ obeys the commutation relations usual for the angular momentum components

$$(3.29) \quad [L_\alpha, L_\beta] = \epsilon_{\alpha\beta\gamma} L_\gamma$$

by this justifying the name of quasi-rotation angles for the $\vec{\omega}_l$ variables set. Meanwhile, the set of pseudomomentum components B_γ (in fact these infinitesimal operators differ from the true momentum operators by the multiplier i) is not closed with respect to commutation relations:

$$(3.30) \quad [B_\alpha, B_\beta] = \epsilon_{\alpha\beta\gamma} L_\gamma$$

The pseudomomentum components also do not commute with the angular (quasi-) momentum components [8] since these commutators are equal to the respective pseudomomentum components:

$$(3.31) \quad [B_\alpha, L_\beta] = [L_\alpha, B_\beta] = \epsilon_{\alpha\beta\gamma} B_\gamma$$

This suggests that the coordinate map eq. (3.25) reparametrizing the $SO(4)$ manifold is not a very good one as it does not permit to recognize the fundamental fact about the $SO(4)$ group that it is a direct product of two $SO(3)$ subgroups ($SO(4) = SO(3) \times SO(3)$). In terms of the infinitesimal operators the direct product structure of a group means that the infinitesimal operators of each subgroup multiplier commute with each other. It is not necessarily so for all coordinate maps introduced into the group, but it is possible to select some special coordinate maps where it is so. For the $SO(4)$ group the required commutation relations are achieved by making a coordinate transform:

$$(3.32) \quad \omega_{\gamma\pm} = \epsilon_{\alpha\beta\gamma} \omega_{\alpha\beta} \pm \omega_{s\gamma}$$

leading to a new set of infinitesimal operators:

$$(3.33) \quad F_\gamma = \frac{1}{2}(L_\gamma + B_\gamma) = \frac{\partial h(\vec{\omega})}{\partial \omega_{\gamma+}}, \; G_\gamma = \frac{1}{2}(L_\gamma - B_\gamma) = \frac{\partial h(\vec{\omega})}{\partial \omega_{\gamma-}}$$

Their commutation relations reveal the direct product structure of the $SO(4)$ group, as one can check:

$$(3.34) \quad [F_\alpha, G_\beta] = 0, \; \forall \alpha, \beta, \gamma$$

In other words, the vector operator $\vec{F}$ commutes with the vector operator $\vec{G}$ and each of them forms a basis in the tangent space to the corresponding $SO(3)$ subgroup of the $SO(4)$ group of interest:

$$(3.35) \quad [F_\alpha, F_\beta] = \epsilon_{\alpha\beta\gamma} F_\gamma, \; [G_\alpha, G_\beta] = \epsilon_{\alpha\beta\gamma} G_\gamma, \forall \alpha, \beta, \gamma$$

which means that both vector operators $\vec{F}$ and $\vec{G}$ are some momentum operators as their own components conform to the characteristic commutation rules and by this define three-dimensional rotations on the respective angle triples $\vec{\omega}_\pm$. The new set of parameters (angles $\vec{\omega}_\pm$) lacks any clear physical meaning as it represents neither pure rotation nor pure deformation of the system of HOs. Those conceptually important transformations can be recovered either by setting $\vec{\omega}_+ = \vec{\omega}_-$, which results in a pure rotation, or by setting $\vec{\omega}_+ = -\vec{\omega}_-$, which corresponds to a pure deformation.

The coordinate map given by the variables $(\vec{\omega}_+, \vec{\omega}_-)$ is a significant improvement as compared to eq. (3.25). Nevertheless, an explicit expression for an h matrix in its terms is still a clumsy combination of the trigonometric functions of two triples of reparametrizing angles $\vec{\omega}_\pm$. It is known however that in the case of the $SO(3)$ group [8] its quaternion [27] parameterization has the advantage that the matrix elements of $SO(3)$ rotation matrices, when expressed in terms of the components of the normalized quaternion, are quadratic functions of these components.

As quaternions have disappeared from the common chemists' mathematical background, we recall here the basic properties of this beautiful mathematical instrument. Quaternions are objects having four components. The first one can be treated as a real scalar, whereas the other three can be considered components of a three-dimensional vector. These objects are customarily represented in a form similar to that of complex numbers:

$$\mathbf{q} = q_0 + iq_x + jq_y + kq_z = (q_0, \vec{q})$$

with three imaginary units i, j, k taking incidentally parts of the orts of the 3-dimensional space. Two quaternions can be added by adding their corresponding components (scalar and vector ones)

$$(3.36) \quad \mathbf{q} = \mathbf{q}_1 + \mathbf{q}_2$$

$$q_i = q_{1i} + q_{2i}$$

They can be also multiplied generalizing the multiplication rule for complex numbers by the following multiplication table for the orts:

$$(3.37) \quad
\begin{array}{c|cccc}
\diamond & 1 & i & j & k \\
\hline
1 & 1 & i & j & k \\
i & i & -1 & k & -j \\
j & j & -k & -1 & i \\
k & k & j & -i & -1
\end{array}$$

The quaternion

$$\tilde{\mathbf{q}} = q_0 - iq_x - jq_y - kq_z = (q_0, -\vec{q})$$

is called conjugate to $\mathbf{q}$ and the norm of a quaternion is given by:

$$(3.38) \quad \|\mathbf{q}\| = (\mathbf{q} \diamond \tilde{\mathbf{q}})^{\frac{1}{2}} = \sqrt{q_0^2 + q_x^2 + q_y^2 + q_z^2}$$

For a pair of quaternions $\mathbf{a}$ and $\mathbf{b}$: $\mathbf{a} = (a_0, \vec{a}); \mathbf{b} = (b_0, \vec{b})$, not necessarily normalized, the following relation holds:

$$(3.39) \quad \mathbf{a} \diamond \mathbf{b} = (a_0 b_0 - (\vec{a}, \vec{b}), a_0 \vec{b} + b_0 \vec{a} + \vec{a} \times \vec{b})$$

The multiplication $\diamond$ of quaternions is associative and distributive, but not commutative:

$$(3.40) \qquad \mathbf{a} \diamond \mathbf{b} \neq \mathbf{b} \diamond \mathbf{a}$$

After a short reminder of what the quaternions are, we briefly sketch how they can be used to represent 3-dimensional rotations. Due to the obvious importance of the capacity to describe 3-dimensional rotations of usual physical space, numerous techniques or parameterizations for them have been suggested. The most widespread is of course the representation of rotations by 3×3 matrices which, as mentioned above, can be decomposed in a product of rotations around three coordinate axes. Another natural description for a 3-dimensional rotation is by defining a rotation axis and a rotation angle. For the rotation described by the rotation axis and the rotation angle it is easy to construct a representation by a normalized quaternion. If ω is a rotation angle and $\vec{l}$ is a unit vector directed along the rotation axis (the vector of its directing cosines) then the normalized quaternion r representing this rotation is

$$(3.41) \qquad \mathbf{r} = \left(\cos \frac{\omega}{2}, \vec{l} \sin \frac{\omega}{2} \right)$$

Any 3-dimensional vector $\vec{v}$ can be treated as a special case of a quaternion with the zero scalar component

$$(3.42) \qquad \mathbf{v} = (0, \vec{v})$$

The rotation given by the quaternion r has a formal expression:

$$\mathbf{v}' = \tilde{\mathbf{r}} \diamond \mathbf{v} \diamond \mathbf{r}$$

in terms of the quaternion product introduced above. Provided the quaternion r is normalized one can easily see that $\tilde{\mathbf{r}} = \mathbf{r}^{-1}$:

$$\mathbf{v}' = \mathbf{r}^{-1} \diamond \mathbf{v} \diamond \mathbf{r}.$$

One can easily check that performing two sequential rotations described by quaternions $\mathbf{r}_1$ and $\mathbf{r}_2$ is equivalent to performing one rotation described by the product quaternion

$$(3.43) \qquad \mathbf{r} = \mathbf{r}_1 \diamond \mathbf{r}_2$$

Performing the algebra (quaternion multiplications according to the rules eq. (3.39)) one can find that the 3×3 rotation matrix $\mathcal{R}$ corresponding to the quaternion r can be written in the form [8, 27]:

$$(3.44) \qquad \mathcal{R} = \begin{pmatrix} r_0^2 + r_x^2 - r_y^2 - r_z^2 & 2(r_x r_y - r_0 r_z) & 2(r_x r_z + r_0 r_y) \\ 2(r_x r_y + r_0 r_z) & r_0^2 - r_x^2 + r_y^2 - r_z^2 & 2(r_y r_z - r_0 r_x) \\ 2(r_x r_z - r_0 r_y) & 2(r_y r_z + r_0 r_x) & r_0^2 - r_x^2 - r_y^2 + r_z^2 \end{pmatrix}$$

where r_0, r_x, r_y, and r_z are the components of a normalized quaternion r defining the rotation in question. This is the announced rational (in fact quadratic) parameterization of the elements of the $SO(3)$ group ($\mathcal{R} \in SO(3)$) by the components of the normalized quaternion. It is not unique as the quaternion $-\mathbf{r}$ obviously produces the same matrix $\mathcal{R}$ as $-\mathbf{r}$ defines the rotation around the axis with the opposite positive direction and by the angle $\omega \pm 2\pi$. By this the two quaternions define the same rotation $\mathcal{R}$ i.e. the same element of the $SO(3)$. The normalized quaternions obviously form a group $\mathbb{H}^{\diamond}$ as well (if the multiplication eq. (3.39) is considered as the group operation) and eq. (3.44) defines a homomorphism of $\mathbb{H}^{\diamond}$ on $SO(3)$ with $\mathbf{r}$ and $-\mathbf{r}$ defining the same rotation.

Our purpose is to construct an analogous parameterization for the $SO(4)$ group. In order to reach this goal we mention that there exists a similar homomorphism between the $SU(2)$ group of 2×2 unitary matrices with complex elements with the unit determinant and the $SO(3)$ group. The correspondence establishes as follows: for a rotation $\mathcal{R}$ in the three dimensional space one can choose a quaternion representation $\mathbf{r} = (r_0, \vec{r})$. This quaternion is normalized and it defines a 2×2 matrix:

$$(3.45) \qquad \begin{pmatrix} r_0 - ir_z & -ir_x - r_y \\ -ir_x + r_y & r_0 + ir_z \end{pmatrix}$$

It is easy to check that the rows and columns of this matrix are orthogonal and its determinant equals unity. The independent complex matrix elements in eq. (3.45) are known as Cayley-Klein parameters of the rotation group. Also, one can see that for quaternions connected by the relation $\mathbf{r} = \mathbf{r}_1 \diamond \mathbf{r}_2$ the corresponding 2×2 matrices are connected by the same relation with replacement of the quaternion multiplication by the usual matrix product. This establishes isomorphism between the $SU(2)$ group and the group of normalized quaternions $\mathbb{H}^{\diamond}$ which can be continued to the homomorphism on $SO(3)$.

Now we notice that the (para)rotations by the triples of angles $\vec{\omega}_{\pm}$ can be represented as (para)rotations by angles

$$(3.46) \qquad \omega_{\pm} = \sqrt{\sum_{\gamma} \omega_{\pm\gamma}^2}$$

around axes with the directing cosines $\dfrac{\omega_{\pm\gamma}}{\omega_{\pm}}$, respectively. The normalized quaternions $\mathbf{q}$ and $\mathbf{p}$ corresponding to these pararotations have the following components:

$$(3.47) \qquad
\begin{aligned}
q_0 &= \cos \frac{\omega_+}{2}, \quad q_x = \frac{\omega_{+x}}{\omega_+} \sin \frac{\omega_+}{2}, \quad q_y = \frac{\omega_{+y}}{\omega_+} \sin \frac{\omega_+}{2}, \quad q_z = \frac{\omega_{+z}}{\omega_+} \sin \frac{\omega_+}{2} \\
p_0 &= \cos \frac{\omega_-}{2}, \quad p_x = \frac{\omega_{-x}}{\omega_-} \sin \frac{\omega_-}{2}, \quad p_y = \frac{\omega_{-y}}{\omega_-} \sin \frac{\omega_-}{2}, \quad p_z = \frac{\omega_{-z}}{\omega_-} \sin \frac{\omega_-}{2}
\end{aligned}$$

By these quaternions $\mathbf{q} = \mathbf{q}(\vec{\omega}_+)$ and $\mathbf{p} = \mathbf{p}(\vec{\omega}_-)$ two $SU(2)$ matrices can be constructed. Each of these matrices acts in a two-dimensional space. Using a physical analogy they can be considered as respective configuration spaces for two particles with spin 1/2 each (like an electron) [28, 29]. The basis states in these two-dimensional spaces can be thought as states with the spin projections $\pm\frac{1}{2}$ on a fixed axis. For example the $SU(2)$ matrix defined by the quaternion $\mathbf{q} = \mathbf{q}(\vec{\omega}_{+})$ acts as:

$$(3.48) \qquad \begin{pmatrix} q_0 - iq_z & -iq_x - q_y \\ -iq_x + q_y & q_0 + iq_z \end{pmatrix} \begin{pmatrix} \left|+\tfrac{1}{2}\right\rangle \\ \left|-\tfrac{1}{2}\right\rangle \end{pmatrix}$$

Constructing an $SO(4)$ matrix in terms of two $SU(2)$ matrices parametrized by $\mathbf{q}$ and $\mathbf{p}$ is done as follows: each of the $SU(2)$ matrices corresponding to $\mathbf{q}$ and $\mathbf{p}$, respectively, acts in a separate space of states of two particles with $\frac{1}{2}$-spins [28, 29]. Since the $SO(4)$ group is a direct product of two $SO(3)$ (or of $SU(2)$ locally isomorphous to $SO(3)$) groups the matrix representing an element of $SO(4)$ is the direct (Kronecker) product of two $SU(2)$ matrices. The space in which it acts is a direct product of two spaces spanned by the basis states $\left\{\left|+\tfrac{1}{2}\right\rangle, \left|-\tfrac{1}{2}\right\rangle\right\}$ each. The configuration space for the pair of spins $\frac{1}{2}$ is spanned by four product functions:

$$(3.49) \qquad \left|+\frac{1}{2}; +\frac{1}{2}\right\rangle ; \left|+\frac{1}{2}; -\frac{1}{2}\right\rangle ; \left|-\frac{1}{2}; +\frac{1}{2}\right\rangle ; \left|-\frac{1}{2}; -\frac{1}{2}\right\rangle$$

The direct (Kronecker) product of the $SU(2)$ matrices representing the $\mathbf{q}$- and $\mathbf{p}$-pararotations acts in this space with the notion that the $\mathbf{q}$-dependent matrix eq. (3.48) acts on the states of the first particle and the $\mathbf{p}$-dependent one on the states of the second particle in the product state. Then we form linear combinations of the above states, which correspond to specific values of the total spin and desired spatial symmetry. The combination which corresponds to the zero total spin of two particles transforms as a scalar i.e. (singlet) s-function. Those which correspond to the total spin equal to unity form the basis in the three-dimensional (triplet) space of p-functions. The coordinate (x-, y-, and z-) functions are obtained as the following combinations of the states with the definite spin projections (the above product states):

$$
\begin{aligned}
|s\rangle &= \tfrac{1}{\sqrt{2}}\left(\left|+\tfrac{1}{2}; -\tfrac{1}{2}\right\rangle - \left|-\tfrac{1}{2}; +\tfrac{1}{2}\right\rangle\right) \\
|x\rangle &= \tfrac{i}{\sqrt{2}}\left(\left|+\tfrac{1}{2}; +\tfrac{1}{2}\right\rangle - \left|-\tfrac{1}{2}; -\tfrac{1}{2}\right\rangle\right) \\
|y\rangle &= \tfrac{1}{\sqrt{2}}\left(\left|+\tfrac{1}{2}; +\tfrac{1}{2}\right\rangle + \left|-\tfrac{1}{2}; -\tfrac{1}{2}\right\rangle\right) \\
|z\rangle &= -\tfrac{i}{\sqrt{2}}\left(\left|+\tfrac{1}{2}; -\tfrac{1}{2}\right\rangle + \left|-\tfrac{1}{2}; +\tfrac{1}{2}\right\rangle\right)
\end{aligned}
\qquad (3.50)
$$

Transforming the Kronecker product of the $SU(2)$ matrices defined by $\mathbf{q}$ and $\mathbf{p}$ to the basis of spatial s- and p-functions in eq. (3.50) yields the required $SO(4)$ matrix:

$$h = \begin{pmatrix} q_0p_0 + q_xp_x + q_yp_y + q_zp_z & q_0p_x - q_xp_0 - q_yp_z + q_zp_y \\ -q_0p_x + q_xp_0 - q_yp_z + q_zp_y & q_0p_0 + q_xp_x - q_yp_y - q_zp_z \\ -q_0p_y + q_xp_z + q_yp_0 - q_zp_x & q_0p_z + q_xp_y + q_yp_x + q_zp_0 \\ -q_0p_z - q_xp_y + q_yp_x + q_zp_0 & -q_0p_y + q_xp_z - q_yp_0 + q_zp_x \end{pmatrix}$$

$$\begin{pmatrix} q_0p_y + q_xp_z - q_yp_0 - q_zp_x & q_0p_z - q_xp_y + q_yp_x - q_zp_0 \\ -q_0p_z + q_xp_y + q_yp_x - q_zp_0 & q_0p_y + q_xp_z + q_yp_0 + q_zp_x \\ q_0p_0 - q_xp_x + q_yp_y - q_zp_z & -q_0p_x - q_xp_0 + q_yp_z + q_zp_y \\ q_0p_x + q_xp_0 + q_yp_z + q_zp_y & q_0p_0 - q_xp_x - q_yp_y + q_zp_z \end{pmatrix}$$

(3.51)

in terms of two normalized quaternions defined by eq. (3.47). This formula has also a technical advantage in that it is as required a rational function of its arguments – the quaternions' components. Its disadvantage is that the number of variables here equals eight although the number of independent angular variables is only six.[2]

The number of variables can be easily reduced in the vicinity of the unity matrix, which is of real interest to us, without the loss of the rational character of dependence on variables. The $SO(4)$ matrix H close to the unity matrix I has an expansion:

$$(3.52) \quad H = I + \delta^{(1)}H + \delta^{(2)}H$$

The unity matrix obviously corresponds to a pair of quaternions with:

$$(3.53) \quad q_0 = p_0 = 1; \vec{q} = \vec{p} = \vec{0}$$

The matrix close to the unity is characterized by quaternions with small vector parts:

$$(3.54) \quad |\vec{q}|, |\vec{p}| \ll 1$$

The normalization condition for the two quaternions involved allows us to write:

$$(3.55) \quad q_0 = \sqrt{1 - q_x^2 - q_y^2 - q_z^2} \approx 1 - \frac{1}{2}|\vec{q}|^2 = 1 - \frac{1}{2}q^2$$

$$p_0 = \sqrt{1 - p_x^2 - p_y^2 - p_z^2} \approx 1 - \frac{1}{2}|\vec{p}|^2 = 1 - \frac{1}{2}q^2$$

$$q^2 = \sum_\gamma q_\gamma^2, \ p^2 = \sum_\gamma p_\gamma^2, \ \gamma = x, y, z$$

By singling out the contributions eq. (3.53) up to the second order with respect to the (small) components of the vector parts of the quaternions **q** and **p**, we obtain the first order correction:

$$(3.56) \quad \delta^{(1)}H = \begin{pmatrix} 0 & p_x - q_x & p_y - q_y & p_z - q_z \\ q_x - p_x & 0 & -q_z - p_z & q_y + p_y \\ q_y - p_y & q_z + p_z & 0 & -q_x - p_x \\ q_z - p_z & -q_y - p_y & q_x + p_x & 0 \end{pmatrix}$$

---

[2]This beautiful formula is apparently known in the community, but the author failed to find an adequate reference.

which is an antisymmetric matrix and the second order correction:

$$\delta^{(2)}H = -\tfrac{1}{2}(q^2 + p^2)I +$$

$$(3.57) \quad +\begin{pmatrix} q_xp_x + q_yp_y + q_zp_z & -q_yp_z + q_zp_y & q_xp_z - q_zp_x & -q_xp_y + q_yp_x \\ -q_yp_z + q_zp_y & q_xp_x - q_yp_y - q_zp_z & q_xp_y + q_yp_x & q_xp_z + q_zp_x \\ q_xp_z - q_zp_x & q_xp_y + q_yp_x & -q_xp_x + q_yp_y - q_zp_z & q_yp_z + q_zp_y \\ -q_xp_y + q_yp_x & q_xp_z + q_zp_x & q_yp_z + q_zp_y & -q_xp_x - q_yp_y + q_zp_z \end{pmatrix}$$

which is a symmetric matrix. These expressions suffice for our purposes of constructing a mechanistic model of PES, which will be done below.

3.2.3. Quaternion form of the hybrid orbitals and hybridization tetrahedra

In the previous section we used quaternions to construct a convenient parameterization of the hybridization manifold, using the fact that it can be supplied by the $SO(4)$ group structure. However, the strictly local HOs allow for the quaternion representation for themselves. Indeed, the quaternion was previously characterized as an entity comprising a scalar and a 3-vector part: $\mathbf{h} = (h_0, \vec{h}) = (s, \vec{v})$. This notation reflects the symmetry properties of the quaternion under spatial rotation: its first component $h_0 = s$ does not change under spatial rotation i.e. is a scalar, whereas the vector part $\vec{h} = \vec{v} = (h_x, h_y, h_z)$ expectedly transforms as a 3-vector. These are precisely the features which can be easily found by the strictly local HOs: the coefficient of the s-orbital in the HO's expansion over AOs does not change under the spatial rotation of the molecule, whereas the coefficients at the p-functions transform as if they were the components of a 3-dimensional vector. Thus each of the HOs located at a heavy atom and assigned to the m-th bond can be presented as a quaternion:

$$(3.58) \quad \mathbf{h}_m = (s_m, \vec{v}_m), \quad s_m^2 + |\vec{v}_m|^2 = 1$$

The HOs and thus the quaternions representing them are subject to the normalization condition. This allows us to construct a visual picture of hybridization by using four vector parts $\vec{h}_m = \vec{v}_m$ of HO quaternions residing at a given atom. Directions of the vectors forming the latter coincide with those of the HOs themselves, the angles between the vectors coincide with the interhybrid angles, and the lengths of the vectors are square roots of the weights of the p-states. These vectors can be assumed to have a common origin coincident with the position of the atom (nucleus). This set of vectors forms a *hybridization tetrahedron*. The formal operation giving it is the projection of the set of HOs by:

$$(3.59) \quad I - |s\rangle\langle s|$$

which cuts out the scalar part (the s-orbital component) of each HO $(s_m, \vec{v}_m)$ in the quaternion representation. It can be easily recovered from the normalization condition.

Although they seem to be very flexible objects (the lengths of the vectors and intervector angles are likely to be variable) the hybridization tetrahedra are in fact subject to very strict conditions due to the orthonormality of the HOs at each given atom. Only a three-dimensional manifold of the possible forms of the hybridization tetrahedra spanned by the triple $\vec{\omega}_b^A$ of the Jacobi angles is available. This considerably reduces the freedom of choice of the shapes of the hybridization tetrahedra. As one can easily check, the standard sp^3-hybridization is naturally represented by a perfect tetrahedron with $|\vec{v}_m| = \sqrt{3}/2$; the sp^2-hybridization is represented by a trigonal pyramid with one of the vectors (aligned to its height) having a unit length representing the π-orbital, and three others representing the sp^2-hybrids lying in the plane with $|\vec{v}_m| = \sqrt{2/3}$; finally, the nonhybridized atom is represented by a tetrahedron formed by three perpendicular unit vectors, while the fourth one representing the pure s HO is a zero vector. All the intermediate forms of hybridization tetrahedra are covered by the triples $\vec{\omega}_b$ of Jacobi angles. By this a coordinate map in the space of possible hybridizations of the atoms with sp-valence shell is introduced.

Quaternion representation allows us to easily formulate various important facts about the geometry of the systems of HOs, which become the facts concerning the geometry of hybridization tetrahedra. The simplest is the definition of the interhybrid angles or, in other words, of the shapes of the hybridization tetrahedra. The interhybrid angles $\theta_{mm'}$ relate to the coefficients of the s-function in the corresponding HOs through the orthonormalization conditions:

$$(3.60) \qquad s_m s_{m'} + (\vec{v}_m, \vec{v}_{m'}) = \delta_{mm'}$$

by the following formula:

$$(3.61) \qquad \cos\theta_{mm'} = -\frac{s_m}{\sqrt{1 - s_m^2}}\frac{s_{m'}}{\sqrt{1 - s_{m'}^2}}$$

A very elegant statement concerning the properties of hybridization tetrahedra belongs to Kennedy and Schäffer [30]: in any hybridization tetrahedron two planes formed by any two pairs of HOs are orthogonal. It can be easily proven using the quaternion representation: for the scalar product of the vectors normal to two said planes, the following chain of equalities holds (numeration is obviously arbitrary):

$$(\vec{v}_1 \times \vec{v}_2, \vec{v}_3 \times \vec{v}_4) =$$

$$(3.62) \qquad = (\vec{v}_1, \vec{v}_3)(\vec{v}_2, \vec{v}_4) - (\vec{v}_1, \vec{v}_4)(\vec{v}_2, \vec{v}_3) =$$

$$= s_1 s_3 s_2 s_4 - s_1 s_4 s_2 s_3 = 0$$

General linear relations between the elements of the HOs residing on a heavy atom as taken in the quaternion form represent some interest. The orthonormality condition for the HOs written in the quaternion form allows us to establish the shape of the hybridization tetrahedra through eq. (3.61). On the other hand, the 4×4 matrix formed by HOs expansion coefficients is orthogonal not only with respect to rows, each representing one HO, but also with respect to columns, so that:

$$(3.63) \qquad \sum_m h_{m\alpha} h_{m\alpha'} = \delta_{\alpha\alpha'}$$

From this general relation one easily derives the linear dependence condition for the vectors $\vec{v}_m$ forming hybridization tetrahedra:

$$(3.64) \qquad \sum_m s_m \vec{v}_m = 0$$

Applying the orthogonality relation eq. (3.63) to the vector parts allows us to write:

$$(3.65) \qquad \sum_m \vec{v}_m \otimes \vec{v}_m = \mathcal{I}$$

The quaternion representation of the HOs is useful also for analysis of the symmetry properties of the energy components arising within the SLG-based semiempirical theories. Using the quaternion notation eq. (3.58) we get for the one-center molecular integrals:

$$(3.66) \qquad \begin{aligned} U_m^t &= s_m^2(U_s - U_p) + U_p \\ (t_m t_m \mid t_m t_m) &= C_1 + C_2 s_m^2 + C_3 s_m^4 \\ g_{t_k t_m'} &= C_4 + C_5[s_m^2 + s_k^2] + C_3 s_m^2 s_k^2 \end{aligned}$$

where the combinations C_n, $(n = 1 \div 5)$ of the Slater-Condon parameters [31] are defined by eq. (2.72). From the above one can easily conclude that the matrix elements entering eq. (2.88) are either invariant with respect to basis transformations (the interatomic Coulomb interaction γ_{AB}) or can be uniquely expressed through contributions of s-AO to the HOs (the one-center matrix elements). The only class of molecular integrals depending on the whole structure of the HOs (including directions) within the MINDO/3 type of parameterization of the Hamiltonian is that of the resonance integrals.

In the quaternion representation one can find that simple expressions for the first-order variation of the structure of the HOs are derived from small quasi- and pseudorotations $\delta\vec{\omega}_l$ and $\delta\vec{\omega}_b$ applied to the set of HOs at a given atom:

$$(3.67) \qquad \begin{aligned} \delta^{(1)}s &= -(\delta\vec{\omega}_b, \vec{v}) \\ \delta^{(1)}\vec{v} &= s\delta\vec{\omega}_b + \delta\vec{\omega}_l \times \vec{v} \end{aligned}$$

where $\times$ stands for the vector product of 3-vectors. Finally the formula eq. (3.51) for the system of HOs as expanded over AOs in terms of two quaternions $\mathbf{q}$ and $\mathbf{p}$ beautifully condenses to two quaternion multiplications [32]:

$$(3.68) \qquad \mathbf{h}' = \mathbf{q} \diamond \mathbf{h} \diamond \mathbf{p}^{-1}$$

which can be checked by directly performing the necessary algebra.

3.3. DEDUCTIVE MOLECULAR MECHANICS: FAMILY OF APPROXIMATIONS[3]

At this point we have two main prerequisites for constructing a mechanistic model of PES on the basis of the SLG-based semiempirical model of molecular electronic structure. We have performed an analysis of the ESVs used in it to describe

---

[3]Reprinted from A.L. Tchougréeff. *J. Mol. Struct.* (THEOCHEM), **630**, 243, 2003 with permission from Elsevier.

molecular electronic structure – those related to the geminal amplitudes/density matrices and those describing the shapes and orientations of the sets of HOs – the hybridization tetrahedra – in whose basis the densities are written and are shown to be transferable. These two ESV subsets are rather different in nature and this assumes a different treatment for these two groups in what follows. The geminal amplitudes/density matrices have been shown to be transferable within different orders of magnitude in small parameters. This allows for a variety of approximations where the corrections to the transferable values to the different orders of magnitude in different parameters are taken into account. As for the HO related ESVs, only very general relations have been established so far.

The constructions of different approximations will be done in the sections that follow on the basis of the variational principle for molecular electronic energy in the SLG-based approximation. We shall demonstrate that this treatment leads to a mechanistic model which can in a sense be considered a "generic" or "deductive" form of MM. It means that although the simple "balls-and-springs" model can hardly be justified from any general point of view, it does not mean that any other mechanistic model cannot be justified at all. And that is what we shall provide.

The related history can be dated back to the beginning of the last century. As it is indicated in [33] as early as in the year 1901 a certain company advertised a model set of wooden (i.e. rigid) tetrahedra which were designed to represent the forms (mutual spatial arrangement) of atoms in organic molecules. A classical (mechanistic) description, as opposed to the contemporary MM (if someone had pursued this direction) could then naturally arise in terms of parameters characterizing such tetrahedra, rather than ball sizes and spring elasticities known nowadays. Such a description in principle should not be worse than a conventional "balls-and-springs", MM at least for the reason that atoms with bonds are to the same extent similar to balls with springs (or sticks) as they are to the wooden tetrahedra. The QM model based on the SLG trial wave function allows us, as it will be shown, to substantiate the "tetrahedral" form of the MM. Indeed, a mathematical object referred to earlier as the "hybridization tetrahedron" and used so far to visualize the shape of the HO system residing at a given atom, can be used also for representing the energy functional in its terms. Of course, tetrahedral shapes have been previously used in the literature many times to *visualize* atoms. It is enough to mention the textbook [34]. However, as far as we know, nobody has tried to proceed further and to ascribe any definite energy significance to the form and relative orientation of these tetrahedra, though the qualitative considerations of Pauling, leading to the maximal hybrid strength or the maximal overlap [35] principles, are well known.

We are going to deduce a mechanistic model for molecular energy from the SLG-based QM method described in Section 2.4. We shall perform transformations and approximations, following the line mentioned above and arrive naturally at a "tetrahedral" representation of heavy atoms consistent with facts known from stereochemistry and usually interpreted in the VSEPR framework. The mechanistic model of PES will be derived in terms of these objects.

The derivation is based on the variational principle for energy and it naturally starts from writing it down. The analysis of the properties of the ESVs pertinent to the SLG approximation performed in Section 3.2.2 allows us to rewrite the energy eq. (2.88) using our current knowledge of the transferability of the density related ESVs as follows:

$$
\begin{aligned}
(3.69)\quad E = \sum_m & \left[\left(2U_m + \frac{\Delta\gamma_m}{2} - 2\beta^{R_m L_m}_{r_m l_m} \right) + \frac{1}{2} \sum_{k<m} \sum_{tt' \in \{r,l\}} \delta_{T_k T'_m} g^{T_k}_{t_k t'_m} \right] + & (a)\\[2mm]
& + \sum_{A<B} Z_A Z_B D_{AB} + & (b)\\[2mm]
& + \sum_m \sum_{t \in \{r,l\}} \left[t\langle \hat{\tau}_{zm}\rangle (U^t_m + \frac{1}{2}(t_m t_m | t_m t_m)^{T_m}) + \right. & (c)\\[2mm]
& \left. + \frac{1}{2} \sum_{k<m} \sum_{tt' \in \{r,l\}} \delta_{T_k T'_m} g^{T_k}_{t_k t'_m} (t\langle \hat{\tau}_{zk}\rangle + t'\langle \hat{\tau}_{zm}\rangle) \right] + & (d)\\[2mm]
& + \sum_m \left[\sum_{t \in \{r,l\}} \frac{1}{2}(t_m t_m | t_m t_m)^{T_m} - \gamma_{R_m L_m} \right] \delta\langle \hat{\tau}^2_{zm}\rangle + & (e)\\[2mm]
& + \sum_{A<B} Q_A Q_B \gamma_{AB} + \sum_m \gamma_{R_m L_m} \langle \hat{\tau}_{zm}\rangle^2 + & (f)\\[2mm]
& + \frac{1}{2} \sum_{k<m} \sum_{tt' \in \{r,l\}} \delta_{T_k T'_m} g^{T_k}_{t_k t'_m} tt' \langle \hat{\tau}_{zk}\rangle \langle \hat{\tau}_{zm}\rangle + & (g)\\[2mm]
& - \sum_m \beta^{R_m L_m}_{r_m l_m} \delta\langle \hat{\tau}_{+m} + \hat{\tau}_{-m}\rangle & (h)
\end{aligned}
$$

The above expression is based on the MINDO/3 parameterization. If the NDDO type of parameterization is used for the two-center Coulomb integrals the rows (e) and (f) have to be modified accordingly (see below). The representation eq. (3.69) allows the following family of approximate treatments for the energy. If the geminal *amplitude*-related ESVs (expectation values of the pseudospin operators or one- and two-electron density matrix elements) are *fixed* at their transferable values – the corresponding approximation is termed the FA i.e. the fixed amplitudes – the energy eq. (3.69) reduces to the lines (a) and (b), which yield an expression dependent on the molecular geometry and the hybridization ESVs only. The other lines in eq. (3.69) reappear if corrections to the amplitude related ESVs are taken into account. This corresponds to the *tuning* of the geminal *amplitudes* in response to the geometry variations or chemical substitution and this family of approximations is thus called the TA, i.e. the *tuned amplitudes* approximation. The simplest one in this type is that which retains only the terms linear in $\langle \hat{\tau}_{zm}\rangle \sim \mu_m$ (lines (c) and (d)) thus allowing the bond asymmetry (polar bonds and nonvanishing effective charges). Including the terms linear in ζ_m^{-1} provides the possibility to take into account the variations of two-electron density matrices in response to geometry and environment variations (line (e)), but including only the quadratic terms yields corrections to the bond orders (line (h)). Further in this work we mean by the TA approximation its simplest (μ_m-linear) version. Since the corrections to the bond orders appear only in the second order in small parameters μ_m and ζ_m^{-1} the FA and the simplest TA approximations may profit from the transferable values of the bond orders.

Further components of the derivation of the mechanistic model relate to the HOs. The latter enter the theory through the Hamiltonian matrix elements in the HO basis. As it has been shown, the HO related ESVs also fall into two subclasses: the angular variables defining the shapes of the systems of HOs – hybridization tetrahedra – centered at a given atom and those responsible for the orientation of this system (or of the tetrahedron representing it) in the space. This latter set of variables (the Jacobi angles $\vec{\omega}_l^A$) are very flexible – their actual values depend on the orientation of the molecule and thus can never be fixed. As for the HO related ESVs defining the shapes of the hybridization tetrahedra (the Jacobi angles $\vec{\omega}_b^A$) they can in principle be fixed at certain values and then the hybridization tetrahedra so defined can be considered as rigid bodies. This type of approach will be referred as one with the *fixed* hybrid *orbitals* (FO). This setting is used as a starting point for further analysis of the MM atomic types, which may be considered as atoms in different hybridization states characterized by the s-/p-weights ratios of their HOs, ultimately defined by the $\vec{\omega}_b^A$ triples – coordinates of the shapes of the hybridization tetrahedra. Alternatively one can think about *tuning* the shapes of the hybrid *orbitals* (TO) and thus of those of the hybridization tetrahedra. More precisely, we have to establish the relation between the geometry and composition variations and those of the hybridization tetrahedra deformations and rotations.

In what follows we shall consider a variety of approximate treatments of eq. (3.69) in more detail, by allowing the fixation or tuning of each of the specified classes of ESVs, each leading to a specific mechanistic description.

3.3.1. Fixed amplitudes fixed orbitals (FAFO) model

This is the simplest possible mechanistic model of the PES, derived from an approximate treatment of energy according to eq. (3.69). The FA type of treatment implies that the geminal amplitude-related ESV eqs. (2.78) and (2.81) are fixed at their invariant values eq. (3.7). This corresponds clearly to a simplified situation where all bonds are single ones. Within such a picture, the dependence of the energy on the interatomic distance reduces to that of the matrix elements of the underlying QM (MINDO/3 or NDDO) semiempirical Hamiltonian.

The FO type of treatment for the HOs implies that the weights of the s- and p-components of the HO system at each heavy atom are fixed as the interhybrid angles are fixed by the s-weights of HOs. This also means that the shapes of the hybridization tetrahedra remain fixed. Selection of the s-weights or equivalently of the $\vec{\omega}_b$ triples of angles can be done in a variety of ways, each potentially producing a specific implementation of the FAFO model. The simplest way is to fix them at the standard sp^n hybridizations with integer values of $n = 1 \div 3$. Alternatively one may produce a series of hybridization tetrahedra by fitting the experimental data. Other methods may also be invented. Below we analyze theoretical arguments in favor of that or another type of deformation of a prototypical sp^3 hybridization tetrahedron in different situations. In any case, the tetrahedral shapes once found are fixed in the FO setting and interact with each other (and with the "spheres" representing the hydrogen atoms). The number of bonding interactions each tetrahedron is allowed to take part in equals to four minus the number of lone pairs residing on it i.e. the number of bonds is determined by the usual valence rules.

Analysis of the general energy expression eq. (3.69) shows that for the MINDO/3 Hamiltonian the only HO orientation dependent contribution to the energy is the resonance energy of the two center bonds. In the NDDO approximation there are the orientation-dependent Coulomb contributions, but they are much less important and we consider them separately later.

The resonance interaction can be recast in the form of interaction between the hybridization tetrahedra, which in its turn depends on the distance between the centers of the tetrahedra, on their mutual orientation, and on their orientation with respect to the bond axis – that connecting the centers of the tetrahedra involved (the nuclei). The latter can be proven by the following construction: consider the m-th two-center bond and the 4×4 matrix of the resonance integrals between the AOs in the diatomic coordinate frame (DCF) which is defined by setting its z-axis to be directed along the $R_m L_m$ two center bond (the bond axis):

$$(3.70) \quad B^{R_m L_m} = \begin{pmatrix} \beta_{\sigma\sigma}^{R_m L_m} & 0 & 0 & \beta_{\sigma\zeta}^{R_m L_m} \\ 0 & \beta_{\pi\pi}^{R_m L_m} & 0 & 0 \\ 0 & 0 & \beta_{\pi\pi}^{R_m L_m} & 0 \\ \beta_{\zeta\sigma}^{R_m L_m} & 0 & 0 & \beta_{\zeta\zeta}^{R_m L_m} \end{pmatrix}$$

Elements of this matrix depend only on the $R_m L_m$-interatomic separation. The resonance integral $\beta_{r_m l_m}^{R_m L_m}$ for the m-th bond (geminal) can be written in a concise form:

$$(3.71) \quad \beta_{r_m l_m}^{R_m L_m} = \tilde{\mathbf{h}}_m^{R_m} B^{R_m L_m} \mathbf{h}_m^{L_m}$$

where the HOs centered on nonhydrogen atoms are taken in the DCF as well. To get rid of the relation with the DCF we notice that the only necessary components of the vector parts of the HO quaternions i.e. their ζ-components, have a representation independent of the coordinate frame according to:

$$(3.72) \quad v_{m\zeta}^{T_m} = (\vec{v}_m^{T_m}, \vec{e}_{R_m L_m})$$

Using the latter, the resonance integral can be rewritten in the form:

$$(3.73) \quad \begin{aligned} \beta_{r_m l_m}^{R_m L_m} &= \beta_{\sigma\sigma}^{R_m L_m} s_m^{R_m} s_m^{L_m} + \beta_{\sigma\zeta}^{R_m L_m} s_m^{R_m} (\vec{v}_m^{L_m}, \vec{e}_{R_m L_m}) + \\ &+ \beta_{\zeta\sigma}^{R_m L_m} (\vec{v}_m^{R_m}, \vec{e}_{R_m L_m}) s_m^{L_m} + \beta_{\pi\pi}^{R_m L_m} (\vec{v}_m^{R_m}, \vec{v}_m^{L_m}) + \\ &+ (\beta_{\zeta\zeta}^{R_m L_m} - \beta_{\pi\pi}^{R_m L_m})(\vec{v}_m^{R_m}, \vec{e}_{R_m L_m})(\vec{v}_m^{L_m}, \vec{e}_{R_m L_m}) \end{aligned}$$

which is already coordinate frame-independent. Another shorthand form of the above resonance integral is:

$$(3.74) \quad \beta_{r_m l_m}^{R_m L_m} = (s_m^{R_m}, \vec{v}_m^{R_m}) B^{R_m L_m} \begin{pmatrix} s_m^{L_m} \\ \vec{v}_m^{L_m} \end{pmatrix}$$

where usual matrix multiplications are assumed and the resonance integrals' $B^{R_m L_m}$ matrix is rewritten according to the quaternion representation of HOs, possessing scalar and vector parts:

$$(3.75) \quad B^{R_m L_m} = \begin{pmatrix} \beta_{\sigma\sigma}^{R_m L_m} & \beta_{\sigma\zeta}^{R_m L_m} \vec{e}_{R_m L_m} \\ \beta_{\zeta\sigma}^{R_m L_m} (\vec{e}_{R_m L_m})^\dagger & \beta_{\pi\pi}^{R_m L_m} \mathcal{I} + \\ & + (\beta_{\zeta\zeta}^{R_m L_m} - \beta_{\pi\pi}^{R_m L_m}) \\ & \vec{e}_{R_m L_m} \otimes \vec{e}_{R_m L_m} \end{pmatrix}$$

Here $\mathcal{I}$ stands for the 3×3 unit matrix acting (as does the 3×3 diadic product $\vec{e}_{R_m L_m} \otimes \vec{e}_{R_m L_m}$) on the vector parts of HOs in the quaternion form.

Multiplying the resonance integral by the quadrupled transferable spin bond order $P_{0m}^{rl} = \frac{1}{2}$ eq. (3.14) results in the resonance energy of the m-th bond which is the only nontrivial contribution to the molecular energy at this (FAFO) level of approximate treatment of the MINDO/3 Hamiltonian using the SLG trial wave function. Within this picture the hybridization tetrahedra interact and the interaction energy depends on separations between centers of the tetrahedra, their mutual orientation, with respect to the bond axis.

Going to the NDDO type of the Hamiltonian, parameterization brings additional energy terms dependent on the mutual orientation of hybridization tetrahedra. When the NDDO parameterization scheme for the Hamiltonian is used, the two-center Coulomb integrals become dependent on the magnetic quantum numbers of the AOs for which they are defined. This brings also the dependence of the two-center Coulomb integrals in the HOs basis on the shape and the orientation of the HOs involved. The strictly local character of one-electron basis functions forming carrier spaces for geminals allows us to introduce point multipoles describing charge distributions for atoms defined in terms of the electron densities located on the HOs. For convenience, we write them in the units $-e$. The atomic charge (monopole) is then defined according to eq. (2.79). Using this definition the HOs' populations P_m^{tt} can be written as:

$$(3.76) \quad P_m^{tt} = \frac{Q_A + Z_A}{8} + \delta' P_m^{tt} \text{ for } t_m \in A$$

In this expression (in variance with the transferable value of the one-electron density $\frac{1}{2}$) the average spin density $\dfrac{Q_A + Z_A}{8}$ takes into account the net electron transfer to the atom A. For this reason the deviations $\delta' P_m^{tt}$ at each given atom satisfy the condition:

$$(3.77) \quad \sum_m \delta' P_m^{tt} = 0$$

The details of electron distribution in atoms characterized by higher multipoles (dipoles and quadrupoles) are defined by the deviations $\delta' P_m^{tt}$. As it has been already mentioned the vector parts of the HOs centered at each given atom $\vec{v}_m^A$ transform as 3-vectors under the molecule/space rotations, and the hybrid densities $s_m^A \vec{v}_m^A$ transform as 3-vectors as well. On the other hand, the diadic products $\vec{v}_m^A \otimes \vec{v}_m^A$ under 3-dimensional rotations transform as a sum of a scalar and of the rank two tensor of the 3-dimensional space. The values of these momenta are obtained by averaging their standard definitions:

$$(3.78) \quad \vec{\mu} = e\vec{r}$$
$$\mathcal{D} = e(3\vec{r} \otimes \vec{r} - r^2 \mathcal{I})$$

over the electron density distribution around each given atom ($\vec{r}$ have to be understood as coordinates of electrons relative to the nucleus at hand). In the SLG approximation, the latter is described by the HOs and their populations. Using the representation of an HO in the quaternion form of eq. (3.58), we obtain the dipole moment of the charge distribution on atom A to be:

$$(3.79) \quad \vec{\mu}^A = 2d_1^A \sum_{t_m \in A} P_m^{tt} s_m^A \vec{v}_m^A$$

Inserting the HOs' populations expressed in terms of the effective charge of the atom and HO specific deviations, and taking into account the condition eq. (3.64) we obtain for the dipole moment:

$$(3.80) \quad \vec{\mu}^A = 2d_1^A \sum_{t_m \in A} \delta' P_m^{tt} s_m^A \vec{v}_m^A$$

Analogously the quadrupole moment on atom A is:

$$(3.81) \quad \mathcal{D}^A = 6(d_2^A)^2 \sum_{t_m \in A} P_m^{tt}(\vec{v}_m^A \otimes \vec{v}_m^A - (v_m^A)^2 \mathcal{I})$$

which reduces to:

$$(3.82) \quad \mathcal{D}^A = 6(d_2^A)^2 \sum_{t_m \in A} \delta' P_m^{tt}(\vec{v}_m^A \otimes \vec{v}_m^A - (v_m^A)^2 \mathcal{I})$$

due to the linear condition eq. (3.65). In the above expressions the characteristic lengths d_1^A and d_2^A define the radial extent of the corresponding quantities and are expressed through the orbital (Slater) exponents [36] (n stands for the principal quantum number of the orbitals under consideration):

$$(3.83) \quad
\begin{aligned}
d_1 &= \frac{2n+1}{\sqrt{3}} \frac{(4\zeta_{ns}\zeta_{np})^{n+1/2}}{(\zeta_{ns} + \zeta_{np})^{2n+2}} \\[2mm]
d_2 &= \sqrt{\frac{(2n+1)(2n+2)}{20}} \, \zeta_{np}^{-1}
\end{aligned}$$

As it has been shown long ago in [37, 38] the molecular two-electron integrals can be represented with high accuracy as energies of interactions of the corresponding multipole momenta. In the MNDO context, the potentials acting between the multipoles and representing the two-center two-electron integrals are not those known from electrostatics, but semiempirical interaction potentials selected to flow to the one-center values of the respective integrals when the interatomic separation vanishes. This does not seem to be very practical as noticeable deviations from the multipole based estimates appear at the unphysical internuclear separations which are significantly smaller than the characteristic bond lengths. For all physically meaningful internuclear separations, the integrals can be approximated by multipole expansions, thus yielding for the two-center contribution to the Coulomb energy:

$$\begin{aligned}
E_{\text{Coul}}^{AB} = {} & Q^A G^{00} Q^B + Q^A G^{01} \vec{\mu}^B - \vec{\mu}^A G^{10} Q^B - \\
& - \vec{\mu}^A G^{11} \vec{\mu}^B + Q^A G^{02} \mathcal{D}^B + \mathcal{D}^A G^{20} Q^B - \\
& - \vec{\mu}^A G^{12} \mathcal{D}^B + \mathcal{D}^A G^{21} \vec{\mu}^B + \mathcal{D}^A G^{22} \mathcal{D}^B
\end{aligned} \tag{3.84}$$

where

$$G_{\alpha\beta\ldots\mu}^{ll'} = \underbrace{\nabla_\alpha \nabla_\beta \ldots \nabla_\mu}_{l+l'} R_{AB}^{-1} \tag{3.85}$$

are the derivatives of the Coulomb potential with respect to the Cartesian components $\alpha, \beta, \ldots, \mu$ taken l times with respect to the components of the radius vector of atom A and l' times with respect to the components of the radius vector of atom B necessary to describe the interactions of respective multipoles (for details see [39–41]). This replaces the Coulomb interaction of effective atomic charges in row (f) in eq. (3.69) for the nonbonded atoms, A and B. For the (singly) bonded atoms, the two-electron two-center matrix element involving HOs centered at two centers of the bond contributes its product by the two-electron density matrix element rather than the product of two one-electron density matrix elements. So the corresponding contribution must be taken away from the multipole-multipole contribution eq. (3.84) and added bond-energy term.

3.3.1.1. Local equilibrium conditions for hybridization tetrahedra and quasitorques

In the FAFO picture, when the form of the HOs is fixed, the equilibrium condition for the hybridization tetrahedron can be written as the equilibrium condition for the orientation of the latter. Due to the angular character of the variables involved, the corresponding set of the energy derivatives with respect to the $\vec{\omega}_i^A$ components can be thought to be a (quasi)torque (here the prefix *quasi* as previously refers to the fact that no rotation of any physical body is involved in its definition rather that of a fictitious hybridization tetrahedron). As one can check, each (m-th) bond, incident to the given atom A, contributes to the quasitorque the following increment:

$$\begin{aligned}
\vec{K}_m^{R_m L_m} = {} & -4 P_{0m}^{rl} \left\{ \{ \vec{e}_{R_m L_m} \times \vec{v}_m^{R_m} \left[\beta_{\zeta\sigma}^{R_m L_m} s_m^{L_m} + \right. \right. \\
& \left. + (\beta_{\zeta\zeta}^{R_m L_m} - \beta_{\pi\pi}^{R_m L_m})(\vec{v}_m^{L_m}, \vec{e}_{R_m L_m}) \right] + \\
& \left. + \beta_{\pi\pi}^{R_m L_m} \vec{v}_m^{L_m} \times \vec{v}_m^{R_m} \right\}
\end{aligned} \tag{3.86}$$

Assuming to simplify the notations that for all the incident bonds the atom A is the right-end atom ($A = R_m$) we obtain the overall quasitorque acting upon the hybridization tetrahedron centered on the atom A and the corresponding energy minimum conditions with respect to orientations of all hybridization tetrahedra in the molecule:

$$(3.87) \quad \begin{aligned} \vec{K}_A &= \sum_m \vec{K}_m^{R_m L_m} \\ \forall A, \ \vec{K}_A &= 0 \end{aligned}$$

These equilibrium conditions are completely analogous to the equilibrium conditions for a system of rigid bodies [42] which requires evanescence of all (quasi)torques.

Though the equilibrium conditions eq. (3.87) require that a sum of the contributions eq. (3.86) vanishes, it is of interest to consider archetypal situations when some of these contributions vanish separately. These situations are twofold as two vector terms eq. (3.86) sum up to give a quasitorque contribution. The first one, proportional to $\vec{e}_{R_m L_m} \times \vec{v}_m^{R_m}$, vanishes if the HO on the right-end atom and the bond vector are collinear. If the same holds also for the left-end atom, one can see that the vector parts of both HOs ascribed to the bond under consideration are collinear so that the second vector term proportional to $\vec{v}_m^{L_m} \times \vec{v}_m^{R_m}$ also vanishes. This clearly corresponds to the equilibrium condition for two singly σ-bonded hybridization tetrahedra. A quasitorque appears if an HO ascribed to the bond under consideration is not collinear with the bond axis it is ascribed to and the quasitorque tends to align them.

An alternative equilibrium condition is possible only for a pair of HOs with vanishing s-contributions. For two pure p-orbitals residing on the right- and left-end atoms of the bond, the numerical coefficients at the first vector terms vanish if the HOs are perpendicular to the bond axis. In this case the second vector term vanishes if two vectors representing the pure p-orbitals are parallel. This clearly corresponds to the picture of a π-bond between the hybridization tetrahedra. A quasitorque then appears, tending to orient two hybridization tetrahedra in such a way that the two heights of the unit length of two hybridization tetrahedra are parallel.

The above considerations are valid for the MINDO/3 type of the Hamiltonian parameterization. In the NDDO setting the intrabond contribution to the quasitorque must be modified to take into account the dependence of the two-center Coulomb interaction integrals on the shape and orientation of the HOs on both ends of the bond. Moreover, additional contributions to the quasitorque acting upon the hybridization tetrahedron residing at a given atom come from all nonbonding Coulomb interactions eq. (3.84). Their form can be easily figured out from the multipolar representation of the nonbonded Coulomb interactions. Indeed, under the action of a 3-dimensional rotation described by (small) angles $\delta\vec{\omega}_i^A$ the dipole moment centered on atom A acquires a correction of the form:

$$(3.88) \quad \delta\vec{\mu}^A = \delta\vec{\omega}_i^A \times \vec{\mu}^A$$

and the quadrupole momentum acquires a correction of the form:

$$\delta \mathcal{D}^A = 6(d_2^A)^2 \sum_{t_m \in A} \delta' P_m^{tt}((\delta\vec{\omega}_l^A \times \vec{v}_m^A) \otimes \vec{v}_m^A + \vec{v}_m^A \otimes (\delta\vec{\omega}_l^A \times \vec{v}_m^A)) =$$

$$(3.89)$$

$$= \delta\Omega_l^A \mathcal{D}^A - \mathcal{D}^A \delta\Omega_l^A$$

where $\delta\Omega_l^A$ stands for the 3×3 matrix representing the vector multiplication by $\delta\vec{\omega}_l^A$ from the left.

Then the contributions to the quasitorques can be found. The simplest one is the contribution coming as an effect of the orientation of the dipole moment $\vec{\mu}^A$. As the energy contribution of all terms involving the dipole can be written as

$$(3.90) \quad -(\vec{\mu}^A, \vec{E}^A)$$

where $\vec{E}^A$ is the electric field vector on the atom A from all sources, inserting the dipole moment variation $\delta\vec{\mu}^A$ gives:

$$(3.91) \quad -(\delta\vec{\mu}^A, \vec{E}^A) = -(\delta\vec{\omega}_l^A \times \vec{\mu}^A, \vec{E}^A) = (\delta\vec{\omega}_l^A, \vec{\mu}^A \times \vec{E}^A)$$

and the contribution to the quasitorque is precisely $\vec{\mu}^A \times \vec{E}^A$. The situation with the quadrupole is only slightly more complex and can be treated according to [43].

3.3.1.2. Global equilibrium conditions for hybridization tetrahedra

In the previous subsection we formulated the equilibrium conditions for the hybridization tetrahedra which follow from the FAFO approximation for the molecular energy eq. (3.69). They do not seem to be practical for performing calculations as they require tedious recalculations on the scalar and vector parts of the HOs after a step along the energy gradient eq. (3.87) is performed. An alternative would be to use eq. (3.25) with fixed pseudorotation angles $\vec{\omega}_b^A$ or in a more geometric formulation with the fixed shapes of the hybridization tetrahedra. These latter are possibly identifiable with specific atomic "types", which produce the matrix $H(\vec{\omega}_b^A)$ with the columns corresponding to the system of HOs at a given atom. If these HOs are treated as quaternions, their vector parts $\vec{v}_m^{A(0)}$ form the hybridization tetrahedron at atom A. The actual orientations of these tetrahedra are defined by the interactions of each hybridization tetrahedron with its neighbors: either other tetrahedra or spheres, representing hydrogen atoms. For each atom the orientation of its system of HOs is given by a rotation matrix $R(\vec{\omega}_l^A)$ eq. (3.25) according to:

$$(3.92) \quad \begin{pmatrix} s_m^A \\ \vec{v}_m^A \end{pmatrix} = R^A \begin{pmatrix} s_m^A \\ \vec{v}_m^{A(0)} \end{pmatrix}$$

The 4×4 rotation matrices $R^A = R(\vec{\omega}_l^A)$ have the following structure:

$$(3.93) \quad R = \begin{pmatrix} 1 & 0 & 0 & 0 \\ 0 & & & \\ 0 & & \mathcal{R} & \\ 0 & & & \end{pmatrix}$$

ensuring the invariance of the scalar parts of the HOs. The vector parts of all HOs residing at a given atom transform according to:

$$(3.94) \quad \vec{v}_m^A = \mathcal{R}^A \vec{v}_m^{A(0)}$$

which gives the actual orientation of the hybridization tetrahedra in the molecule. The resonance energy thus becomes a quadratic function of the components of the normalized quaternions $\mathbf{r}^A$ used to parametrize matrices $\mathcal{R}^A$. The equilibrium orientation of these tetrahedra satisfies the energy minimum condition. Taking derivatives with respect to the components of $\mathbf{r}^A$ and including the normalization conditions $\|\mathbf{r}^A\| = 1$ by using the Lagrange multipliers ξ^A results in a set of 4-dimensional linear eigenvalue problems:

$$(3.95) \quad \Xi^A \mathbf{r}^A = \xi^A \mathbf{r}^A$$

which must be solved self-consistently for all nonhydrogen atoms, as matrices Ξ^A depend on orientation of the hybridization tetrahedra of the atoms bonded to atom A. The eigenvector $\mathbf{r}^A$ corresponding to the lowest eigenvalue ξ^A must be taken throughout the iteration process.

The matrix elements of Ξ^A can be evaluated using a fundamental fact concerning quaternions: the rotation of the vector part of an HO according to eqs. (3.92), (3.93), and (3.94) in the quaternion representation can be written as:

$$(3.96) \quad \mathbf{h}_m^A = \mathbf{r}^A \diamond \mathbf{h}_m^{A(0)} \diamond \tilde{\mathbf{r}}^A$$

Then, performing the necessary algebra, we arrive at a pair of 4×4 matrices:

$$(3.97) \quad C_m = \begin{pmatrix} 0 & \frac{1}{2}\vec{v}_m^{R_m}(0) \times \vec{e}_{R_m L_m} \\ \frac{1}{2}(\vec{v}_m^{R_m}(0) \times \vec{e}_{R_m L_m})^\dagger & \begin{aligned} &(\vec{v}_m^{R_m}(0), \vec{e}_{R_m L_m})\mathcal{I} + \\ &+ \frac{1}{2}(\vec{v}_m^{R_m}(0) \otimes \vec{e}_{R_m L_m} + \\ &+ \vec{e}_{R_m L_m} \otimes \vec{v}_m^{R_m}(0)) \end{aligned} \end{pmatrix}$$

$$(3.98) \quad D_m = \begin{pmatrix} 0 & \frac{1}{2}\vec{v}_m^{R_m}(0) \times \vec{v}_m^{L_m} \\ \frac{1}{2}(\vec{v}_m^{R_m}(0) \times \vec{v}_m^{L_m})^\dagger & \begin{aligned} &(\vec{v}_m^{R_m}(0), \vec{v}_m^{L_m})\mathcal{I} + \\ &+ \frac{1}{2}(\vec{v}_m^{R_m}(0) \otimes \vec{v}_m^{L_m} + \vec{v}_m^{L_m} \otimes \vec{v}_m^{R_m}(0)) \end{aligned} \end{pmatrix}$$

In their terms the sought symmetric matrix Ξ^A acquires the form:

$$(3.99) \quad \Xi^A = 4 \sum_m P_m^{rl} \left[\left(\beta_{\zeta\sigma}^{R_m L_m} s_m^{L_m} + \right. \right.$$

$$\left. + (\beta_{\zeta\zeta}^{R_m L_m} - \beta_{\pi\pi}^{R_m L_m})(\vec{v}_m^{L_m}, \vec{e}_{R_m L_m}) \right) C_m +$$

$$\left. + \beta_{\pi\pi}^{R_m L_m} D_m \right]$$

This comprises the global equilibrium condition for hybridization tetrahedra in the FAFO model. Their direct relation with the molecular shape which enters in an invariant manner through the bond vectors $\vec{e}_{R_m L_m}$ is remarkable. As previously, the above formulae apply only if the MINDO/3 type of parameterization is assumed for the SLG theory underlying the DMM description. If the NDDO parameterization is employed, the corrections due to electrostatic interactions between the multipoles residing on heavy atoms, whose orientations are dependent on those of the hybridization tetrahedra, must be included.

3.3.1.3. Librations of hybridization tetrahedra

In the previous subsections we considered the equilibrium conditions for the hybridization tetrahedra in molecules, which represent the orientation of the systems of HOs at each heavy atom with fixed weights of the s- and p-functions in each HO or equivalently with fixed shapes of hybridization tetrahedra. In order to have a description of the energy in the vicinity of the equilibrium, the second order corrections to it are necessary. The terms of interest are of two types. First, these are the terms of the second order with respect to variations of quasirotation angles $\delta\vec{\omega}_l^A$ at each given atom, which describe the energy variation when the hybridization tetrahedron of the atom at hand slightly rotates (librates) while all surrounding hybridization tetrahedra for heavy atoms and spheres representing hydrogens remain in their equilibrium positions. Second, there are terms of the overall second order bilinear with respect to $\delta\vec{\omega}_l^{R_m}$ and $\delta\vec{\omega}_l^{L_m}$. These terms describe the contribution to the energy which appears when hybridization tetrahedra residing on two bonded atoms librate simultaneously.

Due to the FAFO type of approximations used, only the resonance energy is affected by the librations of the hybridization tetrahedra. The terms of the first type may be obtained by inserting the second order correction $(\delta^{(2)}s_m^{R_m}, \delta^{(2)}\vec{v}_m^{R_m})$ for the right-end HOs into the expression for the resonance integral eqs. (3.73), (3.74). In the FO approximation $\delta^{(2)}s_m^{R_m}$ naturally vanishes. Inserting the second order corrections for the HOs eqs. (3.73), (3.74) results in the second order correction to the resonance integrals. The latter must be habitually multiplied by the quadrupled spin-bond orders for the corresponding bonds and summed up. This procedure has been performed in [44] and [45] for the sp^3 hybridized atom with four symmetric substituents. In this case, the energy correction is a diagonal quadratic form in $\delta\vec{\omega}_l^{R_m}$ with three degenerate eigenvalues:

$$
\begin{aligned}
&\delta^{(2)}_{\omega_l\omega_l} E = 4P_0^{rl}(\delta\vec{\omega}_l^{R_m} | \mathcal{G}_{ll}^{R_m R_m} | \delta\vec{\omega}_l^{R_m}) \\
&\mathcal{G}_{ll}^{R_m R_m} = \tfrac{4}{\sqrt{3}}(\beta_{\zeta\sigma}^{R_m L_m} s_m^{L_m} - \beta_{\zeta\zeta}^{R_m L_m}\sqrt{1 - (s_m^{L_m})^2})\mathcal{I}
\end{aligned}
\tag{3.100}
$$

The corrections of the second type can be easily obtained if one inserts the first order corrections $(\delta^{(1)}s_m^{R_m}, \delta^{(1)}\vec{v}_m^{R_m})$ and $(\delta^{(1)}s_m^{L_m}, \delta^{(1)}\vec{v}_m^{L_m})$ to eqs. (3.73), (3.74). As previously $\delta^{(1)}s_m^{R_m} = \delta^{(1)}s_m^{L_m} = 0$ due to the FO approximation. After some algebra we get:

$$
\delta^{(2)}_{\omega_l\omega_l} E = 4P_{0m}^{rl}\big(\delta\vec{\omega}_l^{R_m} | \mathcal{G}_{ll}^{R_m L_m} | \delta\vec{\omega}_l^{L_m}\big), \text{ where}
$$

$$
\begin{aligned}
\mathcal{G}_{ll}^{R_m L_m} = {}& \beta_{\pi\pi}^{R_m L_m}\big((\vec{v}_m^{R_m} \otimes \vec{v}_m^{L_m}) - (\vec{v}_m^{R_m}, \vec{v}_m^{L_m})\mathcal{I}\big) + \\
& + (\beta_{\zeta\zeta}^{R_m L_m} - \beta_{\pi\pi}^{R_m L_m})(\vec{v}_m^{R_m} \times \vec{e}_{R_m L_m}) \otimes (\vec{v}_m^{L_m} \times \vec{e}_{R_m L_m})
\end{aligned}
\tag{3.101}
$$

The terms of this type must be summed over all bonds between the heavy atoms. The formulae eqs. (3.100), (3.101) represent the potential energy of the molecular system as a quadratic function on small variations of the variables $\vec{\omega}_l^A$. This may be used either in a frame of a linear response analysis of reaction of the system of hybridization tetrahedra to various perturbations or (if the hybridization tetrahedra are supplied by fictitious inertia momenta) as potential energy of the system of the tetrahedra in a frame of a Car-Parinello-like [46] procedure.

The interaction of the neighbor hybridization tetrahedra is particularly simple if the tetrahedra involved correspond to the sp^3 hybridized atom with equivalent bonds. In this case the HOs in the equilibrium are collinear with the bond vectors so that the 3×3 matrix in eq. (3.101) becomes:

$$(3.102) \quad \mathcal{G}_{ll}^{R_m L_m} = \frac{3}{4}\beta_{\pi\pi}^{R_m L_m}\left(\left(\vec{e}_{R_m L_m} \otimes \vec{e}_{R_m L_m}\right) - \mathcal{I}\right)$$

Numerical estimates for the libration force constant can be easily done. For the methane molecule only the term eq. (3.100) appears. With the MINDO/3 parametrization at the equilibrium geometry of methane, it amounts to $17.19\,\mathrm{eV/rad}^2$. For neopentane under the same conditions the diagonal libration force constant is $21.15\,\mathrm{eV/rad}^2$, whereas the coefficient at the off-diagonal 3×3 matrix block responsible for coupling of librations of two neighbor carbon hybridization tetrahedra is only $2.02\,\mathrm{eV/rad}^2$.

The above formulae are modified in the NDDO parametrization where the librations of the hybridization tetrahedra of nonbonded atoms also couple through the electrostatic forces acting between corresponding multipoles.

3.3.2. Fixed amplitudes tuned orbitals (FATO) model

The deductive mechanistic model for molecular PES proposed in the previous subsection corresponds to the picture of the rigid ("wooden") hybridization tetrahedra. Within such a picture, whatever perturbations happen to a molecule may result only in variations of the orientation of the hybridization tetrahedra representing the systems of HOs residing at each heavy atom. Meanwhile, the proposed treatment of the hybridization manifold locally using its $SO(4)$ group structure may be used to construct another, somewhat wider (but also deductive) mechanistic representation of molecular energy where heavy atoms are depicted as flexible ("rubber") tetrahedra. Analysis of results of semiempirical calculations performed by the SLG-MINDO/3 method underlying our derivation, done in [44] and above in Section 3.2.1, shows that the hybridization related ESVs are much more sensitive to any perturbation affecting the molecule than the ESVs related to the geminal amplitudes. This puts into the agenda developing an approximation which allows first of all the adjustment of the shapes of the HOs (or equivalently of the hybridization tetrahedra) to various perturbations. The geminal related ESVs may still be considered to be fixed at their transferable values.

Specificity of any semiempirical parametrization is that in the FA approximation the one-center energies E_A eq. (2.88) related to the carbon atom remain hybridization-independent (see below and [44] and [45]). This result which ultimately comes from the fact that in carbon the valence shell is half filled, distinguishes carbon among other elements. For that reason (in the FA approximation) only the resonance contribution to the total energy depends both on orientation (as in the FAFO model) and on the form of the hybridization tetrahedra. This considerably simplifies the derivation in the case of carbon atoms. For that reason we consider it separately.

3.3.2.1. FATO molecular mechanics of sp^3 carbons

Pseudotorques and Local equilibrium conditions for sp^3 carbons. As mentioned previously, the only hybridization-dependent contribution to the total energy in the case of carbon atom in the FATO approximation is still the resonance energy. So the equilibrium conditions with respect to the shape and orientation of the hybridization tetrahedra representing the system of HOs residing at a carbon atom A reduce to a requirement of evanescence of the first derivatives of the resonance energy with respect to pseudo- and quasirotation angles $\vec{\omega}_b^A$ and $\vec{\omega}_l^A$ of eq. (3.25). Using the expansion for the resonance energy up to linear terms [45] in small pseudo- and quasirotations ($\delta\vec{\omega}_b^A$ and $\delta\vec{\omega}_l^A$) results in the equilibrium conditions:

$$(3.103) \quad \vec{N}^A = \nabla_{\vec{\omega}_b^A} E = \vec{0}; \; \vec{K}^A = \nabla_{\vec{\omega}_l^A} E = \vec{0}$$

for all atoms A, where the quasitorque $\vec{K}^A$ is defined by eq. (3.86), whereas the pseudotorque $\vec{N}^A$ is:

$$\vec{N}^A = -4 \sum_{m \in A} P_{0m}^{rl} \{ \beta_{\sigma\sigma}^{R_m L_m} \vec{v}_m^{R_m} s_m^{L_m} + \beta_{\sigma\zeta}^{R_m L_m} \vec{v}_m^{R_m} (\vec{v}_m^{L_m}, \vec{e}_{R_m L_m}) -$$

$$(3.104) \qquad - \beta_{\zeta\sigma}^{R_m L_m} s_m^{R_m} s_m^{L_m} \vec{e}_{R_m L_m} - \beta_{\pi\pi}^{R_m L_m} s_m^{R_m} \vec{v}_m^{L_m} -$$

$$- (\beta_{\zeta\zeta}^{R_m L_m} - \beta_{\pi\pi}^{R_m L_m}) s_m^{R_m} (\vec{v}_m^{L_m}, \vec{e}_{R_m L_m}) \vec{e}_{R_m L_m} \}$$

As previously, here, for the sake of simplicity of notation, we assume that for all bonds incident to atom A, this atom is a "right-end" atom of the bond. Though the equilibrium conditions are rather cumbersome, for symmetric cases they can be solved yielding obvious answers: the carbon atom in symmetric tetrahedral environment acquires the sp^3 hybridization with the HOs collinear to the bonds, etc.

The analog of the global equilibrium conditions eq. (3.95) for the FAFO model can be obtained in the FATO setting as well. To do so we notice that inserting the $SO(4)$ matrix parametrized by a pair of quaternions eq. (3.53) yields the resonance energy as a bilinear function in each of the normalized quaternions $\mathbf{q}^A$ and $\mathbf{p}^A$ describing together the shape and orientation of the hybridization tetrahedron on atom A. Combining this bilinear form with the normalization conditions for the quaternions:

242   *Andrei L. Tchougréeff*

$\|\mathbf{q}^A\| = \|\mathbf{p}^A\| = 1$ taken into account with use of the Lagrange multipliers θ^A and v^A results in a system of pairs of coupled linear equations:

$$(3.105) \quad \Theta^A \mathbf{q}^A = \theta^A \mathbf{p}^A; \, \Upsilon^A \mathbf{p}^A = v^A \mathbf{q}^A$$

which must be solved consistently for all A. We do not give the explicit form of matrices Θ^A and Υ^A both because they are too cumbersome and this treatment of the global equilibrium cannot be generalized to atoms other than carbon, as for other atoms one-center energies involve higher powers of the quaternion components.

Second order corrections to the energy of sp^3 carbon atom. In order to construct the required mechanistic picture, the estimate of the restoring force which opposes both the quasi- and pseudorotation (deformation) of the hybridization tetrahedra is necessary. That can be obtained by a linear response procedure. For the sp^3 carbon atom in the symmetric tetrahedral environment, the related resonance energy is a diagonal quadratic form with respect to small quasi- and pseudorotations together with triply degenerate eigenvalues [44, 45]:

$$\delta_{\omega\omega}^{(2)} E = 4 P_{0m}^{rl} \left((\delta\omega_b^{R_m} | \mathcal{G}_{bb}^{R_m R_m} | \delta\omega_b^{R_m}) + \right.$$
$$\left. + (\delta\omega_l^{R_m} | \mathcal{G}_{ll}^{R_m R_m} | \delta\omega_l^{R_m}) \right),$$

with

$$(3.106)$$

$$\mathcal{G}_{bb}^{R_m R_m} = 2 \left[\left(\beta_{\sigma\sigma}^{R_m L_m} + \frac{1}{\sqrt{3}} \beta_{\zeta\sigma}^{R_m L_m} \right) s_m^{L_m} - \right.$$
$$\left. - \left(\beta_{\sigma\zeta}^{R_m L_m} + \frac{1}{\sqrt{3}} \beta_{\zeta\zeta}^{R_m L_m} \right) \sqrt{1 - (s_m^{L_m})^2} \right] \mathcal{I}$$

and where $\mathcal{G}_{ll}^{R_m R_m}$ is given by eq. (3.100). This is used to obtain the response of the shape and orientation of the hybrids ($\delta\vec{\omega}_b$ and $\delta\vec{\omega}_l$) to various perturbations (see below). Analogous expressions can be obtained using eq. (3.53) for arbitrary hybridization.

Further terms are necessary to describe the interaction between the shape and orientation modes of two bonded tetrahedra, which appears when either of them is quasi- or pseudorotated in the vicinity of the equilibrium. These formulae can be obtained by considering those cross terms in the resonance integral expansion which are bilinear in $\delta\vec{\omega}^{R_m}$ and $\delta\vec{\omega}^{L_m}$, respectively. This result can be represented as a matrix element:

$$(3.107) \quad \delta_{\omega\omega}^{(2)} E = 4 P^{rl} \left(\delta\omega_b^{R_m}, \delta\omega_l^{R_m} \right) \begin{pmatrix} \mathcal{G}_{bb}^{R_m L_m} & \mathcal{G}_{bl}^{R_m L_m} \\ \mathcal{G}_{lb}^{R_m L_m} & \mathcal{G}_{ll}^{R_m L_m} \end{pmatrix} \begin{pmatrix} \delta\omega_b^{L_m} \\ \delta\omega_l^{L_m} \end{pmatrix}$$

of a 6×6 off-diagonal block where the 3×3 subblocks $\mathcal{G}_{bb}^{R_m L_m}, \mathcal{G}_{bl}^{R_m L_m}, \mathcal{G}_{lb}^{R_m L_m}$ are:

$$\mathcal{G}_{bb}^{R_m L_m} = \beta_{\sigma\sigma}^{R_m L_m} \vec{v}_m^{R_m} \otimes \vec{v}_m^{L_m} - \beta_{\sigma\zeta}^{R_m L_m} s_m^{R_m} \vec{v}_m^{R_m} \otimes \vec{e}_{R_m L_m} -$$

$$- \beta_{\zeta\sigma}^{R_m L_m} s_m^{R_m} \vec{e}_{R_m L_m} \otimes \vec{v}_m^{L_m} + \beta_{\pi\pi}^{R_m L_m} s_m^{R_m} s_m^{L_m} \mathcal{I} +$$

$$+ (\beta_{\zeta\zeta}^{R_m L_m} - \beta_{\pi\pi}^{R_m L_m}) s_m^{R_m} s_m^{L_m} \vec{e}_{R_m L_m} \otimes \vec{e}_{R_m L_m}$$

$$(3.108) \quad \mathcal{G}_{bl}^{R_m L_m} = -\beta_{\sigma\zeta}^{R_m L_m} \vec{v}_m^{R_m} \otimes (\vec{v}_m^{L_m} \times \vec{e}_{R_m L_m}) - \beta_{\pi\pi}^{R_m L_m} s_m^{R_m} \mathcal{V}_m^{L_m} +$$

$$+ (\beta_{\zeta\zeta}^{R_m L_m} - \beta_{\pi\pi}^{R_m L_m}) s_m^{R_m} \vec{e}_{R_m L_m} \otimes (\vec{v}_m^{L_m} \times \vec{e}_{R_m L_m})$$

$$\mathcal{G}_{lb}^{R_m L_m} = -\beta_{\zeta\sigma}^{R_m L_m} (\vec{v}_m^{R_m} \times \vec{e}_{R_m L_m}) \otimes \vec{v}_m^{L_m} + \beta_{\pi\pi}^{R_m L_m} s_m^{L_m} \mathcal{V}_m^{R_m} +$$

$$+ (\beta_{\zeta\zeta}^{R_m L_m} - \beta_{\pi\pi}^{R_m L_m}) s_m^{L_m} (\vec{v}_m^{R_m} \times \vec{e}_{R_m L_m}) \otimes \vec{e}_{R_m L_m}$$

Here $\mathcal{V}_m^{R_m}$ stands for the 3×3 matrix representing the vector multiplication by $\vec{v}_m^{R_m}$: $\mathcal{V}_m^{R_m} \vec{x} = \vec{v}_m^{R_m} \times \vec{x}$; and the $\mathcal{G}_{ll}^{R_m L_m}$ subblock is defined by eq. (3.101). These subblocks couple small pseudo- and quasirotations of the hybridization tetrahedra corresponding to the right- and left-end atoms of the bond (in the specified order) in a bilinear fashion. Their form particularly simplifies for the sp^3 carbon atom in a symmetric environment for which we have:

$$\mathcal{G}_{bb}^{R_m L_m} = \tfrac{1}{4} \beta_{\pi\pi}^{R_m L_m} \mathcal{I} - \tfrac{1}{4} \left[3\beta_{\sigma\sigma}^{R_m L_m} + \sqrt{3}\beta_{\sigma\zeta}^{R_m L_m} - \sqrt{3}\beta_{\zeta\sigma}^{R_m L_m} - \right.$$

$$(3.109) \qquad \left. - (\beta_{\zeta\zeta}^{R_m L_m} - \beta_{\pi\pi}^{R_m L_m}) \right] \vec{e}_{R_m L_m} \otimes \vec{e}_{R_m L_m}$$

$$\mathcal{G}_{bl}^{R_m L_m} = \mathcal{G}_{lb}^{R_m L_m} = \tfrac{1}{2} \beta_{\pi\pi}^{R_m L_m} \mathcal{E}_{R_m L_m}$$

where $\mathcal{E}_{R_m L_m}$ stands for the 3×3 matrix representing the vector multiplication by $\vec{e}_{R_m L_m}$: $\mathcal{E}_{R_m L_m} \vec{x} = \vec{e}_{R_m L_m} \times \vec{x}$.

For the diagonal restoring force constants related to the deformations of the hybridization tetrahedra the following estimates can be obtained: 30.58 eV/rad^2 in CH$_4$; 34.49 eV/rad^2 in neopentane. This estimate indirectly explains the observation extracted from the numerical experiments: the shapes of the hybridization tetrahedra are more stable than their orientations as the diagonal force constant for deformations is 1.5 times larger than that for rotations of the hybridization tetrahedra.

3.3.2.2. FATO molecular mechanics of nitrogen atom. Model "ammonia" molecule.

As it has already been shown, within the FA approximation, the form of the HOs on four-coordinated carbon atom is ultimately defined by the two-center resonance interactions. It is the $SO(4)$ group structure of the hybrid manifold that restricts the capacity of the HOs residing on the atom to adjust themselves to the arrangements of the surrounding atoms (groups). In this subsection, we apply the linear response method to estimate the shape of the hybridization tetrahedron and to analyze the stereochemistry and molecular mechanics of the triply bonded nitrogen atom. Even in the FA picture, the presence of a lone pair on the nitrogen atoms results in a significant hybridization dependence of the one-center energy contributions eq. (2.88) which cannot be considered a small perturbation. We consider the hybridization dependent parts of molecular energy in order to extract information on equilibrium shapes of the corresponding hybridization tetrahedra. For the sake of simplicity, we restrict ourselves with one in the ammonia molecule as this model problem retains all the characteristic features of the general case. In order to study the ammonia molecule we consider it as maintaining its C_{3v} symmetry with the 3rd order axis directed along the z-axis of the coordinate frame. The geometry is then characterized by the pyramidalization angle δ equal to zero for the planar structure. The overall resonance energy of three N-H bonds is then a function of only one of the three pseudorotation angles ω_{sz} and of the pyramidalization angle δ:

$$(3.110) \quad -\sqrt{3}P^{rl}\left[\beta_{\sigma\sigma}^{R_m L_m}\cos\omega_{sz} + \beta_{\zeta\sigma}^{R_m L_m}\sin\delta\sin\omega_{sz} + \sqrt{2}\beta_{\zeta\sigma}^{R_m L_m}\cos\delta\right]$$

It is easy to see that the minimum of the above expression with respect to both its arguments is reached precisely for the planar configuration and for the sp^2 hybridization ($\delta = 0$, $\omega_{sz} = 0$). This result comes from the (two-center) resonance energy only.

The hybridization dependent part of the one-center energy of the nitrogen atom is:

$$(3.111) \quad \left[(U_s - U_p) + \frac{1}{4}(3C_2 + 2C_3 + 4C_5)\right]\sin^2\omega_{sz} + \frac{1}{4}C_3\sin^4\omega_{sz}$$

with obvious extrema: a minimum at $\omega_{sz} = \frac{\pi}{2}$ (no hybridization) and a maximum at $\omega_{sz} = 0$ (sp^2 hybridization). Characteristic values of the atomic parameters [47] show that the contributions depending on Coulomb integrals can provide the total variation in energy less than 0.8 eV whereas the difference of the core attraction parameters U_s and U_p results in a huge amount of about 10 eV. Thus the nontrivial equilibrium in such a system is only possible if the strong deforming potential exerted by the contribution eq. (3.111) and tending to no hybridization is counterpoised by other contributions. Within the FA approximation, the only counterpoise is the resonance energy considered here. By this, we arrive at a very simple (but internally consistent) picture of hybridization/stereochemistry of the nitrogen atom. There exist two contributions to the energy. One (eq. (3.111)) tends to keep the valence angles at 90°, while another (eq. (3.110)) tends to place all substituents at the nitrogen atom on one plane with the latter. The observed pyramidal form is a result of the interplay between these two contributions. A pyramidalization (inversion) potential in which no kind of interbond interaction is involved (see below) comes from the same source.

In the present setting, the equilibrium shape of the nitrogen hybridization tetrahedron is given by the value of the pseudorotation angle ω_{sz} only. In the vicinity of the equilibrium it is reasonable to assume that the latter value arises as a result of an action of a deforming force exerted due to the resonance interaction with hydrogens on the otherwise nonhybridized nitrogen atom. The nonplanar form is maintained by the reaction of the one-center energy terms proportional to the second derivative of eq. (3.111) taken in the minimum, corresponding to nonhybridized atomic orbitals. The total pseudotorque exerted by three symmetrical bonds equals to the derivative of eq. (3.110) with respect to ω_{sz} at the point corresponding to the minimum of eq. (3.111). Finally the correction to the pseudorotation angle is:

$$(3.112) \quad \delta\omega_{sz} = \frac{\sqrt{3}P_0^{rl}\beta_{\sigma\sigma}^{R_m L_m}}{2[(U_s - U_p) + \frac{1}{4}(3C_2 + 2C_3 + 4C_5)]}$$

which defines to a first approximation the shape of the hybridization tetrahedron of the nitrogen atom. That raw estimate results in the numerical value of $\delta\omega_{sz}$ of only *ca.* 0.38 rad. The equilibrium value of this pseudorotation angle is *ca.* 0.95 rad. That large discrepancy is clearly due to the pseudotorque as the tetrahedron shape is determined by eq. (3.112) by the resonance interactions of the nitrogen's *s*-orbital only and for that reason does not depend on actual pyramidalization angle δ. The equilibrium pseudorotation angle in its turn is estimated at the equilibrium geometry, which appears as a result of taking into account the additional deforming force exerted upon the system of the nitrogen HOs.

Minimization of the sum of resonance eq. (3.110) and one-center eq. (3.111) energies on ω_{sz} allows us to determine the optimal value of this pseudorotation angle as a function of the angle δ. This minimization is possible analytically, but results in an equation of the fourth degree in $\tan\omega_{sz}$. So, we tried to find realistic estimates for the dependence $\omega_{sz}(\delta)$. First of all, we can determine the value of the pyramidalization angle δ optimal for a given value of pseudorotation angle ω_{sz}. It can be easily done by taking the derivative of the resonance energy with respect to δ since the lone pair contribution does not depend on δ explicitly. It should be noted that the relatively small specific correction to the core-core interaction adopted in semiempirical schemes is pyramidalization angle-dependent (the H-H nonbonding interactions depend on δ). Also, small polarization of the N-H bonds within the TA approximation framework modifies the energy expression. We neglect the effect of these contributions, which allows us to obtain a simple relation:

$$(3.113) \quad \sin\omega_{sz} = \sqrt{2}\tan\delta$$

which holds exactly for all geometry critical points – minima, maxima, and saddle points and can be used as an interpolation formula for the intermediate geometries. The relation eq. (3.113) also assumes that the HOs are directed along the bond vectors. Inserting the interpolation formula eq. (3.113) into molecular energy immediately results in a pyramidalization potential of rather nontrivial form:

$$2(U_s - U_p)\tan^2\delta + \left(\tfrac{3}{2}C_2 + C_3 + 2C_5\right)\tan^2\delta - C_3\tan^4\delta -$$

$$(3.114) \quad -4P^{rl}\sqrt{3}\left[\beta_{\sigma\sigma}^{R_m L_m}\sqrt{1 - 2\tan^2\delta} + \sqrt{2}\beta_{\zeta\sigma}^{R_m L_m}\sin\delta\tan\delta + \right.$$

$$\left. + \sqrt{2}\beta_{\zeta\sigma}^{R_m L_m}\cos\delta\right]$$

where the combinations C_n of the Slater-Condon parameters have been defined previously (eq. (2.72)). By this the existence of the pyramidalization potential is proven by sequential derivation from a QM expression for the energy eq. (3.69) rather then decided on a "school-wise basis" [3]. Formally, the source of this potential is a purely quantum mechanical requirement of mutual orthogonality of HOs centered on the nitrogen atom. Its physical nature may be characterized as the energy of excited configurations of the nitrogen atom admixed to its ground state by the perturbation induced by the resonance interaction with surrounding bonded atoms. The admixture coefficients (weights) of the excited atomic configurations appear as functions of hybridization parameters, which can be explicitly written according to [16]. Neither of these sources has anything to do with interpair Coulomb interactions.

3.3.2.3. *FATO molecular mechanics of oxygen atom. Model "water" molecule.*[4]

As mentioned previously, the qualitative difference between atomic types in the deductive MM scheme reduces to the number of lone pairs they bear. In the previous section we derived the existence of the pyramidalization potential at the sp^3-nitrogen atom analyzing the interplay between the deforming contribution (eq. (3.111)) produced by the lone pair and the hybridizing contribution (eq. (3.110)). In this section we address the model "water" molecule in a similar manner in order to get the deductive molecular mechanics of the oxygen atom. The properties of the oxygen atom with two covalent bonds in the SLG picture are determined by the interplay of two energy contributions similar to the case of nitrogen: (i) the one-center energy of the atom and (ii) the resonance energy of the two covalent bonds it forms. The hybridization/density dependent part of the one-center energy eq. (2.88) for such a model reads:

$$
\begin{aligned}
E' = {} & 2\sum_m (s_m^2 (U_s - U_p) + U_p)\delta P_m + \\
& + \sum_m (C_1 + C_2 s_m^2 + C_3 s_m^4)\delta\Gamma_m + \\
& + \sum_{k\neq m} (C_4 + C_5[s_m^2 + s_k^2] + C_3 s_m^2 s_k^2) \\
& \left(\delta P_k \delta P_m + \tfrac{1}{2}\delta P_k + \tfrac{1}{2}\delta P_m\right)
\end{aligned}
\tag{3.115}
$$

where we assume that the corrections δP_m and $\delta\Gamma_m$ are counted from their invariant values characteristic for covalently bonding geminals (eq. (3.12)). In the case of bonding geminals with $m = 3, 4$ $\delta P_m = \delta P$ and $\delta\Gamma_m = \delta\Gamma$ correspond to their characteristic values in oxygen compounds. These are controlled by the μ_0 parameters for the bonds formed by the oxygen atom and by their variations around this value which are, however, negligibly small for the purpose of our current analysis. For the lone pairs, $m = 1, 2$ and $\delta P_m = \dfrac{1}{2}$ and $\delta\Gamma_m = \dfrac{3}{4}$ (see above). The symmetry condition for the shapes of the HOs assigned to the covalent bonds $s_3 = s_4 = s$ also holds. The one center energy, however, depends on the overall weight of the atomic s-orbital in two lone pairs $(s_1^2 + s_2^2)$ rather than on the specific distribution of the s-character between them. This degeneracy leads to certain complications later. Adding the resonance energy of two covalent bonds we obtain the energy of a doubly bonded oxygen atom:

---

[4]Reprinted from A.L. Tchougréeff. *J. Mol. Struct.* (THEOCHEM), **632**, 91, 2003 with permission of Elsevier.

$$E = C_3(\tfrac{1}{2} - \delta P)^2[1 - 2s^2]^2 + [2[U_s - U_p](\tfrac{1}{2} - \delta P) +$$
$$+ 2C_3(\tfrac{1}{4} - \delta P^2) + 2C_5(\tfrac{3}{4} - \delta P - \delta P^2) +$$
$$+ C_2(\tfrac{3}{4} - \delta\Gamma)][1 - 2s^2] - 2C_3(\delta P + \delta P^2 - \delta\Gamma)s^4 +$$
$$- 4\left(2\beta_{\sigma\sigma}^{OH}s + \beta_{\zeta\sigma}^{OH}[(\vec{v}_3,\vec{e}_3) + (\vec{v}_4,\vec{e}_4)]\right)P_{OH}^{rl}$$

(3.116)

This expression contains all the deductive molecular mechanics of a doubly bonded oxygen (at least as it comes from the semiempirical SLG scheme). As in the case of ammonia, the minimum of the one-center energy corresponds to $s = 0$, which refers to no hybridization. The reason, as previously, is that the eq. (3.116) for the one-center energy is dominated by the $U_s - U_p$ difference, which amounts to -12 eV, whereas the contributions from intraatomic Coulomb terms (both of second and fourth order in s_m's) entering with different signs (as in the case of nitrogen) cover the overall variation of no more than 2 eV. Taking into account that the range of variation of the variable s is restricted by the inequality $1 - 2s^2 \geq 0$ due to the normalization conditions allows us to conclude that the one-center energy is most probably a function with a single minimum at $s = 0$.

Inserting eq. (3.67) into the expression for resonance energy we find that the minimum of the resonance part alone is reached when:

(3.117) $\beta_{\sigma\sigma}^{OH}(\delta\vec{\omega}_b, \vec{v}_3 + \vec{v}_4) + \beta_{\zeta\sigma}^{OH}s(\delta\vec{\omega}_b, \vec{e}_3 + \vec{e}_4) = 0$

so that the linear configuration with:

(3.118) $\vec{v}_3 = -\vec{v}_4; \vec{e}_3 = -\vec{e}_4$

is obviously a solution. Thus, as in the case of an "ammonia" molecule modeling the sp^3 hybridized nitrogen atom, the actual (bent) form of the model "water" molecule is a result of an interplay between the one- and two-center contributions to the energy. The remarkable difference from the nitrogen case is the degeneracy i.e. the fact that the energy depends only on the sum of the s-weights residing either in the bonding HOs or in the lone pairs, so that the weight $1 - 2s^2$ of the s-orbital falling to two lone pairs on oxygen is arbitrarily distributed between them. Thus the shape of the hybridization tetrahedron on the oxygen atom remains undefined.

The form of the bonding HOs may be, nevertheless, specified on the basis of the orthonormality relations for the HOs, which are consequences of the group $SO(4)$ structure of the hybridization manifold. According to eq. (3.61) the interhybrid angle for the bonding HOs is given by:

(3.119) $\cos\theta_{34} = -\dfrac{s^2}{1 - s^2}$

For symmetry reasons, the p_y-AO of the oxygen atom does not contribute to the lone pair of HOs and thus its weight is equally distributed between the bonding HOs:

(3.120) $v_{3y} = -v_{4y} = \dfrac{1}{\sqrt{2}}$

Since the overall p-weight residing in each bonding HO equals $1 - s^2$ the x-components of the bonding HOs become:

$$(3.121) \quad v_{3x} = v_{4x} = -\sqrt{\frac{1}{2} - s^2}$$

This comprises the description of the system of the HOs centered at the doubly bonded oxygen atom.

3.3.3. Tuned amplitudes fixed orbitals (TAFO) model

A further step in developing the different possible approximation procedures for treating the energy eq. (3.69) would be one which allows us to tune the ESVs related to the geminal amplitudes (TA) but keeps intact the shapes of hybridization tetrahedra (FO). This results in a TAFO approach which *a priori* seems an acceptable option for constructing a classical (mechanistic) scheme for the molecular PES. However, a simple analysis of the above expansions allows one to conclude that the sensitivity of two subsets of the ESVs characteristic for the underlying SLG-MINDO/3 QM method, opposes this approximation scheme. In fact, whatever perturbation affects the HO-related ESVs is much stronger than the geminal related ones. Thus, when the geminal amplitudes are expected to be affected by the environment, the HOs are affected much more. The opposite may happen only if some very rare special perturbation (such as completely symmetric deformation of the carbon tetrahedron) takes place. For that reason we do not discuss this approximation further.

3.3.4. Tuned amplitudes tuned orbitals (TATO) model

The mechanistic model of the PES closest to the underlying QM procedure (and eventually coincident with the latter [44]) is of course that where both classes of ESVs are adjusted to each other and to the geometry variations. It can be shown that the corrections to the invariant (transferable values) of the geminal related ESVs are small, though not negligible. As for the HO related ESVs they remain as much sensitive to whatever perturbation as in the FATO class of approximations.

In this section we consider the effects of small variations of the geminal related ESVs upon the shapes of the hybridization tetrahedra. We take into account only the corrections to the ESVs of the first order with respect to ζ^{-1} and μ. According to this assumption the bond orders remain invariant, which considerably simplifies the whole treatment. As the bond orders in the TATO model are kept at their transferable values, taking into account the geminals' tuning affects first of all the one-center contributions to the molecular energy. The one-center energy E_A eq. (2.88) can be rewritten:

$$E_A = E^{(0)} + E' = E_1 + E_2 + E_3, \text{ where } E_i = E_i^{(0)} + E_i', \text{ and}$$

$$E_1^{(0)} = \sum_{t_m \in A} U_m^t, E_1' = 2 \sum_{t_m \in A} U_m^t \delta P_m^{tt},$$

(3.122)

$$E_2^{(0)} = \frac{1}{4} \sum_{t_m \in A} (t_m t_m \mid t_m t_m)^{T_m}, E_2' = \sum_{t_m \in A} (t_m t_m \mid t_m t_m)^{T_m} \delta \Gamma_m^{tt},$$

$$E_3^{(0)} = \frac{1}{4} \sum_{k \neq m} \sum_{tt'} g_{t_k t_m'}^{T_k}, E_3' = \sum_{k \neq m} \sum_{tt'} g_{t_k t_m'}^{T_k} (\delta P_k^{tt} + \delta P_m^{t't'})$$

in a form where the transferable expressions marked with the (0) superscript are separated from the bond-specific contributions. We also dropped the second order term $\delta P_k^{tt} \delta P_m^{t't'}$ from E_3' which is the one-center contribution to the energy of Coulomb interaction of electrons residing in different geminals.

The one-center energy components have no clear correspondence in the standard MM setting. In our approach the one-center contributions E_i' arise due to deviations of the geminal amplitude related ESVs (δP_m^{tt} and $\delta \Gamma_m^{tt}$) from their transferable values. These deviations interfere with hybridization. The derivatives of E_i''s with respect to the angles $\vec{\omega}_b$ and $\vec{\omega}_l$, taken at the values characteristic for the stable hybridization tetrahedra shapes which appear in the FATO model, yield quasi- and pseudotorques acting upon the hybridization tetrahedron. In evaluating these quantities we notice that all the hybridization dependence which appears in the one-center terms is that of the matrix elements of eq. (2.71). In the latter, the only source of the hybridization dependence is that of the second and fourth powers of the coefficients of the s-orbital in the HOs. Since they do not depend on the orientation of the hybridization tetrahedra, we immediately arrive at the conclusion that no quasitorques caused by the variation of electron densities appear in the TATO setting:

(3.123) $\quad \vec{K}_i' = 0.$

For the pseudotorques the situation is different and we get:

$$\vec{N}_1' = -4(U_s - U_p) \sum_{t_m \in A} \delta P_m^{tt} s_m \vec{v}_m$$

(3.124)

$$\vec{N}_2' = -2 \sum_{t_m \in A} \delta \Gamma_m^{tt} (C_2 + 2C_3 s_m^2) s_m \vec{v}_m$$

$$\vec{N}_3' = -2 \sum_{k \neq m} \sum_{tt'} (C_5(s_m \vec{v}_m + s_k \vec{v}_k) +$$

$$+ C_3 s_m s_k (s_k \vec{v}_m + s_m \vec{v}_k)) (\delta P_k^{tt} + \delta P_m^{t't'})$$

These general expressions must be evaluated at characteristic points. To evaluate the effect of single substitution at the sp^3 carbon atom, the choice of a symmetric set of hybrids is an appropriate zero approximation. We have $\forall m \ s_m = \frac{1}{2}; \vec{v}_m = \frac{\sqrt{3}}{2}\vec{e}_{R_m L_m}$ which results in the following pseudotorque contributions:

$$\vec{N}_1' = -\sqrt{3}(U_s - U_p)\sum_{t_m \in A}\delta P_m^{tt}\vec{e}_{R_m L_m}$$

$$(3.125) \quad \vec{N}_2' = -\frac{\sqrt{3}}{2}\left(C_2 + \frac{1}{2}C_3\right)\sum_{t_m \in A}\delta\Gamma_m^{tt}\vec{e}_{R_m L_m}$$

$$\vec{N}_3' = -\frac{\sqrt{3}}{2}\left(C_5 + \frac{1}{4}C_3\right)\sum_{k \neq m}\sum_{tt'}(\vec{e}_{R_m L_m} + \vec{e}_{R_k L_k})(\delta P_k^{tt} + \delta P_m^{t't'})$$

In the linear response approximation, the above pseudotorques give the following pseudorotations of the hybridization tetrahedron on the atom under consideration:

$$(3.126) \quad \delta\vec{\omega}_{bi} = -\frac{\vec{N}_i'}{8P_{0m}^{rl}\mathcal{G}_{bb}^{R_m R_m}}; \delta\vec{\omega}_b = \sum_i \delta\vec{\omega}_{bi}$$

These quantities ultimately define what can be related to the atom types of the standard MM setting. Indeed, the atom types in the MM differ among other features by their preferable valence angles. In the deductive MM setting, the counterpart for the preferred valence angles are the interhybrid angles. If a small pseudorotation $\delta\vec{\omega}_b$ is applied to a hybridization tetrahedron the variation of the interhybrid angles is:

$$(3.127) \quad \delta\theta_{mm'} = -\frac{1}{\sqrt{1-s_m^2-s_{m'}^2}}\left(s_{m'}(\delta\vec{\omega}_b, \vec{v}_m)\sqrt{\frac{1-s_{m'}^2}{1-s_m^2}} + \right.$$
$$\left. + s_m(\delta\vec{\omega}_b, \vec{v}_{m'})\sqrt{\frac{1-s_m^2}{1-s_{m'}^2}}\right)$$

For the symmetric sp^3 tetrahedron the above expression simplifies to:

$$\delta\theta_{mm'} = -\sqrt{\frac{3}{8}}(\delta\vec{\omega}_b, \vec{e}_{R_m L_m} + \vec{e}_{R_{m'} L_{m'}})$$

For the atoms of the second row the pseudotorque $\vec{N}_1'$ coming from the nonuniform distribution of electronic density in the bonds dominates the whole picture due to the magnitude of $(U_s - U_p)$. Assuming that only one HO acquires a density correction δP_1^{rr} we get as a first approximation:

$$\delta\theta_{1m} = -\frac{1}{4\sqrt{2}}\frac{U_s - U_p}{\mathcal{G}_{bb}^{RR}}\delta P_1^{rr} \ \forall m \neq 1$$
$$(3.128)$$
$$\delta\theta_{mm'} = -\delta\theta_{1m} \ \forall mm' \neq 1$$

Since $U_s - U_p < 0$ the density increase $(\delta P_1^{rr} > 0)$ at the 1st HO results in an increase of the incident interhybrid angles and in equal decrease of the angles between otherwise nonperturbed HOs. All this, of course, is in agreement with the analysis performed in [48] with the difference that HOs here are not arbitrarily assumed to be collinear with the bonds. Numerical estimates are as follows. The calculation on the CH_3F molecule results in the value of δP_1^{rr} for the C-F bond geminal of -0.13 (the carbon atom is meant). Using the estimate of $\mathcal{G}_{bb}^{RR}$ performed for the methane molecule, we get $\delta\theta_{1m} \approx -4°$ which is in perfect agreement with the precise SLG-MINDO/3 calculation. With this, one can try to introduce a certain systematization in the currently chaotic picture of atomic types employed in MM. It is reasonable to assume that the atomic type is defined by the atom's environment and thus by the polarities of the incident bonds or equivalently by the parameters μ_{0m} of these bonds. According to eq. (3.19) they define the populations of the HOs δP_m^{rr} assigned to each of the bonds and the ideal valence angles for types so defined.

This result allows us to readdress the Nyholm-Gillespie [17–19] idea of considering the electron pair Coulomb repulsion in the valence shell as a reason for the observed stereochemistry. According to these authors, the interpair repulsion energies conform to the rule that the more populated the bond, the stronger it repels the bonds incident to the same atomic vertex, which within the limit, results in the rule that a lone pair repels other bonds and the corresponding valence angles are smaller than the ideal tetrahedral ones. We have already shown that this result appears without any relation to the Coulomb repulsion while analyzing the source of the pyramidalization potential of nitrogen. Here as well, we see that an infinitesimal increase of electron population at one of the HOs makes others increase the interhybrid angles with the more populated HO. Though this is in perfect agreement with the Nyholm-Gillespie rules, the real source of the effective interhybrid interaction has nothing to do with the Coulomb repulsion of electron pairs.

The above consideration is in agreement also with the well known qualitative Bent's rules [49] which state that the weight of the s-AO increases in the HO, which is involved in bonding with a more electropositive substituent. Indeed, electropositive substituents would lead to the positive values of δP_1^{rr} and after using eqs. (3.126), (3.67) the variation for the s-coefficient becomes:

$$(3.129) \quad \delta s_1 = -\frac{3}{8}\frac{U_s - U_p}{\mathcal{G}_{bb}^{RR}}\delta P_1^{rr}$$

which is positive for the second row atoms ($U_s - U_p < 0$). The latter formula shows that the Bent's rule validity depends crucially on the sign of the $U_s - U_p$ difference. If for any reasons the opposite sign of the above factor occurs or the effect of the $\vec{N}_1'$ pseudotorque is superseded by that of $\vec{N}_2'$ (it has an opposite sign, but according to our estimate is much smaller for the atoms of the second row, which is likely to change for heavier elements) the inversion of the Bent's rule takes place, and its modification proposed by Frenking [50] on the base of analysis of numerical experiments acquires a theoretical explanation.

As it has been mentioned several times, the derivations presented in this and earlier sections are largely based on the MINDO type of the parametrization of the underlying SLG-based semiempirical method. In the NDDO parametrization, the multipole-multipole electrostatic interactions between hybridization tetrahedra yield not only the dependence of the energy on the mutual orientations of the hybridization tetrahedra centered on nonbonded atoms, but also on the shapes of the tetrahedra which define the magnitudes of the multipoles. This produces the contributions to the pseudotorques. As in the case of quasitorques, the simplest is the one from the dipole, residing at a given atom A. As one can check, its variation under the deformation of the hybridization tetrahedron is given by:

$$(3.130) \quad \delta \vec{\mu}^A = 2d_1 \sum_{t_m \in A} \delta' P_m^{tt} \left[s_m^2 \mathcal{I} - \vec{v}_m^A \otimes \vec{v}_m^A \right] \delta \vec{\omega}_b^A$$

Inserting this into the energy, yields the following contribution to the pseudotorque acting on the hybridization tetrahedron of the atom A:

$$(3.131) \quad \vec{N}_{\text{Coul/dipole}}^A = -2d_1 \sum_{t_m \in A} \delta' P_m^{tt} \left[s_m^2 \mathcal{I} - \vec{v}_m^A \otimes \vec{v}_m^A \right] \vec{E}^A$$

where $\vec{E}^A$, as previously, is the electric field from all sources acting at the point where atom A is located. Further contributions (those from the variations of effective charges – atomic monopoles – and quadrupoles) also can be written straightforwardly.

3.3.5. Relation between DMM and standard MM

The content of the deductive molecular mechanics (DMM) as formulated in [51] and in the earlier text, is a description of the molecular energy in the form of eq. (3.69) as a function of shapes and mutual orientations of the hybridization tetrahedra and of geometry parameters. On the other hand, the standard MM briefly reviewed in Section 2.5 can qualify as a scheme directly parametrizing molecular energy as a function of molecular geometry (nuclear coordinates) only. At the same time, the MM theory implies the SLG type of the molecular electronic structure as an assembly of almost independent two-electron two-center bonds. From the traditional MM point of view the angular variables $\vec{\omega}_b$, $\vec{\omega}_l$ describing the shapes and orientations of hybridization tetrahedra are superfluous and must be excluded. This can be done by finding the response of the corresponding ESVs to the variations of bond lengths and valence angles using linear response relations between different subsets of variables pertinent to the DMM picture. To do so, let us consider a minimum of the energy with respect to both geometry and the ESVs. In the vicinity of an energy minimum x_0, q_0 it can be expanded up to second order with respect to nuclear displacements $q - q_0$ and variations of the ESVs $x - x_0$:

$$E = E_0 + \frac{1}{2}(x - x_0 | \nabla_x \nabla_x E | x - x_0) + (x - x_0 | \nabla_x \nabla_q E | q - q_0) +$$

$$(3.132) \qquad + (q - q_0 | \nabla_q \nabla_x E | x - x_0) + \frac{1}{2}(q - q_0 | \nabla_q \nabla_q E | q - q_0)$$

where linear terms disappear due to minimum conditions. For the sake of definiteness we restrict ourselves to the FA picture. Alternatively we can think that the density matrix elements in the TA family of approximations are calculated by the explicit formulae eqs. (3.16) and (3.17) or (3.14), which excludes the amplitude related ESVs from consideration. Then the only remaining ESVs are the $\vec{\omega}_b$, $\vec{\omega}_l$ angles describing the shapes and orientations of the hybridization tetrahedra sensitive to whatever variations of molecular composition and/or geometry. Minimizing the energy eq. (3.132) with respect to x for a given value of q leads to basic linear response relation between the ESVs and the geometry distortions:

$$(3.133) \quad x - x_0 = -(\nabla_x \nabla_x E)^{-1} \nabla_x \nabla_q E | q - q_0)$$

which formally represents the response of ESVs to the geometry variation $q - q_0$.

3.3.5.1. Linear response relations for hybridization ESVs

The main use of eq. (3.133) is for exclusion of the angular variables describing the hybridization tetrahedra from the DMM mechanistic picture and for going by this to a more standard classical MM-like description of the PES. However, before doing that, we have to estimate the precision of the linear response relations eq. (3.133) between geometry and hybridization variations themselves by numerical study. This has been done in [26] on the example of elongation of C-H bonds and deformations of valence angles in the methane molecule.

In the tetrahedral methane molecule (its parameters then correspond to subscript 0 in eqs. (3.132), (3.133)), we notice that the $\nabla_x \nabla_x E$ matrix further simplifies as $s_m^{L_m} = 1$ and, therefore, simple analytical expressions become possible. Also, we notice that the FA approximation is adequate here as, for example, even very large elongation of one C-H bond by 0.1 Å leads to changes of the bond geminal amplitudes u, v, and w not exceeding 0.003. The same applies to the expectation values of the pseudospin ($\hat{\tau}$) operators representing the one- and two-electron density matrix elements.

Linear response of hybridization to bond elongation. First the relation between hybridization and elongation of the C-H bond is considered. For this we need the mixed second order derivatives coupling the bond stretching with the hybridization ESVs. For every C-H bond in methane we can introduce diatomic coordinate frame with the z-axis directed along the bond and express the resonance integral related to this bond as:

$$(3.134) \quad \beta_{r_m l_m}^{CH_m} = \beta_{\sigma\sigma}^{CH} s_m + \beta_{\zeta\sigma}^{CH} v_{m\zeta}$$

where the subscript m enumerates the C-H bonds. Changing the bondlength causes the response of the vector $\delta\vec{\omega}_l$ to be exactly zero (vector product of collinear vectors) since the directions of the chemical bonds and the HOs coincide in the reference structure of methane. Thus the variation of the resonance integral under the elongation reads:

$$(3.135) \quad \beta_{r_m l_m}^{CH_m} = (\theta_{\sigma\sigma}^{CH} s_m - \theta_{\zeta\sigma}^{CH} v_{m\zeta})\delta r_m$$

where the derivative of $\beta_{\mu\nu}^{CH}$ with respect to the interatomic distance is $\theta_{\mu\nu}^{CH}$ and δr_m is the variation of the length of the m-th C-H bond. In the case of methane $s_m = \frac{1}{2}$ and $v_{m\zeta} = \frac{\sqrt{3}}{2}$. Thus the response of the shape of the hybridization tetrahedron represented by the vector $\delta\vec{\omega}_b$ is nonvanishing and can be written as [52]:

$$(3.136) \quad \delta\vec{\omega}_b = -\frac{\sqrt{3}}{4} \cdot \frac{\sqrt{3}\theta_{\sigma\sigma}^{CH} - \theta_{\zeta\sigma}^{CH}}{\sqrt{3}\beta_{\sigma\sigma}^{CH} + \beta_{\zeta\sigma}^{CH}} \vec{e}_m \delta r_m$$

where $\vec{e}_m$ is the unit vector directed along the m-th C-H bond. The numerator corresponds to the block of the $\nabla_x \nabla_q E$ matrix where q is the bondlength r_m and x is $\vec{\omega}_b$. The denominator in this expression is nothing but the eigenvalue of the $\nabla_x \nabla_x E$ matrix given by eq. (3.106) referring to variation of $\vec{\omega}_b$. Formula eq. (3.136) gives the analytical expression for the coupling between the bond elongation and variation of pseudorotation angles $\delta\vec{\omega}_b$ in methane. As one can see, the vector of angles' variation $\delta\vec{\omega}_b$ is collinear to the unit vector directed along this bond. Therefore the considered distortion produces the following form for the matrix of small transformation of the system of HOs (4×4 matrix H in eq. (3.27)):

$$(3.137) \quad \begin{pmatrix} 1 & -\delta & -\delta & -\delta \\ \delta & 1 & 0 & 0 \\ \delta & 0 & 1 & 0 \\ \delta & 0 & 0 & 1 \end{pmatrix}$$

The linear response estimate for δ is:

$$(3.138) \quad \delta = C_1 \cdot \frac{\delta r}{\sqrt{3}}$$

where the numerical value of the coefficient C_1 is $0.2764 \, \text{rad} \cdot \text{Å}^{-1}$. This form of the transformation matrix is perfectly reproduced numerically both in the FA and TA pictures. The linear response is a good estimate even for large distortions (deviation from linearity is only about 1% for the bond stretching of 0.05 Å). It also turns out that the error of the linear response approximation itself depends linearly on the variation of the bond length. This indirectly indicates that the second order estimate in principle suffice to perfectly describe the δ parameter obtained variationally.

Linear response of hybridization to valence angle deformation. The linear response relations between the molecular shape and the shape of hybridization tetrahedron are rather tricky due to the complex structure of the hybridization manifold. The molecular shape can be characterized by a group of unit vectors with a common origin at an atom under consideration pointing to the atoms bonded to the central one. In the case of methane, the deformations of the coordination polyhedron defined thus are small rotations of unit vectors $\vec{e}_{R_m L_m}$ pointing to hydrogen atoms. The valence angle bending can be described by introducing small rotation vectors $\delta\vec{\varphi}_m$, which after applying them to vectors $\vec{e}_{R_m L_m}$ lead to new (distorted) coordination tetrahedron:

$$(\vec{e}_{R_m L_m})' = \vec{e}_{R_m L_m} + \delta\vec{\varphi}_m \times \vec{e}_{R_m L_m} +$$

$$(3.139) \qquad + \tfrac{1}{2}(\delta\vec{\varphi}_m \otimes \delta\vec{\varphi}_m - \mathcal{I}\delta\vec{\varphi}_m^2)\vec{e}_{R_m L_m} =$$

$$= \vec{e}_{R_m L_m} + \delta\vec{\varphi}_m \times \vec{e}_{R_m L_m} + \tfrac{1}{2}(\delta\vec{\varphi}_m(\delta\vec{\varphi}_m, \vec{e}_{R_m L_m}) - \delta\vec{\varphi}_m^2 \vec{e}_{R_m L_m})$$

The small rotations $\delta\vec{\varphi}_m$ form an 8-dimensional space[5] which decomposes to a direct sum of two subspaces: one 3-dimensional, corresponding to rotations of the molecule as a whole, and another 5-dimensional, corresponding to independent variations of valence angles. The former is precisely mapped on the 3-dimensional space of quasirotations $\delta\vec{\omega}_l$ while the latter (5-dimensional) must be mapped on the 3-dimensional space of pseudorotations $\delta\vec{\omega}_b$ corresponding to changes of the shape of the hybridization tetrahedron [52]. General theorems of linear algebra [53] stipulate that under these conditions there exists a two-dimensional kernel in the space of deformations of molecular shape which maps to the vanishing deformation of hybridization tetrahedron. In [52] the term "hybridization incompatible" has been coined for the deformations from this kernel. The structure of deformations lying in the kernel is quite simple: they are produced by equal variations of opposite (spiro) valence angles. In contrast, the variations which correspond to the increase of one valence angle by $\delta\chi$ and decrease of its spiro counterpart by the same value, fall into "coimage" of this mapping i.e. to the subspace which one-to-one maps to the space of pseudorotations $\delta\vec{\omega}_b$. The deformations in the coimage can be called "hybridization compatible". It is clear that only these latter variations should be considered. In general, an arbitrary deformation $\{\delta\vec{\varphi}_m | m = 1 \div 4\}$ decomposes into a sum of the pure rotation, hybridization compatible and hybridization incompatible contributions so that both the orientation and the shape of the hybridization tetrahedron must be adjusted.

First we consider the geometry issues. As mentioned, only the hybridization compatible deformations of geometry affect the shape of the hybridization tetrahedron. On the other hand one can easily see that variation of the valence angle $\chi_{mm'}$ with $m < m'$ reduces to rotations of the involved bond vectors $\vec{e}_{R_m L_m}$ and $\vec{e}_{R_{m'} L_{m'}}$ around the axis orthogonal to the both coordination tetrahedron vectors:

$$(3.140) \quad \delta\vec{\varphi}_m = -\frac{\delta\chi_{mm'}}{2} \frac{\vec{e}_{R_m L_m} \times \vec{e}_{R_{m'} L_{m'}}}{|\vec{e}_{R_m L_m} \times \vec{e}_{R_{m'} L_{m'}}|}; \quad \delta\vec{\varphi}_{m'} = -\delta\vec{\varphi}_m$$

The reaction of hybridization tetrahedron on the changes of local geometry can be considered in the linear response approximation eq. (3.133). It is clear that any variation of the valence angle is a sum of equal amounts of hybridization- compatible and hybridization- incompatible deformations. The denominator in the linear response relation eq. (3.133) is the same as for eq. (3.136) while the relevant block of the $\nabla_x \nabla_q E$ matrix (with q taken as a difference of two opposite valence angles) is proportional to $\beta_{\zeta\sigma}^{CH}$. For the methane molecule with the carbon atom put in the origin of the coordinate frame, substitution of matrices of second derivatives for the energy

---

[5]Not 12-dimensional, as the components $\delta\vec{\varphi}_m \parallel \vec{e}_{R_m L_m}$ do not affect the result and can be safely set equal to zero.

into eq. (3.133) gives the reaction of the form of hybridization tetrahedron on the angular distortions of molecular geometry in the form:

$$(3.141) \quad \delta\vec{\omega}_b = -\frac{\beta_{\zeta\sigma}^{CH}}{\sqrt{2}(\sqrt{3}\beta_{\sigma\sigma}^{CH} + \beta_{\zeta\sigma}^{CH})}(\delta\chi_{12}\vec{k} + \delta\chi_{13}\vec{j} + \delta\chi_{14}\vec{i})$$

provided the parameters $\delta\chi_{1m}$ describe the hybridization compatible deformations of the coordination tetrahedron. Taking the "hybridization compatible" variation of two appropriate spiro valence angles, we obtain for example:

$$(3.142) \quad \delta\vec{\omega}_b = -\frac{\beta_{\zeta\sigma}^{CH}}{\sqrt{2}(\sqrt{3}\beta_{\sigma\sigma}^{CH} + \beta_{\zeta\sigma}^{CH})}\delta\chi\vec{k}$$

This is the coupling between the change of the pseudorotation vector and totally hybridization compatible deformation of valence angles. The considered distortion produces the following HO transformation matrix (matrix H in eq. (3.27)):

$$\begin{pmatrix} 1 & 0 & 0 & -\delta \\ 0 & 1 & 0 & 0 \\ 0 & 0 & 1 & 0 \\ \delta & 0 & 0 & 1 \end{pmatrix}$$

where the linear response estimate for δ is:

$$(3.143) \quad \delta = C_2 \cdot \delta\chi$$

The numerical value of the coefficient C_2 is -0.20734 for the equilibrium inter-atomic distance in methane. The above form of the HO transformation matrix is perfectly confirmed by our numerical experiments performed within the FA picture even for very large distortions, which is a consequence of the mathematical structure of the hybridization manifold described above. The numerical data show that the linear response estimate performs very well up to improbably large distortions (the deviation from the linear response estimate is smaller than 0.25% for the distortion of 0.3 rad (about 17°)).

The smallness of the coupling coefficient C_2, even for the hybridization compatible deformations, allows us to qualitatively understand some features of the electronic structure of strained organic molecules as it appears in the numerical experiments performed by the SLG-based methods. In cyclopropane, a very large distortion of the C-C-C valence angle from the tetrahedral one to the 60° one leads only to a relatively small distortion of the corresponding interhybrid angle. We model this process by strongly deforming the methane molecule. The simple estimate is based on eq. (3.141) and runs as follows. The valence angle variation, when going from methane to cyclopropane, is 49.5° (=109.5° − 60°); only one half of it is hybridization compatible; after multiplying by C_2 this yields the value of the interhybrid angle between the HOs corresponding to the "untouched" C-H bonds of 114.5°

(i.e. the angle variation amounts to only $5°$). From the energy minimum condition for hydrides it follows that the C-H bonds must follow the directions of the HOs. Numerical experiments performed using the SLG-MINDO/3 method show that if one of the H-C-H valence angles is fixed at the cyclopropane value of $60°$, then the energy minimum is reached when its spiro counterpart equals $115°$. This result can be directly compared with the experimental value of the H-C-H valence angle in the cyclopropane molecule which equals $115.1°$.

An analogous estimate can be applied to cyclobutane. In this case we consider the distorted methane molecule with one of the valence angles fixed at $90°$. In our model, the response of the HOs to the deformation is proportional to the deviation of the valence angle from the tetrahedral one. The deviation of the C-C-C angle from the tetrahedral one in cyclobutane ($19.5°$) amounts to 40% of that in cyclopropane. Therefore, we can expect that about the same ratio will be observed for the deviations of the H-C-H valence angle from the tetrahedral one in the cyclobutane and cyclopropane molecules. In fact, the ratio got in the SLG-MINDO/3 numerical experiment is about 39%.

3.3.5.2. Estimates of parameters of the standard MM force fields based on DMM

In the previous section we demonstrated numerically the validity of the linear response approximation for the hybridization tetrahedra. Now we can use these relations to perform the announced transition from the DMM model of molecular PES to a model dependent on molecular geometry. It is formally obtained by inserting eq. (3.133) into eq. (3.132) which yields:

$$(3.144) \quad \begin{aligned} E = {} & E_0 + \tfrac{1}{2}(q - q_0 | \nabla_q \nabla_q E | q - q_0) - \\ & - \tfrac{1}{2}(q - q_0 | \nabla_q \nabla_x E (\nabla_x \nabla_x E)^{-1} \nabla_x \nabla_q E | q - q_0) \end{aligned}$$

and the check of the validity of the linear response performed above is necessary to be sure that this substitution is not merely a formal trick. Formula eq. (3.144) apparently consists of two contributions: (i) the leading one – the second derivatives of the energy eq. (3.69) with respect to the geometry parameters q, and (ii) a correction appearing as a result of projecting out the ESVs related to the hybridization tetrahedra. Due to the SLG form of the wave function, eq. (3.69) is naturally represented as a sum of atom and bond increments (as in eq. (2.124)). Staying for the sake of simplicity within the FA picture, we can conclude that the only geometry-dependent contribution is that proportional to the resonance integrals of respective bonds. These contributions depend on the natural nuclear coordinates [55]: bond lengths and valence angles entering the definition of the standard MM force fields. This makes it sensible to consider the geometry dependence of individual bond energies. For a pair of singly bonded atoms, the corresponding energy is:

$$(3.145) \quad E_{R_m} + E_{L_m} + E_{R_m L_m}^{bond} + E_{R_m L_m}^{nonbond}$$

which for a symmetric bond may be rewritten as:

$$(3.146) \quad \begin{aligned} E = {} & 2(U_m - \gamma_{R_m L_m}) + \frac{g_m}{2}\left(1 - \frac{1}{\Gamma(\zeta_m)}\right) + \\ & + \frac{\gamma_{R_m L_m}}{2}\left(1 + \frac{1}{\Gamma(\zeta_m)}\right) - 2\beta_{r_m l_m}^{R_m L_m}\frac{\zeta_m}{\Gamma(\zeta_m)} + Z_{R_m}Z_{L_m}D_{R_m L_m} \end{aligned}$$

where U_m is the mean arithmetic value of the one-center electron core attraction parameters U_m^r and U_m^l.

The energy curve for the C-H bond corresponding to the sp^3 hybridization of the carbon atom and to the symmetric TA picture (eq. (3.12)) has the correct qualitative behavior for all interatomic separations. The minimum depth on this curve is approximately -0.23 a.u. and can be considered as the "pure" energy of the C-H bond. It is generally accepted in the literature that the energy of the C-H bond is approximately 0.15 a.u. The latter value is a thermodynamic one, while our value is obtained by extracting the contributions to the energy intrinsic to this bond and excluding the interaction between the bonds. The difference between the thermodynamic value for the bond energy and that obtained from the SLG energy in the FA picture can be explicitly written in a quite simple form:

$$(3.147) \quad 1/4[U_s(C) + 3U_p(C) + 3g_{tt'}^C + 6D_{HH} - E_A(C)]$$

where $E_A(C)$ is the energy of the non-hybridized carbon atom. Adding this value to the minimum of the energy curve for the C-H bond gives the value close to the thermodynamic one, as the heats of formation are rather well reproduced within the SLG-MINDO/3 method.

Comparing the form of the bond energy curve with the Morse potential is done by approximating the curve of eq. (3.146) by the Morse function $D_0[1 - \exp(-a(r - r_e)/r_e)]^2$ minimizing the area between the two curves in the interval from 0.72 Å to 2.50 Å. With the parameters D_0 and r_e fixed at the values equal to the minimum depth and position on the curve (0.2295 a.u. and 1.078 Å) the optimal value of parameter a is then 2.306, but with these parameters, the two curves are in fact quite different (the area between them is almost 11% of the area between the bond energy curve and the abscissa). If we optimize all three parameters of the Morse curve, they slightly modify: $D_0 = 0.2333$ a.u., $r_e = 1.045$ Å, and $a = 2.295$. This reduces the area between the curves by 30%. It should be concluded that the energy profile in the TA approximation is not particularly well reproduced by any Morse curve. On the other hand, the analytical expressions for the molecular integrals entering eq. (3.146) are not much more complex than the exponentials entering the Morse formula, so one can employ this expression as one for the bond-stretching energy force field.

To estimate the parameters of the harmonic force fields that may appear from the SLG-based semiempirical treatment, we consider the symmetric correlated single bond, where the energy can be obtained without any reference to its environment. In our case the derivative of the bond energy with respect to a geometry parameter q has the form:

$$(3.148) \quad \frac{\partial E_m}{\partial q} = Z_{R_m} Z_{L_m} \frac{\partial D_{R_m L_m}}{\partial q} - 2 \frac{\zeta_m}{\Gamma(\zeta_m)} \frac{\partial \beta^{R_m L_m}_{r_m l_m}}{\partial q} -$$

$$- \frac{1}{2}\left(1 - \frac{1}{\Gamma(\zeta_m)}\right) \frac{\partial \gamma_{R_m L_m}}{\partial q}$$

where the derivatives of different ESVs with respect to geometry variables exactly cancel each other so that the final expression for the energy derivative acquires the form expected from the perturbative analysis of the ESVs given above. This is quite an expectable result as the Hellmann-Feynmann theorem is valid for the functions of the SLG approximation and for this reason the derivative of the energy equals the expectation value of the Hamiltonian derivative with respect to the geometry parameter calculated over the density matrix elements. If q is the interatomic distance, setting the derivative equal to zero yields the equation for determining the minimum position. Results are given in Table 3.4. In the limit $\zeta_m \gg 1$ we recover the equilibrium geometry condition for the FA picture. The meaning of other notation in Table 3.4 is as follows. The TA_{symm} refers to a TA estimate for the symmetric bond (bond asymmetry/polarity terms omitted), while TA_{pert} refers to the perturbative inclusion of the bond asymmetry effects to the TA picture using the μ_0 parameter. All estimates appear quite reasonable. At the same time, the latter one looks more promising because it corresponds to the equilibrium interatomic separation in methane and other hydrocarbons.

The same concepts can be used to determine the elasticity constant for bond stretching by taking the second derivative of the energy with respect to the bond length. In the FA picture we get:

$$(3.149) \quad k_{R_m L_m} = \left(Z_{R_m} Z_{L_m} \frac{d^2 D_{R_m L_m}}{dr^2_{R_m L_m}} - 2 \frac{\partial^2 \beta^{R_m L_m}_{r_m l_m}}{\partial r^2_{R_m L_m}} - \right.$$

$$\left. - \frac{1}{2} \frac{d^2 \gamma_{R_m L_m}}{dr^2_{R_m L_m}} \right)_{r^0_{R_m L_m}}$$

We see from Table 3.4 that at least for one of the "experimental" estimates for the stiffness of the C-H bond [54] the agreement is quite acceptable.

The deviation from other cited values may be understood as the bond stretching parameters fitted by other authors in the context of the structure oriented MM schemes are implicitly loaded by the average effects of surrounding atoms and other force fields including nonbonding ones. This disagreement can be partially ascribed to the effects of the projected out Jacobi angles describing the variation of the shape of the hybridization tetrahedron. Indeed, the off-diagonal constant coupling the stretchings of two incident C-H bonds in the methane molecule can be written as:

$$(3.150) \quad K_{\text{off}} = \frac{1}{4\sqrt{3}} \frac{(\sqrt{3}\theta^{CH}_{\sigma\sigma} - \theta^{CH}_{\zeta\sigma})^2}{\sqrt{3}\beta^{CH}_{\sigma\sigma} + \beta^{CH}_{\zeta\sigma}}$$

The only reason why this term appears is the deformation of the carbon hybridization tetrahedron eq. (3.136) effectively coupling stretchings of two C-H bonds. Its estimated magnitude is only 0.120 mdyn/Å which is in agreement with its estimated 0.03 mdyn/Å [55] coming from fitting infrared spectra within the order of magnitude. This estimate establishes the scale of the corresponding effects. One can see that the DMM specific corrections to the bare estimates of the harmonic stretching constants eq. (3.150) are small. Nevertheless, in the cases when the bare harmonic constant vanishes (as the off-diagonal constant does) the corrections to the shape variation allow us to solve the question of the presence of the off-diagonal terms on purely theoretical basis.

Now we consider the valence angle bending force field as it appears from the DMM picture. For this end the geometry variation given by the vectors eq. (3.139) must be inserted in eq. (3.72) and the required elasticity constant can be obtained by extracting the second order contribution in vectors $\delta\vec{\varphi}_m$. In the case of hydride:

$$\delta^{(2)}_{\vec{\varphi}_m\vec{\varphi}_{m'}}E = -2\delta_{mm'}P_m^{rl}\beta_{\zeta\sigma}^{CH}\times$$

$$(3.151)$$

$$\times\left\{(\vec{e}_{R_mL_m},\delta\vec{\varphi}_m)(\delta\vec{\varphi}_m,\vec{v}_m^C) - \delta\vec{\varphi}_m^2(\vec{v}_m^C,\vec{e}_{R_mL_m})\right\}$$

This formula is quite remarkable as it shows that in the FO picture there are no contributions to the bending which can be attributed to any kind of interbond interaction. The bending force field is produced solely by energies of separate chemical bonds.

Typically, in the MM framework, the increment from the bending is considered a quadratic function of valence angles. The formula for bending eq. (3.151) can be rewritten in this form. This is obtained by substituting eq. (3.140) to the second order expansion eq. (3.151) and significant simplifications based on vector algebra. After that we see that the bending force field constant can be written as:

$$(3.152)\quad k_{HCH} = \beta_{\zeta\sigma}^{CH}\left\{P_m^{rl}(\vec{v}_m^C,\vec{e}_{R_mL_m}) + P_{m'}^{rl}(\vec{v}_{m'}^C,\vec{e}_{R_{m'}L_{m'}})\right\}$$

i.e., as a sum of two separate single bond contributions. Moreover, it can be proven [52] that adjusting the hybridization tetrahedron to the geometry change does not modify eq. (3.152) in the FO picture. Inserting the values characteristic for the sp^3-hybridization yields the bare estimate for the harmonic bending constant in the form:

$$k_{HCH} = \frac{\sqrt{3}}{2}\beta_{\zeta\sigma}^{CH}$$

From Table 3.4 one can see that the elasticity constants for bending force fields are in good agreement with the values accepted in the literature.

3.4. WHAT IS DMM?

In the previous sections we performed a sequence of moves intended to bridge the gap between an approximate QM description of molecular electronic structure and a classical representation of the PES of organic molecules suitable for further parametrization and simplifications in order to reach a scheme similar to molecular mechanics i.e. classical force fields. This construct can be qualified as deductive molecular mechanics (DMM) as each of its components has a transparent counterpart in the underlying

QM description and the approximations and simplifications used can be clearly characterized and formulated. From our point of view, this gives a possible explanation for the enormous success both of the MM in describing with considerable precision even tiny details of the molecular geometry of organic compounds and of the VSEPR in explaining and predicting characteristic features of molecular shapes. These two success stories made us consider them as experimental facts which require certain theoretical explanations. We felt that a demand for such an explanation is rather strong, as according to [58] "the situation is scandalous: ... the method [MM] used in thousands of laboratories throughout the world does not have any reliable quantum mechanical derivation". At the same time, the kind of explanation we were looking for fits the remark by Coulson [59], mentioned in the Preface: "... any explanation *why* must be given in terms of concepts which are regarded as adequate or suitable. So the explanation must not be that the electronic computer shows that $D(\mathrm{H} - \mathrm{F}) \gg D(\mathrm{H} - \mathrm{H})$, since this is not an explanation at all, but merely a confirmation of experiment". We can add to this that the result of any calculation is not a theoretical result at all: it is a result of a numerical experiment and the measure of consistency (or inconsistency) between results of different types of experiments (including those performed at the ab initio level) is a subject of separate theoretical consideration. On the other hand, a sequential derivation based on well defined grounds is much more useful for the verification or falsification of a pragmatic model than numerical experiments.

We start the derivation from the proof of transferability of key quantities entering the theory. Despite its long history, the very term "transferability" remains a somewhat vaguely defined synonym of "all the best" in parametrization schemes, referring largely to their capacity to be used without change for any molecule in a sufficiently wide class of similar ones. From the quantically point of view, this concept has received some attention in two related areas. First we mention the estimates of transferability given in [60] where those of the parameters of semiempirical quantum chemical methods have been related to the fact that the corresponding quantities remain the same for all molecules of similar structure up to the second order with respect to overlap integrals between AOs residing at neighbor atoms. That allows one to define the transferability for the quantum chemical parameters (ultimately, for the Hamiltonian matrix elements) as invariance of some quantity to a given order of precision with respect to a small parameter. Analogously, in [61] the problem of constructing transferable dynamic matrices in relation to the analysis of vibrational spectra has been considered. The stability of the dynamic matrix was analyzed with respect to a small parameter of relative mass variation under isotope substitution in a series of related molecules.

The importance of the transferability of the geminals has been pointed out in [62]. It was stated that the assumption of the transferability of the geminal amplitudes is a prerequisite for that of the bond energy. However, in [62] geminal transferability has not been proven and the authors concentrate on the statements equivalent to the transferability of the MM bond stretching force fields. Our proof of course strongly relies on the SLG form of the trial wave function. This may seem to be a very strong restriction on the proposed derivation scheme. However, it is not a restriction at all if correctly understood. In fact, following Ruedenberg we can state that chemical bonds are "observable" objects in chemistry, as their properties are reproducible, follow certain more or less simple laws, etc. Using the SLG form of the wave function simply provides an adequate formal expression for these known facts in terms of quantum mechanics (quantum chemistry).

Under the assumptions given by eqs. (3.6), (3.7) the expectation values of the pseudospin operators (and thus all the amplitude related ESVs) are invariant – transferable – quantities in the sense that they do not depend on the environment of the bond under consideration and even on the particular composition of the bond, i.e. on the nature and the hybridization of the atoms connected by the bond. This is in contrast with the HFR based QC methods where the transferability of bond properties appears as a result of tedious analysis of numerical data. It is important that the invariance (at the established level of precision) of the density matrix elements can be proven only for the basis of the variationally determined HOs – a specific of the SLG proposed approach [9–12]. In the basis of AOs the density matrix elements are not invariant even approximately. The approximation sufficient to obtain formally these invariant results breaks only at large interatomic separations which normally are not covered by any MM-like approximation. This result allows us to pose further questions: to what extent the density ESVs' invariance may stand further improvements of the description and whether it is possible to relate the invariance of the density matrix elements with the transferability of the MM force fields. To answer these questions we notice that the invariant values of ESVs can be improved by perturbative corrections (the TA picture) reflecting all the diversity of chemical compositions and environments the bond may occur in. Despite this, all the variety of perturbations is characterized by two small dimensionless parameters: ζ_m^{-1} eq. (3.13) and μ_m eq. (3.15). Both parameters depend on the atoms connected by the bond, their separation, and their hybridization. The perturbative treatment allows us to estimate the precision of transferability. For example, using eqs. (3.14), (3.16), and (3.17) we conclude that the bond order is the quantity transferable up to second order with respect to both ζ_m^{-1} and μ_m; the ionicity (the total weight of the ionic configurations) is transferable up to second order with respect to μ_m and up to first order with respect to ζ_m^{-1}; the bond polarity is transferable up to first order with respect to both ζ_m^{-1} and μ_m. The second order transferability of bond orders explains the success of the concept of "single bond", the fundamental concept shaping all chemistry and suitable for a large variety of chemical bonds. Note that the second order transferability takes place for bond orders also in the case when we employ the SLG bond wave function with the correct asymptotic behavior despite the fact that the transferable numerical value itself is obtained from the HFR wave function, which does not possess this property. Within this picture, all specific characteristics of the force field can be loaded only onto parameters of the (effective) Hamiltonian, which are either numbers specific for a given atom in certain hybridization state or for a pair of such hybrid states of atoms – ends of the bond. The force fields are basically sums of products of ESVs by matrix elements of molecular Hamiltonian, which are geometry dependent and composition specific. The force fields thus obtained are expected to be the same for the same composition of the bond and to depend on the environment only weakly (to the extent of the variance of the μ_{m1} parameters). These properties are basically much more than necessary for substantiating an MM-like description.

The local character of the orbitals used throughout the derivation is inherent for the suggested approach and the specific form of the orbitals of interest appears as a result of energy minimization procedure, which allows us to avoid a posteriori localizations complemented by poorly defined "tail cutting". The locality of the orbitals used in the SLG picture is in sharp contrast to the standard HFR treatment leading to delocalized orbitals.

Theoretical studies performed in this chapter can be used also for analysis of the problems of traditional MM methods. As described in Section 2.5, the number of parameters in MM force fields may present a considerable problem. Two aspects are important: setting of the atomic types and assigning the force field parameters indexed by pairs, triples, and quadruples of atomic types of the atoms involved. The atomic types in the standard MM setting serve to refine the classification of the point masses involved in the "balls-and-springs" picture as compared to the classification limited to nuclear charge. The first level of this refinement in the standard MM setting refers to the hybridization of atoms understood as standard hybridization types sp^n with integer values of n. Next the immediate neighbors of a given atom may be taken into account to produce further specialization of types. This route is eventually infinite. In the DMM context the hybridization is described using three angles assembled in the $\vec{\omega}_b$ vector. These variables span the whole manifold of possible hybridizations – shapes of the hybridization tetrahedra and eventually the sp^n-ones. The shapes, however, cannot be arbitrarily assigned; as we have seen in Section 3.3.4 the polarities of the bonds incident to the given atom modify the shape of the hybridization tetrahedron following Bent's rule: the more electronegative substituent requires an HO with smaller s-weight. The variation of the resonance integral for the given bond does not affect the bond order in a very wide interval. Nevertheless, the increase of the resonance integral affects the HO composition: the larger resonance integral, similar to the more electronegative atom, requires an HO with a smaller s-weight. Pragmatically one could begin the design of the MM types on the basis of the TATO-DMM treatment from selecting the $sp^n (n = 1 \div 3)$ hybridized atoms as basic types and dividing them into subtypes according to the values of the μ_{0m} parameters of the incident bonds. This method eventually allows one to adequately define the atomic types.

Another application of the DMM theory developed here may be for reducing the number of valence angle bending parameters. The ideal values for the valence angles are naturally identified with the interhybrid angles which are assigned as *one-center* quantities characteristic for a given subtype. The elasticity constants for the bending force fields are, as shown already, assigned, not on the atomic type triples basis, but employing a kind of "combination" of rules yielding the bending elasticity constant as a sum of two contributions indexed by the types of the involved bonds i.e. by atomic type pairs, rather than triples. This allows us to cope with the amount of bending parameters of the order of several 10^4 required by the current force fields.

It remains to find out whether a similar treatment can be developed for the torsion force fields habitually indexed by quadruples of atomic types and thus posing most problems in the prescription of parametrization. One can also think about following problem setting, namely of designing a set of MINDO/3 or NDDO parameters selected for using with the formulae of either of the approximations of the DMM family. In this case, the entire parameter set can be indexed by the only atomic types as no parameters indexed by pairs and even more by triples or quadruples of atomic types are previewed in a semiempirical setting.

The obtained mechanistic picture of molecular potential takes an intermediate position between QM methods and standard MM schemes. Though it can be used as a standalone mechanistic model of molecular PES, the standard MM picture can be derived from it by eliminating the auxiliary (from this point of view) angular variables describing quasi- and pseudorotation of the hybridization tetrahedra. We have provided the exclusion of the angular variables characterizing the shapes and orientations of the hybridization tetrahedra from the mechanistic DMM model of molecular PES. This results in a model announced in Section 3.1, which is similar to the standard MM models but is obtained by the sequential derivation from the QM (SLG) model of molecular electronic structure. As mentioned already the transferability of the ESVs characterizing chemical bonds in molecules and linear response relations for hybridization of ESVs are the main components of deriving the MM theory of molecular PESs from the corresponding QM theory. Both these features have been mathematically derived and numerically checked in Sections 3.2.1 and 3.3.5.1

One of the motives of our analysis was the obvious success of the VSEPR model of stereochemistry [17–19] in systematizing an enormous amount of experimental material. That theory ascribes great significance to Coulomb bond-bond interactions to explain the observed molecular shapes. It is be noticed that in the setting in the DMM, molecular shapes are presented by the unit vectors $\vec{e}_{R_m L_m}$ describing the directions of the bonds. They follow the shapes of the hybridization tetrahedra, but there is surely some misfit due to other contributions to the energy. In the TATO model, that would be interactions between the effective atomic charges. However, even in the TATO model where one could expect a nontrivial effect of electron-electron interactions upon the shape of the hybridization tetrahedra, only the topology of the hybridization manifold assures the latter in carbon atoms. The situation with other organogenic atoms significantly differs from that for carbon. In the case of nitrogen and oxygen atoms even in the FA approximation, the one-center energy is strongly hybridization-dependent, due to the one-electron terms describing the core attraction of electrons in the lone pair and sensitive to the relative weights of the s- and p-AOs in the corresponding HO. The source of this is of course the strong difference between the core attraction parameters in the s- and p-subshells (U_s and U_p) with large preference towards purely s-lone pair for atoms, which has as its source both the Coulomb and kinetic energy of electrons in the atom. In free atoms this immediately results in no hybridization at all for nitrogen and oxygen and in $90°$ valence angles predicted by older theories [16] for water and ammonia with a subsequent need to explain the observed shape of these molecules with the valence angles only slightly smaller than

Table 3.1. DMM based estimates of the MM force field parameters as compared to those accepted in some standard MM parameterizations.

	r_0^{CH} Å	k_{CH} mdyn/Å	k_{HCH} mdyn/deg
FA: 1.069		8.30	0.509
TA$_\text{symm}$: 1.078		7.77	
TA$_\text{pert}$: 1.096		7.17	
	Standard MM:		
[1]: 1.113	[1]: $4.5 \div 4.7$	[56]: 0.549	
[56]: 1.105	[55]: 5.31	[57]: 0.508	
[57]: 1.090	[54]: 7.90	[55]: 0.493	

the tetrahedral ones and both exceeding $100°$. Curiously enough, the authors of the VSEPR model seem to overlook this result, well known for decades, and do not consider it as a starting point and incidentally the limiting case of the electron pair repulsion and started their theory from scratch. If we reside in the FA domain we have to admit that the only source of the observed stereochemistry can be found in the interplay between one-center hybridization dependent terms and resonance energy. This was clear yet to Coulson [16], but seems to acquire a formal proof only in the proposed context.

3.5. TATO-DMM AND INTERSUBSYSTEM FRONTIER

After developing a general theory which describes the transition from a quantum mechanical description of molecular electronic structure and PES based upon the SLG trial wave function to the classical (mechanistic) one built in terms of the hybridization tetrahedra and their interactions, deriving the form of the junction between the subsystems appears to be almost a trivial exercise. As it has been mentioned many times the general setting for dividing a molecular system into parts is to introduce first a basis of orbitals such that these later could be without big doubt ascribed to that or another quantum (R-) or classical (M-) subsystems. Then the frontier between the subsystems, by definition, is formed by the atoms which bear the orbitals which belong to different subsystems. The result of the QM treatment of the quantum system then depends on the form of the orbitals centered on the frontier atoms. It must be noticed that the leading contribution to the interaction energy between the classically and quantally treated parts of the entire molecular system is given by the Coulomb forces between effective charges residing in the subsystems and between the electron densities located on the same frontier atom, but ascribed to different subsystems. These contributions are important in terms of the total energy. However, they turn out to be too symmetric in the sense that they do not strongly depend on or influence the precise form of the orbitals centered on the frontier atoms and these forms (s-/p-weights ratio) are not very sensitive to the charges residing in the subsystems. The shapes of the orbitals on the frontier atoms are by contrast sensitive to the local resonance contributions to the interaction and through it may affect the local interactions in the system responsible for its chemical behavior. Two different effects can be expected from the separation of the system into parts and from treating one of them as being at the MM level in this setting: (i) renormalization of the QM Hamiltonian parameters and (ii) imposing additional forces and torques on the MM subsystem. We exemplify the use of the DMM technique by considering short-range contribution to intersubsystem junction construction. We start from one special but quite characteristic case of a boundary sp^3 carbon atom with one HO pointing to the QM region (we formally assign this HO to the bond with $m = 1$) and others related to the MM one (this case can be classified as the MM boundary atom). The transition to the DMM picture is performed by setting the FA and TO approximations for the HOs centered on this atom. It is to be recalled that the Coulomb interaction between electrons occupying orbitals ascribed to different subsystems reduce to interactions between one-electron densities due to the GF form of the total wave function. The effect of the QM part on the MM part appears due to changes of electron densities in the QM region. These affect both the one-center energy of frontier atom and the resonance energy between the HO $|r_1\rangle$ and all other orbitals in the QM subsystem. The perturbation sets up quasi- and pseudotorques on the hybridization tetrahedra centered at the boundary atom:

$$\vec{K}' = 2\sum_A \{(\delta P_{r_1\sigma}\beta_{\zeta\sigma}^{R_1A} + \delta P_{r_1\zeta}\beta_{\zeta\zeta}^{R_1A})\vec{e}_{R_1A} +$$
$$+ \beta_{\pi\pi}^{R_1A}(\delta P_{r_1\xi}\vec{e}_{R_1A}^{\xi} + \delta P_{r_1\upsilon}\vec{e}_{R_1A}^{\upsilon})\} \times \vec{v}_1^{R_1}$$

$$\vec{N}' = -s_1^{R_1}\vec{v}_1^{R_1}\{(P_1^{rr} - P_1^{ll})[2U_s - 2U_p + C_2 + C_3 + 2C_5] +$$

(3.153)
$$+ (1/2 - \Gamma_1^{rl})[C_2 + 2C_3(s_1^{R_1})^2]\} +$$

$$+ 2\sum_A \{(\delta P_{r_1\sigma}\beta_{\sigma\sigma}^{R_1A} + \delta P_{r_1\zeta}\beta_{\sigma\zeta}^{R_1A})\vec{v}_1^{R_1} -$$
$$- (\delta P_{r_1\sigma}\beta_{\zeta\sigma}^{R_1A} + \delta P_{r_1\zeta}\beta_{\zeta\zeta}^{R_1A})s_1^{R_1}\vec{e}_{R_1A} -$$
$$- \beta_{\pi\pi}^{R_1A}s_1^{R_1}(\delta P_{r_1\xi}\vec{e}_{R_1A}^{\xi} + \delta P_{r_1\upsilon}\vec{e}_{R_1A}^{\upsilon})\}$$

where $\vec{e}_{R_1A}^{\xi}$, $\vec{e}_{R_1A}^{\upsilon}$, and $\vec{e}_{R_1A} = \vec{e}_{R_1A}^{\zeta}$ are the orts of the DCF defined by the R_1A pair of atoms and the quantities C_n are defined by eq. (2.72).

The variations of the one-electron densities $\delta P_{r_1\alpha}$ with $\alpha = \sigma, \xi, \upsilon, \zeta$ and the polarity $(P_1^{rr} - P_1^{ll})$ of the bond with $m = 1$ deserve some discussion. As it is seen from eqs. (3.86), (3.105) each bond incident to an atom contributes an increment to the quasitorque and to the pseudotorque acting upon its hybridization tetrahedron. In the equilibrium these increments separately sum up to zero. We can think that the equilibrium shape and orientation of the hybridization tetrahedron is obtained within a TATO DMM model applied to the entire system. Then, within such a model, there exists an atom corresponding to the left end of the bond with $m = 1$ having number L_1 according to our previous notation. The HOs obtained in this approximation provide an initial guess for HOs in the system including those of the atom R_1, which are of particular interest to us. Going to the hybrid (quantum/classical) description requires the $L_1 R_1$ bond to be broken and all the orbitals centered on the atom L_1 are ascribed to the QM region. The transferable (spin) bond order $\frac{1}{2}$ takes this value only in the basis of the HOs on both ends of this bond. Going to the AO basis on the left end (L_1) produces the initial guess for the one-electron density matrix elements $P_{r_1 \alpha}$ with the subscript $\alpha = \sigma, \xi, \upsilon, \zeta$ running over the AOs of the atom L_1 which are written in the DCF defined by the axis $L_1 R_1$. For all other atoms A in the QM region $A \neq L_1$ the initial guess for the the bond order in the TATO-DMM model is as $P_{r_1 \alpha} = 0$. Then the quantities $\delta P_{r_1 \alpha}$ in eq. (3.153) must be understood as $\delta P_{r_1 \alpha}^{R_1 A}$ – the deviations of the matrix elements of the one-electron density calculated using the QM method chosen for the purpose of treating the QM region of the complete system by a hybrid method under consideration from the guess values defined above. These comprise the contribution of the variation of the two-center elements of the density matrices to quasitorques. The diagonal matrix element of the one-electron density also deviates from its equilibrium value as determined within the TATO-DMM procedure. This value can be safely assumed to be predetermined by the value of the parameter μ_{10}. The quantity δP_1^{rr} entering eq. (3.153) is then deviation of the diagonal matrix element of the one-electron density calculated by the assigned QM method from the value derived from the parameter μ_{10}.

These additional pseudo- and quasitorques produce the pseudo- and quasirotations of the hybridization tetrahedron of the boundary atom R_1. In the linear response approximation, it corresponds to the treatment of the corresponding pseudo- and quasitorques by the $(\nabla_{\vec{\omega}}^2 E)^{-1}$ matrix which is simple (diagonal in the basis of the $\delta \vec{\omega}_b$ and $\delta \vec{\omega}_l$ variables) in the case of symmetric hydride:

$$\delta \vec{\omega}_b = -\frac{\vec{N}'}{4 P^{rl}(\beta_{\sigma\sigma} + \beta_{\zeta\sigma}/\sqrt{3})}$$

$$(3.154) \qquad \delta \vec{\omega}_l = -\frac{\sqrt{3}\vec{K}'}{8 P^{rl}\beta_{\zeta\sigma}}$$

and can be used for estimates also in the general case. Therefore, the hybridization tetrahedron acquires a new shape and orientation which are inconsistent with those of the hybridization tetrahedra centered at the neighbor MM atoms and with their positions. This produces additional classical forces and torques acting on the MM neighbors of the boundary atom at hand. The forces are directed along the $\vec{e}_{R_m L_m}$ vectors and stem from two sources: the variation of the shape of the hybridization tetrahedron:

$$f_{bm} = \frac{1}{2(\beta_{\sigma\sigma} + \beta_{\zeta\sigma}/\sqrt{3})} (\{ -\theta_{\sigma\sigma}^{R_m L_m} s_m^{L_m} \vec{v}_m^{R_m} - \theta_{\sigma\zeta}^{R_m L_m} v_{m\zeta}^{L_m} \vec{v}_m^{R_m} +$$

$$(3.155) \qquad + \theta_{\zeta\sigma}^{R_m L_m} s_m^{R_m} s_m^{L_m} \vec{e}_{R_m L_m} + \theta_{\pi\pi}^{R_m L_m} s_m^{R_m} \vec{v}_m^{L_m} +$$

$$+ (\theta_{\zeta\zeta}^{R_m L_m} - \theta_{\pi\pi}^{R_m L_m}) s_m^{R_m} v_{m\zeta}^{L_m} \vec{e}_{L_m R_m} \}, \vec{N}')$$

and the variation of its orientation:

$$(3.156) \quad f_{lm} = \frac{\sqrt{3}}{4\beta_{\zeta\sigma}}(\{-\theta_{\zeta\sigma}^{R_m L_m} s_m^{L_m} \vec{e}_{R_m L_m} \times \vec{v}_m^{R_m} - \theta_{\pi\pi}^{R_m L_m} \vec{v}_m^{L_m} \times \vec{v}_m^{R_m} -$$
$$- (\theta_{\zeta\zeta}^{R_m L_m} - \theta_{\pi\pi}^{R_m L_m}) v_{m\zeta}^{L_m} \vec{e}_{R_m L_m} \times \vec{v}_m^{R_m}\}, \vec{K}')$$

Analogously, torques acting upon the atoms L_m (the left-end atom for the m-th bond) with $m \geq 2$, which are neighboring to the boundary from the MM side, arise due to variation of the shape of the hybridization tetrahedron:

$$\vec{t}_{bm} = -\frac{1}{2(\beta_{\sigma\sigma} + \beta_{\zeta\sigma}/\sqrt{3})} \{\beta_{\sigma\zeta}^{R_m L_m} (\vec{e}_{R_m L_m} \times \vec{v}_m^{L_m})(\vec{v}_m^{R_m}, \vec{N}') -$$
$$(3.157) \quad - s_m^{R_m} [\beta_{\zeta\sigma}^{R_m L_m} s_m^{L_m} + (\beta_{\zeta\zeta}^{R_m L_m} - \beta_{\pi\pi}^{R_m L_m}) v_{m\zeta}^{L_m}] \vec{e}_{R_m L_m} \times \vec{N}' -$$
$$- (\beta_{\zeta\zeta}^{R_m L_m} - \beta_{\pi\pi}^{R_m L_m}) s_m^{R_m} (\vec{e}_{R_m L_m} \times \vec{v}_m^{L_m})(\vec{e}_{R_m L_m}, \vec{N}')\}$$

and of its orientation:

$$\vec{t}_{lm} = -\frac{\sqrt{3}}{4\beta_{\zeta\sigma}} \{\beta_{\zeta\sigma}^{R_m L_m} s_m^{L_m} (\vec{v}_m^{R_m} \otimes \vec{e}_{R_m L_m} - v_{m\zeta}^{R_m} \mathcal{I}) +$$
$$(3.158) \quad + (\beta_{\zeta\zeta}^{R_m L_m} - \beta_{\pi\pi}^{R_m L_m})[v_{m\zeta}^{L_m} (\vec{v}_m^{R_m} \otimes \vec{e}_{R_m L_m} - v_{m\zeta}^{R_m} \mathcal{I}) +$$
$$+ (\vec{e}_{R_m L_m} \times \vec{v}_m^{L_m}) \otimes (\vec{e}_{R_m L_m} \times \vec{v}_m^{R_m})]\} \vec{K}'$$

The MM subsystem in its turn also affects the parameters of the QM subsystem as any geometry variation in the MM subsystem induces changes in pseudo- and quasirotation angles defining hybridization of the frontier atom. The corrections to the QM one-center Hamiltonian parameters (in the linear approximation) are:

$$(3.159) \quad \begin{aligned} \delta U_1^r &= 2(U_s - U_p) s_1^{R_1} (\delta\vec{\omega}_b, \vec{v}_1^{R_1}) \\ \delta(r_1 r_1 \mid r_1 r_1)^{R_1} &= 2(C_2 s_1^{R_1} + 2C_3 (s_1^{R_1})^3)(\delta\vec{\omega}_b, \vec{v}_1^{R_1}) \end{aligned}$$

The resonance integrals in the QM subsystem are also modified. In the respective DCFs the corrections can be expressed as:

$$(3.160) \quad \begin{aligned} \delta\beta_{r_1\sigma}^{R_1 A} &= \beta_{\sigma\sigma}^{R_1 A} \delta^{(1)} s_1^{R_1} + \beta_{\zeta\sigma}^{R_1 A} \delta^{(1)} v_{1\zeta}^{R_1} \\[4pt] \delta\beta_{r_1\zeta}^{R_1 A} &= \beta_{\sigma\zeta}^{R_1 A} \delta^{(1)} s_1^{R_1} + \beta_{\zeta\zeta}^{R_1 A} \delta^{(1)} v_{1\zeta}^{R_1} \\[4pt] \delta\beta_{r_1\xi}^{R_1 A} &= \beta_{\pi\pi}^{R_1 A} \delta^{(1)} v_{1\xi}^{R_1} \\[4pt] \delta\beta_{r_1\upsilon}^{R_1 A} &= \beta_{\pi\pi}^{R_1 A} \delta^{(1)} v_{1\upsilon}^{R_1} \end{aligned}$$

where the variations of the HOs in response to the geometry variations are given by eq. (3.67).

The formulae given above can be illustrated by numerical estimates of the magnitude of the renormalization of the QM parameters and changes in the DMM forces and torques appearing in the vicinity of an sp^3 carbon atom located on the intersubsystem frontier. The changes in the QM one-center Hamiltonian parameters due to elongation of one of the MM bonds incident to the frontier carbon atom are:

$$(3.161) \quad \frac{\partial U_1}{\partial r_2} = -1.162 \frac{\text{eV}}{\text{Å}}; \quad \frac{\partial (t_1 t_1 | t_1 t_1)}{\partial r_2} = 0.537 \frac{\text{eV}}{\text{Å}}$$

while the effects due to change of the bond angle between two MM bonds incident to the frontier carbon atom are given by:

$$(3.162) \quad \frac{\partial U_1}{\partial \chi_{23}} = 0.755 \frac{\text{eV}}{\text{rad}}; \quad \frac{\partial (t_1 t_1 | t_1 t_1)}{\partial \chi_{23}} = -0.349 \frac{\text{eV}}{\text{rad}}$$

On the other hand the changes in the one- and two-electron densities on the HO centered on the frontier atom lead to the following forces and torques acting on the MM atoms immediately bound to the frontier carbon atom:

$$
(3.163) \quad
\begin{aligned}
f_{bm} &= [-2.225\delta P_m^{rr} + 0.465\delta\Gamma_m^{rr}]\frac{\text{eV}}{\text{Å}} \\
\vec{t}_{bm} &= [2.891\delta P_m^{rr} - 0.604\delta\Gamma_m^{rr}]\frac{\vec{e}_m \times \vec{e}_1}{|\vec{e}_m \times \vec{e}_1|}\frac{\text{eV}}{\text{rad}}
\end{aligned}
$$

The numerical estimates show that in the case of variational determination of orbitals, the effect of the frontier (besides electrostatic, van-der-Waals etc. contributions) can be considered as a relatively weak perturbation.

Similar constructs apply to any kind of possible frontier atom. We consider a special case when the frontier atom serving as the QM/MM junction is the sp^3 nitrogen atom supplying its lone pair to the QM subsystem. Such a setting seems to be quite natural as the basicity or the nucleophilicity functions of the nitrogen atom are both due to interaction of its lone pair with acceptor orbitals. This interaction is naturally to be treated by some kind of QM technique while leaving the rest of the nitrogen neighbors in the MM region.

The one- and two-electron density matrix elements for the QM residing HO are evaluated by the corresponding (QM) procedure, thus invoking the TA type of description for this HO. When the lone pair is involved in the QM subsystem, the corresponding density matrix elements depart from their invariant values ($P_4^{rr} \neq 1$, $\Gamma_4^{rr} \neq 1$). On the other hand, the density matrix elements assigned to HOs of the MM region are fixed at their invariant values according to the FA setting. The nonvanishing intersystem two-electron densities reduce to the products of the corresponding one-electron density matrix elements and the interaction reduces to Coulomb interaction of the densities, a consequence of neglect of one-electron transfers between subsystems. With these assumptions we get the corrected hybridization-dependent one-center energy for the frontier nitrogen atom:

$$
(3.164) \quad
\begin{aligned}
&[(1 + 2\delta P_4^{rr})(U_s - U_p + C_5 + C_3/2) + (3/4 + \delta\Gamma_4^{rr})C_2]\sin^2\omega_{sz} + \\
&+ (1/4 + \delta\Gamma_4^{rr} - \delta P_4^{rr})C_3 \sin^4\omega_{sz}
\end{aligned}
$$

In the case of $\delta P_4^{rr} = \delta\Gamma_4^{rr} = 0$ this expression reduces to eq. (3.111). In practice, variation of the one- and two-electron densities on the lone pair ($\delta P_4^{rr}, \delta\Gamma_4^{rr} < 0$) leads to modification of the nitrogen pyramid. Numerical estimates show that the correction to pyramidalization momentum (the energy derivative with respect to pyramidalization angle δ) is:

$$(3.165) \quad [-45.020\delta P_4^{rr} + 8.424\delta\Gamma_4^{rr}]\frac{\text{eV}}{\text{rad}}$$

The derivatives of the energy correction (of the terms proportional to δP_4^{rr} and $\delta\Gamma_4^{rr}$) with respect to the angles $\vec{\omega}_b$, $\vec{\omega}_l$ yield additional quasi- and pseudotorques ($\vec{K}_4'$ and $\vec{N}_4'$, respectively) acting upon the hybridization tetrahedron of the frontier nitrogen atom:

$$\vec{K}_4' = 0$$

$$(3.166) \quad \vec{N}_4' = -2s_4^N\vec{v}_4^N[\delta P_4^{rr}\{2U_s - 2U_p + C_3 + 2C_5 - 2C_3(s_4^N)^2\} +$$
$$+ \delta\Gamma_4^{rr}\{C_2 + 2(s_4^N)^2\}]$$

The quasitorque induced by the small variations of the one-center ESVs is vanishing, thus resulting in no quasirotation of the hybridization tetrahedron. At the same time the pseudotorque appears due to the involvement of the frontier atom in the density redistribution within the QM part of the complex system. This contribution to the QM induced pseudotorque is collinear to the QM residing HO ($m = 4$).

In the QM part of the system, the variation of the bond orders can also take place. In variance with the pure SLG picture [11, 12] used here as the QM method underlying the MM part of the system, the atoms in the QM part of the combined system may have off-diagonal elements of the one-electron density matrix between orbitals ascribed to the QM subsystem. The latter are obviously the (Coulson) bond orders for the QM part of the system. The corresponding contribution to the energy reads:

$$(3.167) \quad E'_{res} = -2\sum_A \sum_\mu P_{r_4\mu}\beta_{r_4\mu}^{NA}$$

where $P_{r_4\mu}$ are the elements of the one-electron density matrix (spin bond orders) between the r_4-th HO for the lone pair assigned to the QM subsystem and the μ-th AO in the QM system centered on whatever atom A within the latter. The resonance integrals between the lone pair HO residing on the frontier atom and the AOs on any atom in the QM region are functions of six independent angles ($\vec{\omega}_b$ and $\vec{\omega}_l$) defining the shape of the hybridization tetrahedron on the nitrogen atom and its orientation. They take the following form (in the corresponding DCF):

$$\beta_{r_4\sigma}^{NA} = \beta_{\sigma\sigma}^{NA}s_4^N + \beta_{\zeta\sigma}^{NA}v_{4\zeta}^N$$

$$(3.168) \quad \beta_{r_4\zeta}^{NA} = \beta_{\sigma\zeta}^{NA}s_4^N + \beta_{\zeta\zeta}^{NA}v_{4\zeta}^N$$

$$\beta_{r_4\xi}^{NA} = \beta_{\pi\pi}^{NA}v_{4\xi}^N, \quad \beta_{r_4\upsilon}^{NA} = \beta_{\pi\pi}^{NA}v_{4\upsilon}^N$$

Taking into account the values of the components of the vector part of the HO with respect to the DCF:

$$v_{4\zeta}^N = (\vec{v}_4^N, \vec{e}_{NA})$$

$$(3.169) \quad v_{4\xi}^N = (\vec{v}_4^N, \vec{e}_{NA}^\xi)$$

$$v_{4\upsilon}^N = (\vec{v}_4^N, \vec{e}_{NA}^\upsilon)$$

and the expressions (3.67) for the variations of the HO coefficients with respect to pseudo- and quasirotation angles ($\vec{\omega}_b$ and $\vec{\omega}_l$) we get explicit form for the resonance contribution to the pseudo- and quasitorque at the frontier atom:

$$\vec{N}'_{res} = 2\sum_A \{ (P_{r_4\sigma}\beta_{\sigma\sigma}^{NA} + P_{r_4\zeta}\beta_{\sigma\zeta}^{NA})\vec{v}_4^N -$$

$$- (P_{r_4\sigma}\beta_{\zeta\sigma}^{NA} + P_{r_4\zeta}\beta_{\zeta\zeta}^{NA})s_4^N \vec{e}_{NA} -$$

$$(3.170) \qquad - \beta_{\pi\pi}^{NA}s_4^N(P_{r_4\xi}\vec{e}_{NA}^\xi + P_{r_4\upsilon}\vec{e}_{NA}^\upsilon) \},$$

$$\vec{K}'_{res} = 2\sum_A \{ (P_{r_4\sigma}\beta_{\zeta\sigma}^{NA} + P_{r_4\zeta}\beta_{\zeta\zeta}^{NA})\vec{e}_{NA} +$$

$$+ \beta_{\pi\pi}^{NA}(P_{r_4\xi}\vec{e}_{NA}^\xi + P_{r_4\upsilon}\vec{e}_{NA}^\upsilon) \} \times \vec{v}_4^N$$

where $\vec{e}_{NA}^\xi$, $\vec{e}_{NA}^\upsilon$, and $\vec{e}_{NA} = \vec{e}_{NA}^\zeta$ are the orts of the DCF defined by the NA pair of atoms.

The total pseudo- and quasitorques which appear due to quantum behavior of electrons in the QM region then become:

$$\vec{N}' = \vec{N}'_4 + \vec{N}'_{res}$$

$$(3.171)$$

$$\vec{K}' = \vec{K}'_{res}$$

Finally, in the linear response approximation, they produce, after being multiplied by the $(\nabla_{\vec{\omega}}^2 E)^{-1}$ matrix eq. (3.133), the pseudo- and quasirotations of the hybridization tetrahedron on the frontier atom N. The corrections to the pseudo- and quasirotation angles of the hybridization tetrahedron result both in a new form and the orientation of the latter. By this it becomes inconsistent with the positions of the atoms bonded to the frontier atom from the MM side of the system. Multiplying the angular corrections by the $(\nabla_{\vec{\omega}}\nabla_{\vec{\varphi}_m} E)^\dagger$ matrix results in a torque acting upon the T_m atom of the m-th bond incident to the frontier atom N on the MM side. Also, the additional one- and two-electron densities on the frontier atom give additional forces acting upon its MM neighbors. They can be easily obtained if the variations of the quasi- and pseudorotation angles are multiplied by the mixed second derivatives matrix $(\nabla_{\vec{\omega}}\nabla_{r_{NT_m}} E)^\dagger$. These forces are directed along the respective $\vec{e}_{NT_m}$ vectors. This comprises the effect (forces and torques) exerted by the QM subsystem upon the atoms attached to the frontier one on the side of the MM system due to changes of hybridization of the frontier atom.

On the other hand, any deformation in the MM system results in the variation of the pseudo- and quasirotation angles. The shifts of the positions of the MM neighbors of frontier atoms, result in quasi- and pseudotorques acting upon the hybridization tetrahedron of the frontier nitrogen. In its turn, this produces variations of both one-center parameters corresponding to the QM residing HO and of the resonance parameters for the QM residing HO and all other orbitals in the QM region. The variation of the one-center matrix elements of the Hamiltonian corresponding to the QM HO is:

$$\delta U_4^r = 2(U_s - U_p)s_4^N \delta^{(1)} s_4^N = 2(U_s - U_p)s_4^N (\delta\vec{\omega}_b, \vec{v}_4^N)$$

$$(3.172) \quad \delta(r_4 r_4 \mid r_4 r_4)^N = 2C_2 s_4^N \delta^{(1)} s_4^N + 4C_3 (s_4^N)^3 \delta^{(1)} s_4^N =$$

$$= 2[C_2 s_4^N + 2C_3 (s_4^N)^3](\delta\vec{\omega}_b, \vec{v}_4^N)$$

The numerical estimate for the modifications of parameters described by eq. (3.172) are:

$$
\begin{aligned}
(3.173) \quad & \delta U_4^r \simeq -9.53\delta\omega_{sz} \text{ eV} \\
& \delta(r_4 r_4 \mid r_4 r_4)^N \simeq -1.22\delta\omega_{sz} \text{ eV}
\end{aligned}
$$

The modification of the QM resonance integrals to which the HO at hand is involved is somewhat more complex. It nevertheless uses the same (DCF) representation of the resonance integrals as previously:

$$\delta\beta_{r_4\sigma}^{NA} = \beta_{\sigma\sigma}^{NA}\delta^{(1)} s_4^N + \beta_{\zeta\sigma}^{NA}\delta^{(1)} v_{4\zeta}^N$$

$$(3.174) \quad \delta\beta_{r_4\zeta}^{NA} = \beta_{\sigma\zeta}^{NA}\delta^{(1)} s_4^N + \beta_{\zeta\zeta}^{NA}\delta^{(1)} v_{4\zeta}^N$$

$$\delta\beta_{r_4\xi}^{NA} = \beta_{\pi\pi}^{NA}\delta^{(1)} v_{4\xi}^N$$

$$\delta\beta_{r_4\upsilon}^{NA} = \beta_{\pi\pi}^{NA}\delta v_{4\upsilon}^N$$

If these variations are taken into account in the calculations on the QM part of the complex system, the effect of the MM system on the parameters of the effective Hamiltonian for the QM part turns out to be taken into account in the first order. It should be stressed that changes in the hybridization of the frontier atom due to participation of one orbital in the QM subsystem are not taken into account in any of the existing QM/MM schemes. This effect is not very large, so the first-order correction for taking it into account seems to be adequate.

3.6. CONCLUSION

The hybrid QM/MM modeling is quite promising for the study of large molecules especially in the rapidly growing field of computational biochemistry. Covalent bonding between the QM and MM parts is especially important as it arises naturally while modeling enzymatic catalysis. At the same time, this case is the most complex one as the boundary between subsystems cannot be well defined on an intuitive level and the construction of the intersubsystem junction is not straightforward. Many *ad hoc* prescriptions of doing that are proposed in the literature. We have addressed the state-of-the-art in this field in a somewhat critical manner, with special attention to the problems arising during the QM/MM modeling described in Chapter 2. Nevertheless, the sequential derivations of QM/MM junctions have been shown to be possible using simple physical principles which in our opinion govern the sequential construction of all possible hybrid QM/MM schemes. These principles assume the existence of the quantum chemical description underlying the MM one. The SLG trial wave function was taken for constructing the required description. It gives a way of determining the hybrid orbitals centered on the frontier atoms, based on the derivation of the MM description from the QM (SLG) one. The deductive MM derived in this chapter can be of course used as a standalone tool for an economical description of PES of organic molecules, taking into account important details of molecular electronic structure. Its use in the context of hybrid methods allows one to determine the effects of the MM subsystem on the QM one (renormalization of parameters) and of the QM subsystem on the MM one (torques and forces acting on the MM atoms). Explicit expressions are obtained for frontier sp^3 carbon and nitrogen atoms. Numerical estimates obtained illustrate these general points.

References

1. U. Burkert and N.L. Allinger. *Molecular Mechanics*, ACS, Washington, 1982.
2. V.G. Dashevskii. *Conformations of Organic Molecule* [in Russian] Khimiya, Moscow, 1974.
3. A.Y. Meyer. *Theoretical Models of Chemical Bonding, Part 1* ed. by Z.B. Maksić, Springer, Heidelberg, 1989.
4. V.A. Fock. Symmetry of a hydrogen atom, SORENA, **5**, 3, 1935.
5. H. Weyl. *The Theory of Groups and Quantum Mechanics, 1931*, rept. 1950 Dover, NY.
6. H. Weyl. *Classical Groups: Their Invariants and Representations*, Princeton University Press, Princeton, 1939.
7. E. Wigner. *Group Theory and Its Application to the Quantum Mechanics of Atomic Spectra*, Academic Press, New York, 1959.
8. J.P. Elliot and P.G. Dawber. *Symmetry in Physics*, vols. 1, 2, Macmillan, London, 1979.
9. A.M. Tokmachev and A.L. Tchougréeff. *Russ. J. Phys. Chem.*, **73**, 259, 1999.
10. A.M. Tokmachev, A.L. Tchougréeff and I.A. Misurkin. *Russ. J. Phys. Chem.*, **74**, S205, 2000.
11. A.M. Tokmachev and A.L. Tchougréeff. *J. Comp. Chem.*, **22**, 752, 2001.
12. A.M. Tokmachev and A.L. Tchougréeff. *J. Phys. Chem. A.*, **107**, 358, 2003.
13. J.H. van t'Hoff. Archives neerlandaises des sciences exactes et naturelles. **9**, 445, 1874.
14. J.A. LeBel. *Bull. Soc. Chim.*, **22**, 337, 1874.
15. R. Hoffmann. *J. Chem. Phys.*, **39**, 1397, 1963.
16. C.A. Coulson. *Valence*, Oxford University Press, Oxford, 1961.
17. R.J. Gillespie and R.S. Nyholm. *Quart. Rev. Chem. Soc.*, **11**, 339, 1957.
18. R.J. Gillespie. *Angew. Chem. Int. Ed. Engl.*, **6**, 819, 1967; *J. Chem. Educ.*, **47**, 19, 1967; *Molecular Geometry*, Van Nostrand Reinhold, London, 1972.
19. R.J. Gillespie and I. Hargittai. *The VSEPR Model of Molecular Geometry*, Prentice Hall, New Jersey, 1991.
20. J.-P. Daudey, J.-P. Malrieu and O. Rojas. *Localization and Delocalization in Quantum Chemistry, Vol. 1. Atoms and Molecules in the Ground State*, O. Chalvet, R. Daudel, S. Diner and J.-P. Malrieu. eds., Reidel, Dordrecht, 1975.
21. A.L. Tchougréeff. *Chem. Phys. Reports*, **16**, 1035, 1997.
22. P. Surján. *Top. Curr. Chem.*, **203**, 64, 1999.
23. G. Náray-Szabó and P.R. Surján. *Chem. Phys. Lett.*, **96**, 499, 1983.
24. J.M. Parks and R.G. Parr. *J. Chem. Phys.*, **28**, 335, 1958.
25. A.A. Abrikosov, L.P. Gor'kov and I.Y. Dzyaloshinskii. *Quantum Field Theoretical Methods in Statistical Physics*, Pergamon, Oxford, 1965.
26. A.M. Tokmachev and A.L. Tchougréeff. *J. Comp. Chem.*, **26**, 491, 2005.
27. S.L. Altmann. *Rotations, Quaternions and Double Groups*, Oxford Scientific Publications, Clarendon, Oxford University Press, New York, 1986; A.V. Berezin, Y.A. Kurochkin and E.A. Tolkachev. *Quaternions in Relativistic Physics* [in Russian] Nauka i Tehnika, Minsk, 1989.

28. E. Cartan. *Leçons sur la théorie des spineurs.* Hermann, Paris, 1938.
29. R.D. Richtmyer. *Principles of Advanced Mathematical Physics,* vols. 1, 2, Springer, New York, 1981.
30. J.M. Kennedy and C.E. Schäffer. *Inorg. Chem. Acta,* **252**, 185, 1996.
31. E.U. Condon and G.H. Shortley. *Theory of Atomic Spectra,* Cambridge University Press, New York, 1951.
32. I.R. Shafarevich. *Fundamental Concepts of Algebra* [in Russian] R&C Dynamics, Izhevsk, 2001.
33. J. Kozelka. In *Metal Ions in Biological Systems,* vol. 33 (eds. A. Sigel and H. Sigel) Marcel Dekker, New York, 1996.
34. L. Pauling. *General Chemistry,* W.H. Freeman, San Francisco, 1958.
35. G. Del Re. *Theor. Chim. Acta,* **1**, 188, 1963.
36. M.J.S. Dewar and W. Thiel. *J. Am. Chem. Soc.,* **99**, 4899, 1977; M.J.S. Dewar and W. Thiel, *ibid.,* **99**, 4907, 1977.
37. J.F. Mulligan. *J. Chem. Phys.,* **19**, 347, 1951.
38. R.G. Parr and G.R. Taylor. *J. Chem. Phys.,* **19**, 497, 1951.
39. J. Applequist. *J. Math. Phys.,* **24**, 739, 1983.
40. J. Applequist. *J. Chem. Phys.,* **83**, 809, 1985.
41. C.E. Dykstra. *Chem. Rev.,* **93**, 2339, 1993.
42. L.D. Landau and E.M. Lifshits. *Mechanics. Course of Theoretical Physics,* vol. 1, Pergamon, London, 1976.
43. A.D. Buckingham. *In Intermolecular Interactions: From Diatomics to Biopolymers,* ed. by B. Pullman, Wiley Interscience, NY, 1978.
44. A.M. Tokmachev and A.L. Tchougréeff. *Int. J. Quant. Chem.,* **88**, 403, 2002.
45. A.M. Tokmachev and A.L. Tchougréeff. *J. Comp. Meth. Sci. Eng.,* **2**, 309, 2002.
46. R. Car and M. Parinello. *Phys. Rev. Lett.,* **55**, 2741, 1985.
47. R.C. Bingham, M.J.S. Dewar and D.H. Lo. *J. Am. Chem. Soc.,* **97**, 1302, 1307, 1975.
48. I. Mayer. *Struct. Chem.,* **8**, 309, 1997.
49. H.A. Bent. *Chem. Rev.,* **61**, 275, 1961.
50. G. Frenking and N. Frölich. *Chem. Rev.,* **100**, 717, 2000.
51. A.L. Tchougréeff. *J. Mol. Struct. (THEOCHEM),* **630**, 243, 2003.
52. A.L. Tchougréeff and A.M. Tokmachev. *Int. J. Quant. Chem.,* **96**, 175, 2004.
53. A.I. Kostrikin and Y.I. Manin. *Linear Algebra and Geometry* [in Russian] Nauka, Moscow, 1986.
54. R.G. Pearson. *Symmetry Rules for Chemical Reactions,* Wiley, New York, 1976.
55. M.V. Volkenstein, L.A. Gribov, M.A. Elyashevich and B.I. Stepanov. *Molecular Virbations* [in Russian] Nauka, Moscow, 1972.
56. O. Ermer and S. Lifson. *J. Am. Chem. Soc.,* **95**, 4121, 1973; M.J. Hwang, J.P. Stocktisch and A.T. Hagler. *J. Am. Chem. Soc.,* **116**, 2515, 1994.
57. S. Chang, D. McNally, S. Shary-Tehrany, M.J. Hickey and R.H. Boyd. *J. Am. Chem. Soc.,* **92**, 3109, 1970.
58. F. Bernardi, M. Olivucci and M.A. Robb. *J. Amer. Chem. Soc.,* **114**, 1606, 1992.
59. C.A. Coulson. *Rev. Mod. Phys.,* **32**, 170, 1963.
60. I. Fisher-Hjalmars. *In Modern Quantum Chemistry. Istanbul Lectures,* vol. 1, ed. by O. Sinanoğlu, AP, New York, 1965.
61. N.F. Stepanov, G.S. Koptev, V.I. Pupyshev and Y.N. Panchenko. *In Modern Problems of Physical Chemistry,* vol. 11 [in Russian] MSU Publ., Moscow, 1979.
62. T.L. Allen and H. Shull. *J. Chem. Phys.,* **35**, 1644, 1961.
63. A.L. Tchougréeff. *J. Mol. Struct.* (THEOCHEM), **630**, 243, 2003.
64. A.L. Tchougréeff. *J. Mol. Struct.* (THEOCHEM), **632**, 91, 2003.

4

SYNTHESIS: HYBRID MOLECULAR MODELS FOR COORDINATION COMPOUNDS

Abstract   In the last chapter of this book, we employ the general methods developed in the previous chapters in the context of a specific class of molecular systems known as coordination compounds. The simplest statement about this class of systems is negative: it is poorly describable by transitional (classical) MM methods. The reasons are twofold and manifest themselves differently for different subclasses of coordination compounds. The first is the non-directional character of coordination bonds and their unsaturability. This feature is common to all types of coordination centers that manifest a wide variety of coordination numbers and coordination polyhedra. The second important source of problems relates to the coordination compounds of transition metal ions with open d-shells. In this case, the situation is that each of the multiple electronic states available for the open d-shell produces corresponding PES which all lie in a narrow energy interval and may intersect due to their sophisticated dependence on molecular geometry. This produces a picture in which a unique PES usually assumed in the classical MM models does not exist, but a bunch of them is present, all of which must be uniformly treated. As a result of these two sources of problems, the material of this chapter is divided into two parts. In the first we consider the factors responsible for the "unspecific" behavior – a group of electrons occupying three-dimensionally delocalized orbitals of the central atom (ion) of the complex and its close vicinity. This produces a mechanistic picture of PES of coordination compounds capable of reproducing the effects of ligand mutual influence. In the second we address the transition metal complexes with open d-shells. We apply the general hybrid methodology to develop a true QM/MM scheme of including sufficiently quantum subsystem (the d-shell) in the general classical (MM treated) environment.

In this (final) chapter of the book we apply the principles and methods developed in previous chapters for hybrid modeling of the electronic structure of "organic" molecules and methods of constructing intersubsystem junctions to develop analogous constructs for coordination compounds including transition and nontransition metal complexes. Molecular modeling of coordination compounds (CC), reproducing characteristic features of their stereochemistry and electronic structure, is in high demand in the context of studies and development of various industrial and laboratory processes. Ion extraction, ion exchange, isotope separation, and neutralization of nuclear waste are just a few examples. Of particular interest are the structure and reactivity of metal-containing enzymes. Solving these academic and technological problems requires modeling methods allowing massive simulations of PES of CCs in a wide range of molecular geometries including (in the case of, say, coordination processes) internuclear separations, corresponding to dissociation of coordination bonds between central atoms and ligand donor atoms.

Some fundamental features related to the interplay between composition and structure are characteristic of CCs. They are united under the name of ligand influence effects. Only a very superficial description of this circle of phenomena is given here. Let us imagine a symmetric (octahedral) complex of composition ML_6. Its geometrical structure is characterized by the values of the metal-ligand separations. If the ligands L are polyatomic ones, the separation between the central and the ligand donor atoms is meant. Transition from the ML_6 to the ML_5X composition is termed substitution of one of the ligands L by another ligand X (monosubstitution). It obviously reduces the local symmetry from the octahedral one to the tetragonal. The possible geometry variations are limited to potential increase/decrease of the equatorial bond lengths accompanying the decrease/increase of that of the bond in the *trans*-position to the substituent with all possible combinations of the increases and decreases and may be violations of the strictly planar placement of the equatorial ligands around the central atom. Nature appears to be very sophisticated even in this restricted playground. Yet in the 1920s it was observed [1] that in square planar Pt(II) and the octahedral Pt(IV) complexes the *trans*-position (the more remote – axial – one) to the substituent is much more sensitive to the characteristics of the substituent than the *cis*-positions (the closer – equatorial – ones). The only manifestations of this sensitivity known at those earlier times were the rates of the ligand-exchange reactions. It turned out that the ligands occurred in *trans*-positions to the ligands X, which are "stronger *trans*-effectors", exchange much easier than those in the *trans*-positions to the X's, which are weaker *trans*-effectors. It must be remembered that no structural data on bond lengths were available then on the compounds of the complexity at hand. So it was not surprising that any kind of possible *cis*-effect was first assumed as a hypothesis. With the passage of time it has been discovered that (i) the effects of this sort are not specific for the Pt complexes, but occur in a much wider range of the compounds and are not even restricted to the transition metal complexes; (ii) a wider availability of the X-ray techniques has allowed one to establish important structural manifestations of the ligands' influence. The general picture became, however, not clearer but in some respects more obscure. First of all the structural *cis*-effects were unequivocally established. At the same time, no specific *cis*-effectors had been found. It was shown that a stronger *trans*-effector in addition to causing the elongation of the bond in the *trans*-position to the substituent also causes a somewhat smaller shortening of *cis*-bonds and *vice versa*: a weaker *trans*-effector causes elongation of the *cis*-bonds. It was also found that the ligand influence takes place also in the compounds of nontransition metals and even nonmetals like alkali and alkali earth metals, P(V), As(V), Sb(V), S(VI), Se(VI), Sn(IV) etc. In the domain of CCs of nontransition elements the ligand influence turned out to have much more diverse manifestations which depend on the nature of the central atom and the substituents.

Another important manifestation of the ligand influence is the thermochemistry of the isomers of the CCs. The problem of isomerism of the CCs was a crucial point in the whole construction of the coordination theory by Werner in the first two decades of the twentieth century. The possibility of isolating two (and not more than two) distinct species of identical composition ML_4X_2 (together with other observations of this sort) allowed him to argue in favor of the octahedral arrangement of the molecules or atoms nowadays called ligands around the central atom. The heats of formation may differ for the *cis*- and *trans*-ML_4X_2 species, sometimes quite significantly (so that in some cases only one isomer can be observed) and, depending on M, L, and X, either of the *trans*- and *cis*-forms may become more stable.

That wide range of geometry patterns and varieties of the modes of geometry-composition interplays present in CCs could be thought to be inaccessible for any classical (MM) technique in principle. The main concern is the non-monotonic character of the interactions to be assumed at least to somehow mimic the observed features of the ligand influence. However, the practical need for modeling PES of systems with metal ions in an efficient method makes it necessary to formulate a corresponding problem, as that of the search for an effective method of modeling.

In the previous chapters, we developed an approach which can be used to put the process of developing mechanistic descriptions of PES (i.e. of developing MM force fields) on a rational basis. Deductive molecular mechanics [2–4] (DMM) allows us to develop a form of the MM force fields to analyze the form of the electronic wave function relevant to the physical picture of the electronic structure of the considered class of molecules. In this chapter we apply the previously developed DMM approach to analytical derivation of the QM based form of the force fields involving the non-transition metal atoms.

The main advantages of the classical (MM) schemes are their low cost and high efficiency in the prediction of molecular geometry for organic compounds without significant electron correlation. Their principal disadvantage is an intrinsic inability to consider uniformly the situations when electronic structures differ significantly. The above theory explains the reason: estimates or interpolations of ESVs fail if the structure becomes so different that the ESVs useful in one area of the nuclear configuration space become useless in other areas. The QM procedures, if correctly used, are potentially capable of describing different types of electronic structures. The concerted exploitation of the advantages of both the (quantum and classical) approaches can be achieved by hybrid QM/MM schemes, providing another way to bypass the bottleneck of M^n-scalability. The theory of such schemes was the main topic of Chapter 1. The QM/MM schemes describe some relatively small part of the system by an appropriate quantum chemical (QM or QC) method while the rest (relatively inert environment) is covered by classical force fields (molecular mechanics – MM). The practical usefulness and general validity of these approaches is based on the chemically and physically motivated observation: chemical transformations usually affect only a small part of the whole system (reaction center) while the rôle of the surrounding groups and molecules reduces to modification of the PES due for example to some polarization or steric strains. This situation is characteristic for chemical reactions of biological interest (especially, for catalysis by enzymes), when the chemical transformation touches only a restricted region of a molecule so that the electronic structure changes require a thorough correlation account for being adequately described. The situation in TMCs with multiple electronic states accessible in experiment, but localized in the restricted area in the d-shell, calls for a kind of QM/MM technique as well. Of course the TMCs could be treated by some standard QM/MM technique used with the additional prescription – try to extend the quantum subsystem as much as possible – in mind. This is not interesting, however. In what follows we shall try to employ an opposite approach to TMCs: reduce the quantum subsystem as much as possible and on the basis of our previous studies, make steps towards constructing a hybrid method targeted at the TMCs, but taking as much as possible the surrounding of the central (transition metal) atom into the classically treated subsystem. In this way we shall discover how to cope with the problems of the metal targeted MM force fields known from the literature.

The fundamental reasons for the difficulties faced by the MM methods when metal (both transition and nontransition) complexes are involved can be understood if one does not consider the MM as a purely empirical scheme (as it is frequently done), but think about them as of some reflection of specific features of molecular electronic structure, formalized by the form of the trial wave function of that class of compounds where such a parameterization might be possible. As shown in Chapter 3, organic compounds for which the MM methods are known to demonstrate significant successes can be described by the QC method, which directly leads to local and transferable two-center bonds. It is shown in Chapter 3 that the derivation of the MM method from the QC description is possible due to a common background of the MM and SLG description, which consists in the physical presence of two-center, two-electron bonds in organic molecules (in strict terms of Section 1.7 – numbers of electrons in each of the geminals weakly fluctuate).

To cope with the problems of a mechanistic description of CC, we will first analyze three basic questions: the nature of the differences in behavior between central atoms on the one hand and organogenic atoms on the other hand, which results in limitations for the MM techniques when applied to molecules of CC. Getting an idea of the source of these differences tentatively allows us to address further questions: developing an adequate MM-like scheme for CCs of nontransition metals and nonmetals which will be able to reproduce fine structural features of the mutual ligand influence characteristic for this class of molecules. Next we turn to the most complex problem – developing a hybrid modeling technique which would allow us to cover complexes of transition metals with open d-shells.

4.1. CHARACTERISTIC FEATURES OF THE ELECTRONIC STRUCTURE OF COORDINATION COMPOUNDS

The difference in chemical behavior between metals and nonmetals is intuitively clear to any chemist. Theoretical chemistry describes this diversity in terms of different types of chemical bonds. They are portrayed in textbooks as being nonpolar covalent, polar covalent, ionic, dative, donor-acceptor, coordination, and so on. Chemists ascribe specific bonds to the above types without a clear explanation of the grounds used for their classification. (A characteristic example: what is the difference between the polar covalent and donor-acceptor bond; or what is the ionic bond if it is known that the system of charged particles cannot have any equilibrium according to the Earnshow theorem?) In our days, any classification of this type is generally considered obsolete: numbers jumping out by myriads from QC programs, but they do not provide a qualitative understanding of physically different pictures described by theoreticians of previous days. Some solid bond classification may, however, be important not only from the pedagogical point of view: from the MM experience, we know that constructing a mechanistic description is not equally easy for different classes of compounds containing bonds of different types. For purely "organic" molecules with well-defined two-center two-electron bonds, numerous empirical parameterizations have been developed successfully [5–7]. Corresponding efforts, when applied to metal containing compounds and hydrogen bonds, did not give completely satisfactory results until now [8–10]. Of course, a good number of works have appeared, which parametrize PES of some well-defined classes of metal containing compounds, but some questions important from the practical point of view remain unanswered.

A remarkable systematization of chemical bonds is given in [11], which we reproduce here in the form of a Table 4.1.

According to it, the bond types known from theoretical chemistry are placed in relation to characteristics of the electronic structure of different classes of chemical species, and the delocalization pattern of the involved one-electron states is taken to be crucial. The first comment on this classification is based upon our vision of the electronic structure of "organic" compounds. In the Table these bonds are termed as "valence" ones and the corresponding MOs are considered to be localized. If the true MO picture based on the HFR model of electronic structure is employed, the corresponding MOs in CH_4 or NH_4^+ are in fact delocalized at least by symmetry: the

Table 4.1. Chemical bonds classification by electronic structure and properties [11].

Bond type	Electronic structure	Compounds example	Typical properties
Valence	MOs are localized between pairs of atoms and occupied by two paired electrons	CH_4 NH_4^+ Diamond C_2H_4	Distinct character of bond energy, dipole moments, frequencies, polarizabilities, etc.
Orbital	MOs are delocalized in one or two dimensions	Benzene Graphite	There are no distinct characteristics; conductivity, cycles, aromaticity
Coordination	MOs are delocalized in space — three-dimensional	$CuCl_4^{2-}$ $CoCl_2$ (crystal)	There are no distinct characteristics; variable coordination number and magnetic moment, strong mutual influence of ligands

states belonging to one- and three-dimensional representations of the T_d point group are by construction delocalized in three dimensions (*vide infra*). This conclusion is, however, valid only in the frame of the HFR treatment of molecular electronic structure. Alternative approximations, and among them first of all the SLG-based methods described in Section 2.4, restore the local picture of one-electron states and of the bonding itself in the above examples in contrast to that provided by the HFR. With the replacement of the MOs by the local HOs, the bonds of the valence type mentioned in Table 4.1 can be attributed to the situations when the SLG-type wave function gives a dominant contribution to the exact one.

The bonds of the orbital type are attributed in Table 4.1 to the one- and two-dimensional delocalization of MOs. The molecules which can be covered by this description are characterized in chemical terms as polyconjugate organic molecules such as polyenes, aromatic molecules and others of this type (with the limiting case of graphite). Among the characteristic properties of these systems one can mention a far reaching transfer of the substituent influence through the system of delocalized molecular orbitals. This seems to be very reasonable. Finally, in [11] the absence of the characteristicity of the bonds and corresponding vibrational frequencies is related to the three-dimensional delocalization of orbitals in the complexes. These features are exemplified by the metal-ligand bonds in transition metal complexes and the characteristic properties of these bonds mentioned in this respect are the optical spectra and magnetic moment, that distinguish them from all other compounds. These examples deserve some additional comment. Although magnetic and optical properties characteristic of TMCs really require the d- or f-states of the metal atom for their description, the noncharacteristicity of the bonds, their poor directionality and unsaturabilty apply equally to the bonds in CCs of nontransition metals as well (for example, alkali or alkali earth metals). On the other hand, the presence of one-electron states with strong angular dependence (d or f) more likely opposes the nonspecificity or at least poor directionality. In fact the presence of the orbitals with a strong angular dependence leads to quite expectable strong susceptibility of the transition metal ions to the angular characteristics of the coordination polyhedron. However, this susceptibility may have nothing to do with bonding. It is more likely the response of nonbonding electrons in the d-shell to a relatively weak perturbation. Meanwhile, the flexibility of the coordination polyhedra is more easily explained by the dominance of the angle independent s-AOs of the metal ion in forming the metal-ligand bonds. This property of the s-AO does not represent, however, any specificity of transition or rare earth elements.

At first glance, it may even seem that all the characteristics of the complexes listed in Table 4.1 – formation and cleavage of the coordination bonds, mutual influence of the ligands, charge redistribution, dependence of magnetic properties on tiny details of molecular geometry and composition – have too much of the quantum origin so that no mechanistic model of these properties is possible. This point of view seems to be however an opposite extreme. Finally the MM is quite a flexible tool, not limiting in any way the sophistication of the force fields to be used or the number of particles involved in the interaction, or other characteristics of the model. Moreover, as shown previously, it is possible to imagine and successfully construct more general mechanistic models of molecular potentials (PES) than the "balls-and-springs" models accepted in standard MM. The models built remain mechanistic ones, but they naturally take into account important features of the electronic structure, which in a standard formulation would require innumerable parameterizations for more and more tricky force fields, whose form remains without any fundamental basis. Our plan is to construct first a hybrid QM/MM model of CCs and identify its basic features. Then, on the basis of this model, we shall try to separate characteristic situations when quantum description becomes unavoidable from those where one can hope to build some noniterative and more "mechanistic" model. Then we present several such models at different levels of numerical elaboration.

4.2. HYBRID AND CLASSICAL MODELS OF COORDINATION COMPOUNDS OF NONTRANSITION METALS

In this section we apply our methodology of constructing hybrid models of molecular electronic structure to the case of coordination compounds. Our main tool will be the SLG/HFR hybrid scheme described below. It will be used to formalize the difference between the "organic" and "inorganic" parts of the coordination compound molecule. After it is done the ESVs relevant for the most problematic "inorganic" part will be selected and reasonably approximated.

4.2.1. SLG analysis of dative bonding[1]

Before describing the CCs with multiple ligands we consider an intermediate situation between purely "organic" or valence bonding and "coordination" bonding occurring in the "ionic" compounds of metal atoms. It describes so-called "dative bonding" of organic donor molecules with metal ions using lone pairs of their donor atoms and deriving a mechanistic model for this type of interaction. This will be a first step towards understanding the donor-acceptor interactions in CCs with multiple ligands.

The derivation of "organic" DMM performed in the previous chapter used the assumption that the single bond is close to the symmetric one described by the HFR two-electron function. This led to the approximation eq. (3.12) for the density matrix elements of the bond geminals. The infinite bond-length asymptotic wave function of two electrons forming a single bond between two atoms is the singlet with two electrons with equal probability residing one by one on either end of this bond. This describes the homolytic cleavage of a covalent nonpolar σ-bond. Dative bonds in the case of the infinite bond elongation by contrast flow to the ionic limit eq. (2.61). This prevents us from using the approximation eq. (3.12) for the geminal amplitude ESVs since they correspond to the wave function with different asymptotic behavior.

---

[1]Reprinted from A.L. Tchougréeff. Deductive molecular mechanics as applied to develop QM/MM picture of dative and coordination bonds. *J. Mol. Struct. (THEOCHEM)*, **632**, 91, 2003 with permission of Elsevier.

Our immediate purpose is to obtain estimates for the amplitude ESVs in the ionic limit. In a more general context we notice that forming dative bonds with some electron accepting atoms i.e. those which provide an empty orbital with the energy low enough to be partially populated, is the simplest archetypal example of the quantum-classical frontier. Considering these cases in a hybrid perspective, one may think that the lone pair of a donor atom is lent to the QM subsystem of a complex system. All the covalent bonds of the donor atoms, which existed prior to the complexation/donation, remain in the classical domain. For that reason, our immediate purpose is to identify the effect produced by forming a dative bond by a donor atom upon its MM characteristics and more generally the stereochemistry of donor atoms in complexes.

The geminal wave functions in eq. (2.60) in the SLG approximation are by definition obtained by diagonalizing the effective Hamiltonian for the m-th bond. These latter are as previously given by eq. (3.1) which can be recast to the form:

$$(4.1) \quad \begin{aligned} H_m^{\text{eff}} &= H_m^0 + H_m' \\ H_m^0 &= \begin{pmatrix} a & 0 & 0 \\ 0 & b & 0 \\ 0 & 0 & c \end{pmatrix} ; H_m' = d \begin{pmatrix} 0 & 1 & 0 \\ 1 & 0 & 1 \\ 0 & 1 & 0 \end{pmatrix} \end{aligned}$$

with

$$(4.2) \quad \begin{aligned} a &= \mathfrak{R}_m \\ b &= \frac{1}{2}(\mathfrak{R}_m + \mathfrak{L}_m) - \Delta\gamma_m \\ c &= \mathfrak{L}_m \\ d &= -\sqrt{2}\beta_{r_m l_m}^{R_m L_m} \end{aligned}$$

where the quantities $\mathfrak{R}_m$, $\mathfrak{L}_m$, $\Delta\gamma_m$ are defined by eq. (3.2).

The ground state of H_m^{eff} acquired by this bond in the limit of the infinite interatomic separation is controlled by the relative position of the diagonal elements of H_m^0 on the energy scale. If the lowest diagonal matrix element of H_m^0 belongs to the covalent configuration, the asymptotic ground state is also the covalent one (the "organic" case studied previously). An alternative situation occurs if the lowest diagonal matrix element of H_m^0 belongs to one of the ionic configurations. This is possible, provided:

$$(4.3) \quad 0 < \frac{1}{2}(\mathfrak{L}_m - \mathfrak{R}_m) - \Delta\gamma_m$$

In this inequality all quantities depend on interatomic distance. For this reason the character of the bond may (as it is well known, see [12] for the recent analysis of this situation in similar terms) change in the prescription of its formation or cleavage. The critical distance given by the relation:

$$(4.4) \quad \Delta\gamma_m(r_{R_m L_m}^c) = \frac{1}{2}(\mathfrak{L}_m(r_{R_m L_m}^c) - \mathfrak{R}_m(r_{R_m L_m}^c))$$

separates the regions characterized by different physical regimes: at distances shorter than the critical one, independently turning the resonance interaction off yields the wave function coinciding with the covalent configuration; at longer distances, turning the resonance interaction off leads to the ionic wave function. The latter is the lone pair wave function eq. (2.61). This consideration allows us to make precise the definition of dative bonds in the following manner. Instead of speaking about polar covalent *vs.* ionic bonds it might make sense to speak respectively of the (polar) covalent *vs.* dative regimes of maybe the same bond. Then the dative bonds are those whose wave function flows to the single ionic configuration if the resonance is turned off at the actual experimental interatomic separation. For the dative bond geminal namely the ionic function must be used as a zero approximation for constructing the estimates for the density ESVs.

4.2.1.1. Density ESVs in the ionic limit

The implementation of the DMM approach constructed in Sections 3.3.1 and 3.3.2 was based on the fact that for the geminals having the covalent wave function as their asymptotic limit, the related ESVs can be taken as transferable quantities. It is remarkable that the discussed quantities do not themselves pertain to this limit; by contrast, they are better described by the symmetric MO functions, but still the covalent limit wave function is a good starting point for developing the "organic" DMM theory. In its frame it is shown that the (Coulson) bond orders $2P_m^{rl}$ can be set equal to 1 which is equivalent to the well-known picture of the localized single bonds in organic molecules. For a considerable range of the bond-lengths around equilibrium the above transferable value of the bond-order remains invariant (transferable) up to the second order in a small parameter and the interatomic distance dependence of the bond energy is dominated by that of the resonance (one-electron hopping) integrals between the left-end and right-end HOs ascribed to the bond. In the "dative" regime the ionic configuration corresponding to accommodating both bond electrons on (for the sake of definiteness) the right-end atom is the asymptotic wave function for the separated acceptor and donor molecules. This differs significantly from the "organic" situation so that the results of Sections 3.3.1 and 3.3.2 cannot be directly employed.

As mentioned previously, the density ESVs must be obtained from the effective bond Hamiltonian eq. (4.1). In terms of the geminal amplitudes, the ESVs are given by eq. (2.78). To get the required direct estimates of the ESVs, we use again the projection operator technique. In terms of the geminal amplitudes (subject to the normalization condition) the projection operator upon the ground state of a geminal has the form:

$$(4.5) \qquad \mathcal{P} = \begin{pmatrix} u^2 & uz & uv \\ uz & z^2 & zv \\ uv & zv & v^2 \end{pmatrix}$$

The zero approximate operator $\mathcal{P}_0$ projecting on the ionic limit ground state corresponds to $u = 1; z = v = 0$. To ensure the correct (ionic) limit of a general one-dimensional projection operator in a three-dimensional space, we apply the prescription of eqs. (1.107) and (1.104) with the notion that $\dim \mathrm{Im}\, \mathcal{P}_0 = 1$ and

$\dim \mathrm{Im}(1 - \mathcal{P}_0) = 2$. In this case the matrix block V consists of one row and two columns:

$$(4.6) \qquad V = \begin{pmatrix} x, & y \end{pmatrix}$$

so that x and y are two real independent parameters. Clearly the one-times-one matrix can be easily inverted and this results in the following operator projecting to the ground state:

$$(4.7) \qquad \mathcal{P} = \frac{1}{1 + x^2 + y^2} \begin{pmatrix} 1 & x & y \\ x & x^2 & xy \\ y & xy & y^2 \end{pmatrix}$$

The idempotence $\mathcal{P}^2 = \mathcal{P}$ is checked immediately as well as the fact that $\mathrm{Sp}\,\mathcal{P} = 1$. The Schrödinger equation for the projection operator $\mathcal{P}$ eq. (1.95) reads:

$$(4.8) \qquad H\mathcal{P} = \mathcal{P}H$$

can be recast to a system of nonlinear equations for x and y (details are given in [13]):

$$(4.9) \qquad \left.\begin{aligned} x &= \frac{d}{(a-b)}\left(1 + y - x^2\right) \\ y &= \frac{d}{(a-c)} x\left(1 - y\right) \end{aligned}\right\}$$

$$xy = \frac{d}{(b-c)}\left(x^2 - y - y^2\right)$$

The last relation for the product xy is not an independent equation but it must be inserted into that for y and the system becomes one for x and y. Solving this system will be equivalent to solving the original 3×3 eigenvalue problem for the effective bond Hamiltonian. In a perturbative manner we get for the first order approximation:

$$x = \frac{d}{(a-b)}; y = 0$$

Identifying the matrix elements of the projection operators eqs. (4.5) and (4.7) establishes the relation between two parameterizations of the three-dimensional projection operators and produces explicit forms for the ESVs eq. (2.78) in terms of x:

$$(4.10) \qquad \begin{aligned} P_{rr} &\approx 1 - \frac{x^2}{2}; P_{ll} \approx \frac{x^2}{2}; P_{rl} \approx \frac{x}{\sqrt{2}} \\ \Gamma_{rr} &\approx 1 - x^2; \Gamma_{rl} \approx \frac{x^2}{2} \\ \delta P_{rr} &\approx \frac{1}{2} - \frac{x^2}{2}; \delta\Gamma_{rr} \approx \frac{3}{4} - x^2 \end{aligned}$$

The variable x thus describes the deviation of the bond falling in the dative regime from pure ionicity. An alternative to this kind of treatment could be one using LCAOs.

The argument in favor is that limiting wave function for the dative bond is that of the doubly occupied HO – i.e. a HFR one.

4.2.1.2. Bond energy in the ionic limit

In the previous section we obtained the geminal amplitude-related ESVs for the bonding geminal in the "dative" regime. We see that the key estimate used previously to construct the "organic" DMM becomes invalid: the geminal related ESVs in the dative regime are not even approximately transferable numbers. By contrast, the values of all density matrix elements are at least linearly dependent on the resonance integral between the donor HO and the acceptor orbital. Together with approximately exponential distance dependence of the resonance integral, this produces a strongly anharmonic potential characteristic for large interatomic separations. The energy of the geminal in the ionic limit (dative regime) can be easily found:

$$(4.11) \quad E_{DA}^{bond} \approx -\frac{8\beta_{DA}^2}{(\epsilon_m^A - \epsilon_m^D) - 2\Delta\gamma_m}\left(1 + \frac{\gamma_{DA}}{(\epsilon_m^A - \epsilon_m^D) - 2\Delta\gamma_m}\right)$$

This is a natural form of the perturbative estimate of the resonance energy known from the theory. An estimate of this type is employed in the SIBFA [14] method.

4.2.1.3. Bonding contribution in the DMM description of dative bonds

As it is seen from the perturbative estimates, the density ESVs are not transferable and are rather sophisticated functions of those ESVs which define the shape and orientation of the hybridization tetrahedron on the donor atom. At this stage, it is possible to develop a mechanistic description which retains the variable x to keep track of details of the electronic structure. This picture is also suitable for constructing the QM/MM junctions. According to the perturbative estimates for the solutions of eq. (4.9) the equilibrium value of the y variable is always by one order of magnitude in β_{DA} smaller than x. Since only the combinations xy and y^2 enter in the expression of the projection operator and thus in that for the energy, we may set $y = 0$ without causing too large an error in energy and by this further simplify the projection operator in eq. (4.7). Using this approximation, the energy of the dative bond becomes:

$$(4.12) \quad E_{DA}^{bond} \approx -\frac{2\sqrt{2}x}{1 + x^2}\left(\beta_{DA} - \gamma_{DA}x\right)$$

where x must be treated as an independent variable. It must be included in the general energy optimization procedure together with other ESVs and geometry variables. This form must be used either instead of the energy of a covalent bond with constant (transferable) spin bond-orders characteristic of the usual covalent bonds or of the perturbative bond energy estimate. Though they are still rather sophisticated functions of the internuclear separation and of the shape and orientation of the hybridization tetrahedra residing on the donor atoms, they can be easily calculated as they require only elementary functions for their evaluation [74] and have correct asymptotic behavior at infinite internuclear separation.

4.2.2. Qualitative picture of the DMM force fields at donor atoms

In the previous section we presented a derivation of the DMM force fields describing the interactions: stretching of a dative bond formed by a donor atom bearing a lone pair with a metal (or hydrogen or more generally – any acceptor) atom. This derivation resulted in rather complicated formulae where all the terms depend on details of hybridization of the donor atom through the resonance integrals β_{DA}. One may foresee further situations: (i) the shape of the hybridization tetrahedron is, to a large extent, defined by the shape of the ligand molecule itself; (ii) the shape of the hybridization tetrahedron is rather flexible and its variation in the prescription of the formation of the dative bond is significant. We shall see that both situations are realized in practice.

4.2.2.1. Donor-acceptor interactions of the model "ammonia" molecule

When the dative bond is formed by an sp^3-nitrogen atom, the shape of its hybridization tetrahedron changes only to a small extent. This can be understood on the basis of the linear dependency condition of eq. (3.64). Three covalent bonds formed by the nitrogen atom largely predefine the norm and the directions of three vector parts $\vec{v}_m$ of the four HOs residing on the atom. The fourth one is fixed by the cited linear condition. In this case the dependence of the dative bond energy on geometry parameters can be obtained from an analysis of the perturbative estimate eq. (4.11). In the latter expression the molecular geometry dependence is dominated by the square of the resonance integral. Assuming, as in the previous section, that the acceptor is represented only by its empty s-orbital we get, as in the case of a hydrogen atom, a significant simplification for the resonance integral responsible for the dative bonding:

$$(4.13) \quad \beta_{DA} = \beta_{\sigma\sigma}^{DA} s_1 + \beta_{\zeta\sigma}^{DA}(\vec{v}_1, \vec{e}_1)$$

The relative orientation of the acceptor atom and the donor atom hybridization tetrahedron enters through the scalar product of the vector part of the lone pair involved in the formation of the dative bond ($\vec{v}_1$) and the dative bond vector ($\vec{e}_1$) which is a unit vector directed along the line connecting the donor and acceptor atoms. The maximum of the above expression is obviously reached when $\vec{v}_1 \parallel \vec{e}_1$ and whatever escape from this line producing a "lone-pair misdirection" can be thought to contribute to the MM force field for the A-D-X angles bending (where X stands for whatever atom *covalently* bound to the donor atom on the ligand – MM treated – side). It must be noticed, however, that the corresponding contribution is by no means the leading one in terms of energetics. Indeed, the contribution of the dative bonding to the "misdirect potential" is scaled down by a small equilibrium value of x eqs. (4.7) and (4.10) — spin bond order of the dative bond. At the same time the general context of the dative bonding allows one to expect that a local dipole moment $\vec{\mu}_D$ resides on the donor molecule and the ion-dipole energy term:

$$(4.14) \quad \frac{Q_A e(\vec{\mu}_D, \vec{e}_1)}{r_{DA}^2}$$

where Q_A is the effective charge of the acceptor atom (close to its formal ionic charge) and r_{DA}, the donor-acceptor interatomic separation, is larger than the dative bonding contribution. For symmetry reasons (we refer here to a model C_{3v} symmetric "ammonia" molecule) one may expect that $\vec{v}_1 \parallel \vec{\mu}_D$ and thus the resulting force field tend to align $\vec{\mu}_D$ and $\vec{e}_1$ and act to prevent the lone-pair misdirection, but do not have too much relation to the direction of the lone pair itself.

4.2.2.2. Donor-acceptor interactions of the model "water" molecule

Now let us consider what happens when the model "water" molecule described in Section 3.3.2.3 forms an additional "dative" bond with an atom containing one empty s-orbital. The linear response approximation previously employed in the DMM framework cannot be used in the present case as the matrix of the energy second derivatives with respect to variations $\delta\vec{\omega}_b$ and $\delta\vec{\omega}_l$ is degenerate (and thus cannot be inverted) due to invariance of the energy with respect to the deformations of the oxygen hybridization tetrahedron between approximate sp^2 and sp^3 hybridization of the lone pairs mentioned in Section 3.3.2.3 Therefore we try to extract the information on the shape of the oxygen hybridization tetrahedron with an extra dative bond directly from the structure of the $SO(4)$ hybridization manifold. This is an analog for the degenerate perturbation theory for the energy considered as a function of the set of hybridization parameters $\vec{\omega}_b$ and $\vec{\omega}_l$.

To start with, we assume that the HO with $m = 1$ will be used for the dative bond. For it we use the estimates eq. (4.10) for the density ESVs. This may be termed as a "harmonic" approximation in terms of the ESV x. The energy of the dative bond is then given by

$$(4.15) \quad E_{OA}^{bond} \approx -2\sqrt{2}\left(\beta_{OA}x - \gamma_{OA}x^2\right)$$

With these definitions we get a correction for the one-center energy of the "water" oxygen atom due to partial electron density transfer to the acceptor orbital:

$$
\begin{aligned}
\Delta E' = &-(s_1^2(U_s - U_p) + U_p + (C_1 + C_2s_1^2 + C_3s_1^4) - \\
&- (C_4 + C_5[1 - 2s^2] + C_3s_1^2[1 - 2s^2 - s_1^2]))x^2 + \\
&+ 2\left(C_4 + C_5[s_1^2 + s^2] + C_3s_1^2s^2\right)\left(\delta P + \frac{1}{2}\right)x^2
\end{aligned}
$$

(4.16)

This contribution lifts the degeneracy of the one-center energy with respect to the distribution of the s-weight between the two lone pairs. Physically $\Delta E'$ is the energy required to extract an amount of electron density proportional to x^2 from the oxygen lone pair with the weight of the s-function equal to s_1^2. The larger it is, the larger is the s-weight of the HO involved. This is in agreement with the estimates of the ionization potential of the water molecule performed both in the semiempirical SLG approximation [15] and with independent ones which clearly indicate that the first ionized state of the water molecule is the $^2\Pi$ state, which corresponds to extracting an electron from the π-orbital with no s-contribution.

The formulae eqs. (4.15) and (4.16) present together a specific DMM force field for the dative bond formed by the doubly covalently bonded oxygen atom (water, alcohol, simple ether). It is so because the cited equations represent the energy components in terms of the parameters of the semiempirical QM Hamiltonian and of the ESVs characterizing the covalent and dative bond on the one hand and the shape and orientation of the hybridization tetrahedron residing on the oxygen atom on the other hand. As a prescription, the sum of eqs. (4.15) and (4.16) must be added to the total DMM energy and the latter must be optimized also with respect to x and s_1 as well as to all other ESVs at each value of the geometry parameters. A simplified treatment with fixed values of δP and s is also possible. A remarkable feature of this approach is that it remains valid both at very large and short separations between the donor and acceptor atoms, which allows it to cover uniformly the regions normally treated by different methods: by QM at short distances and by standard MM force field for nonbonded atoms at longer ones.

Now we are equipped to study the shape of the hybridization tetrahedron on the oxygen donor atom in the presence of the dative bond. The structure of the $SO(4)$ hybridization manifold does not pose enough restriction on its flexibility. We have to remember in this context that in the case of quadruply bonded carbon or triply bonded nitrogen atoms, the structure of the hybridization manifold fixes the tetrahedral form of the model "methane" or "ammonia" molecules through the linear dependence relation eq. (3.64). In the case of the doubly bonded oxygen, the two vector parts ($\vec{v}_3$ and $\vec{v}_4$) defined by the covalent interactions with the hydrogen atoms do not suffice to determine the other two. We assume that the perturbation incurred by the dative bond formation does not change the overall s-weight ($2s^2$) of the covalent bonding HOs since it would result in a too large energy increase of the "water" molecule due to the response of the bonding HOs to the change of the s/p-ratio. Physically it would correspond to an attempt to extract some electron density from an O–H bonding orbital. The energy of the model "water" molecule is independent of the actual value of s_1 which controls the distribution of the s-weight between two lone pairs residing on the oxygen atom. The latter can access only a restricted range of values:

$$(4.17) \quad 0 \leq s_1^2 \leq 1 - 2s^2; \; s_2^2 = 1 - 2s^2 - s_1^2$$

The limiting values of s_1^2 correspond either to the pure p-character of the dative bond HO ($s_1^2 = 0$) pointing normally to the "water" molecular plane or to approximately sp^2 HO directed along the C_2 axis of the "water" molecule. The hybridization manifold structure allows for the determination of the direction of the HO lend for the dative bonding for each specific value of s_1. According to eq. (3.61) we have:

$$(4.18) \quad \cos\theta_{13} = \cos\theta_{14} = -\frac{s_1 s}{\sqrt{1 - s_1^2}\sqrt{1 - s^2}}$$

With the fixed value of s this defines the projection of the dative bonding HO ($m = 1$) on the C_2 axis (x); the rest of the p-weight comes from the z-component of the HO:

$$v_{1x} = -\frac{\sqrt{2}s_1 s}{\sqrt{1 - 2s^2}}; \; v_{1y} = 0; \; v_{1z} = \sqrt{1 - \frac{s_1^2}{1 - 2s^2}}$$

Thus the datively bonding HO expectedly stays in the mirror plane (which is perpendicular to the molecular plane) of the "water" molecule. For the acceptor ion the going out of this plane results in the corresponding restoring force. The values of the components of the vector part of the dative bonding HO thus obtained must be inserted in the eq. (4.13) for the resonance integral. Analysis of the one-center term eq. (4.16) describing the energy response of the oxygen hybridization tetrahedron to the formation of the dative bond leads to the conclusion that for the dative bond the energy minimum is reached if the pure p-HO is involved in its formation. This result can be easily understood as the resistance to the bond formation (the coefficient at the x^2) is larger for the larger s-contribution to the bonding HO. We arrive at an interesting situation, different from that for the sp^3 nitrogen dative bond. In the case of "water" oxygen we expect that the bond-related force field opposes the electrostatic (ion-dipole) forces which tend to place the acceptor atom on the C_2 axis which coincides with the direction of the effective dipole moment of the "water" molecule.

In order to numerically test the above derivations we performed a series of the SLG-MNDO calculations on a simple model of a complex of the Li^+ ion with H_2O, where the cation has been represented by a single s-orbital with the standard MNDO parametrization for lithium. The calculations have been performed for the Li-O separation of 2.14 Å characteristic for Li^+ complexes with ethers [16]. In agreement with the above estimates, we found first of all that the approximations of the ionic limit (dative regime) are valid for this model system. In all cases, the equilibrium value of x does not exceed 0.3, which can be shown to be a safe estimate for the validity of the ionic limit expansions (neglecting terms higher than x^2). This ESV reaches its maximum for the "π"-coordination of the lithium ion to the water molecule. Nevertheless, the energy of this configuration is not minimal, but maximal. By contrast, the minimal energy is reached for the planar configuration at the oxygen atom although the Li-O bond order is minimal for this situation. This demonstrates the dominance of the electrostatic forces in shaping the molecular geometry of the model complex.

All these conclusions may seem to be too exotic. Nevertheless, experimental facts suggest that the above treatment may be valid. For example, the authors of [16] report the possibility of competition between σ- and π-coordination of ether molecules on the basis of the analysis of the structures of crown-ether complexes of alkali and alkali earth ions. It turns out that the singly charged alkali ions acquire a nonplanar coordination geometry at least with one of the ether oxygen atoms in the crown ether complex more easily than the doubly charged alkali earth ions. For the latter, the planar trigonal geometry of the ether donor oxygen is definitely preferred. This is the trend one would expect on the basis of the relative strength of the ion-dipole interactions of singly and doubly charged ions. This trend becomes even more pronounced if the low-charge acceptors are addressed. In the interaction of the water molecule with electroneutral metal clusters, the adsorption geometry varies from σ- to π-coordination depending on the nature of the metal [17].

To conclude this section, we notice that the described discrepancy between the molecular geometry at the oxygen donor atom and the shape of its hybridization tetrahedron is characteristic only for the ionic limit of the dative bond. If the additional ("coordination") bond reaches the covalent regime characteristic, for example, of the protonation of water molecule (formation of the H_3O^+ cation) there is no such uncertainty and as in the isoelectronic case of ammonia, the pyramidal C_{3v} geometry is the equilibrium one and the misalignment between the bond and HO directions does not exceed a couple of degrees as usual [18].

In this section we have analyzed the behavior of the SLG approximation at the frontier of its applicability area. Being originally designed for treating the systems with well-defined localized two-center bonds it is employed here for analysis of dative (donor-acceptor) bonds and CCs of metal ions. The major result acquired in this way is that we have developed a DMM description for the dative bonds formed by amine nitrogen and ether oxygen atoms. It turned out that the DMM of the dative bonds differs from that of the usual covalent bonds in that the ESVs corresponding to the bond orders are strongly distance dependent. Such a situation is sometimes described in the MM context [19] by referring to Pauling's bond-length–bond-order logarithmic relation [20]. Here we propose a sequential description for such a situation, based on the standard QM technique. This treatment allowed us to analyze the known flexibility of the coordination mode of the ether oxygen to acceptors and to rationalize the observed dependence of the coordination trends on the charge of the metal cation.

4.3. QUALITATIVE PICTURE OF BONDING IN METAL COMPLEXES

In the previous section we developed the DMM methodology and derived formulae for the energy (force fields) for interactions of donor atoms with acceptors. A simplified representation of the acceptor with a single s-orbital was used there. Here we consider the metal-ligand interactions from a different point of view – that of the metal. The metal ion in a CC acquires some density not from one but from many lone pairs of the ligating donor atoms. Constructing a mechanistic or at least an economic QM description for such a case would possibly help to rationalize terms of interligand interaction force fields, which are sometimes included in the standard MM picture to assure proper description of the metal CC.

The analysis performed in the previous section was based on a somewhat over-correlated model of the electronic structure of the dative bond. One of the two ionic configurations, namely, the one with two electrons on the acceptor orbital, was suppressed by setting the ESVs y (or v – in the SLG formulation) equal to zero. This is of course an approximation. It, however, allowed us to stress an important feature missing in all mechanistic models of either coordination or dative bonding, namely the off-diagonal elements of the one-electron density matrix (proportional to the ESV x) being primarily responsible for the variation of the shape of the hybridization tetrahedra on the donor atom and by this, for the effect of the quantum system upon the bonding in the classically treated part of the complex molecule. The charge variability turns out to be thus a higher order effect as compared to the chemical bond formation. From the point of view of energy calculation, these higher order effects are very important as they modify the strong and slowly decaying Coulomb interaction although the ESV x is itself exponentially decaying and multiplies by the exponentially decaying of the resonance integral (see above). The general picture thus appears as follows – at larger separations, where the resonance interaction (one-electron hopping) is negligibly weak, one can approximate the only important Coulomb interaction as one between a central atom (ion) bearing its formal (integer) charge with the effective charges (multipoles) in the ligands, which may be modified due to polarization by the field of the central atom. At shorter distances, the resonance (one-electron hopping) becomes significant; it contributes to the energy both directly as the resonance (bonding) energy and indirectly by modifying the effective charges of the donor and central atoms. It must be realized that the "classical" schemes of the charge redistribution like the "electronegativity equilibration" scheme (see above Section 2.5) do not work here since the adequate variables (the off-diagonal density matrix elements) are missing in them. Incidentally, it is reported [21] that these schemes turn out to be numerically unstable when applied to metal CC.

We see that the charge variability appears due to two types of interactions almost equally important in the case of metal ions binding by donor ligands: one is due to polarization of the ligands by the point metal ion and by the charges residing in other ligands as well; another is due to electron transfers from the donor atoms to the ions' empty shells (Lewis acid-base interactions). Remarkably, neither of them has anything to do with the "flow of electrical fluid" tacitly assumed in the electronegativity equilibration models. The importance of the former mechanism has been recently stressed in [22]. The same concept has been used while developing the COSMOS MM force field [19] employed later for analysis of the behavior of the Zn^{2+} compound with nitrogen containing ligands [31]. The authors [22] mention, however, that the charge redistribution due to electron transfers, i.e. resonance, is not important. This may be true for the class of objects the cited authors actually consider: the crown ether complexes of the Cs^+ or Mg^{2+} ions, where the results of quantum chemical analysis reveal a negligibly weak transfer of electronic density from the oxygen donor atoms to the metal ions (though the calculated extent of this transfer is known to be "method dependent"). But there is no contradiction: the second order quantity – the charge transfer – may be small, but the first order quantity – the Coulson bond order – may remain important. In any case, such a picture with a negligibly small charge transfer, is not generally valid for all metals as some of them are much stronger Lewis acids than heavy alkali cations. For example, our older calculations on the charge distribution in the transition metal complexes revealed a general trend that the formally divalent transition metal cations bear an effective charge of about one unit charge, whereas for the trivalent cations, the effective charge is less than two unit charges [23]. Even in the less pronounced case of the Mg^{2+} ions coordinated through oxygens in xylose isomerase, the effective charge obtained on Mg within the PM3 semiempirical calculation [24] is close to unity. A similar picture has been reported in [25] for Zn^{2+} complexes with imidazole. The remarkable role of the charge redistribution in close vicinity of the Ln^{3+} cations, which does not reduce only to the polarization of the surrounding ligands, has been reported in [26]. Thus the overall picture appears to be too confusing to hope that it can be disentangled by a combination of locally successful ad hoc solutions.

Polarization of the surrounding by the metal ion (either formal or effective) charge can be fit into the electronegativity equilibration scheme by ascribing new electronegativities to the atoms in the "external" field induced by the ion. However, it must be realized that neither of the characteristic "quantum" features of polarizability, such as the alternating polarity law [27] can be reproduced by classical methods.

The above qualitative consideration explains to some extent the reasons why metal ions (both with and without open d-shell) stay away from the general MM picture based on the concept of localized transferable two-center bonds. The physical regime prevents metal-ligand bonds from being saturable, directional (see above and [11]), and transferable, which are important components of the standard MM picture. To be more precise, the metal-ligand bonds lack directionality at the metal center, though the directionality at donor atoms exists and the corresponding force fields are known as energies related to "misdirect" of lone pairs. These characteristics represent, in our opinion, the reasons why, despite numerous attempts present in the literature (for review see e.g. [8, 28]) the PES of metal CCs are not easily covered by the MM-like schemes. Describing the contributions to the energy of a CC requires a narrowly targeted model of its electronic structure. The general methodology will however be the same as that accepted previously in Section 3.1.

The physical picture of the metal-ligand bonding given above is presented largely by negative statements. The metal-ligand bonds are nontransferable, nonlocalized, nondirectional at the metal site and directional only on the ligand side. Thus the trial wave function for the metal complexes is generally not that of the SLG eq. (2.63). On the other hand, the SLG form of the wave function seems to be fine for the free ligands. This brings us to the situation we are already familiar with: that which requires different methods of description to be applied to different parts of a molecule. The physically substantiated picture of the metal CC can be formalized by assuming the following (GF) form of the trial electronic wave function:

$$(4.19) \quad \Psi = \Phi_{\mathrm{CLS}} \wedge \Phi_{\mathrm{SLG}}$$

where $\wedge$ stands as previously for the antisymmetrized product of the electronic functions; Φ_{SLG} is the product of the bonding geminals eq. (2.60) in the "organic" part of the complex; and Φ_{CLS} is a group function invoked to describe the metal ion with its closest environment hereinafter referred to as the closest ligand shell (CLS).

A possible approximation to be used for the Φ_{CLS} function can be chosen considering two ideas. In contrast to the directionality and saturability characteristic for "organic" covalent bonds, those formed by metal ions do not possess these properties. Thus there is no need to invoke the HO formation on the metal ion. At the infinite separation limit, the Φ_{CLS} wave function must flow to the antisymmetrized product of the lone pair geminals of eq. (2.61). The latter is in fact a single determinant function with all lone pair HOs doubly filled. With these arguments, we arrive at the conclusion that the single determinant (HFR) wave function is an appropriate form of the Φ_{CLS} function. It is to be constructed in the carrier space spanned by the lone pair HOs, defined by the Jacobi angles on the donor atoms, and by the valence AOs of the central ion. The number of electrons in Φ_{CLS} equals that in all the lone pairs involved, as the central ion is assumed to provide only its vacant orbitals.

The Φ_{CLS} thus constructed was tacitly assumed by Van Vleck and Owen [29, 30, 32]. This allowed them to describe qualitatively the covalency effects in otherwise "ionic" CCs. The MOs of a CCs were constructed as LCAOs of atomic $4s$- and $4p$-orbitals of the central ion (of the fourth row) with the symmetrized functions of the ligands. If one takes only the ligand σ-orbitals directed along the local axes joining the central ion and the ligand donor atoms, then in the octahedral environment one obtains the bonding (ψ^b) and antibonding (ψ^a) MOs of the a_{1g} and t_{1u} symmetries and nonbonding MOs of the e_g symmetry with respect to the O_h group [30]:

$$
\begin{aligned}
\psi^a(e_{gc}) &= \frac{1}{\sqrt{12}}(2\chi_z + 2\chi_{-z} - \chi_x - \chi_{-x} - \chi_y - \chi_{-y}) \\[6pt]
\psi^a(e_{gs}) &= \frac{1}{2}(\chi_x + \chi_{-x} - \chi_y - \chi_{-y}) \\[6pt]
\psi^a(a_{1g}) &= -x_{a_{1g}}\phi(4s) + \frac{y_{a_{1g}}}{\sqrt{6}}(\chi_x + \chi_y + \chi_z + \chi_{-x} + \chi_{-y} + \chi_{-z}) \\[6pt]
\psi^b(a_{1g}) &= y_{a_{1g}}\phi(4s) + \frac{x_{a_{1g}}}{\sqrt{6}}(\chi_x + \chi_y + \chi_z + \chi_{-x} + \chi_{-y} + \chi_{-z}) \\[6pt]
\psi^a(t_{1u\gamma}) &= -x_{t_{1u}}\phi(4p_\gamma) + \frac{y_{t_{1u}}}{\sqrt{2}}(\chi_\gamma - \chi_{-\gamma}) \\[6pt]
\psi^b(t_{1u\gamma}) &= y_{t_{1u}}\phi(4p_\gamma) + \frac{x_{t_{1u}}}{\sqrt{2}}(\chi_\gamma - \chi_{-\gamma})
\end{aligned}
\tag{4.20}
$$

As in eq. (2.95) here χ_γ, $\chi_{-\gamma}$ are directed along the γ axis ($\gamma = x, y, z$) of the Cartesian coordinate system centered at the metal atom in the positive and negative direction respectively. Coefficients x_Γ and $y_\Gamma = \sqrt{1 - x_\Gamma^2}$ describe the mixing between the metal AOs with the ligand orbitals and are to be determined from the secular equations of the HFR MO LCAO method with the effective Fock operator for the CLS group. This can be applied also for the nontransition elements. In this case the d-functions are not present in the MOs expansion in variance with eq. (2.95). We make an additional reservation concerning the specifics of the electronic structure of the TMCs with open d-shells where unpaired electrons may possibly reside and which are responsible for numerous manifestations of electron correlations in the magnetic properties and optical spectra of these compounds. As mentioned previously, these electrons are better treated as nonbonding ones and for this reason we suggest the d-shells be excluded from the zero-order treatment of the CLS in the TMCs with the open d-shells. This reduces the number of the parameters of the electronic structure of the octahedral complexes to be determined to only two: $x_{a_{1g}}$ and $x_{t_{1u}}$ which turn out to contain all the necessary information. We shall explore this setting first and turn to the complexes with open d-shells in the last section of this chapter.

4.4. HYBRID MODEL FOR COORDINATION COMPOUNDS

4.4.1. Reducing the number of ESVs for CLS

To get an economical description of the ESVs relevant to the Φ_{CLS} function we notice that the standard HFR wave function (without symmetry constraints) implies the MO expansion coefficients over the specified orbital basis set to be the ESVs. This representation may be obtained by diagonalizing the matrix of the effective Fock operator in the specified carrier space. In a QM/MM context, and eventually for constructing a useful DMM description, economical selection of ESVs is desirable. Indeed, the dimension of the carrier subspace we are interested in is $N_M + N_{LP}$, where $N_M(=4)$ is the number of valence orbitals on the metal ion and $N_{LP}(=6)$ (in the simplest case of a single LP per each of six donor atoms in an octahedral complex) is the number of LPs on the attached donor atoms. The number of the ESVs in the MO representation is then $(N_M + N_{LP})^2$, which is the number of the MO expansion coefficients. In the case of complexes with coordination number six, $N_M + N_{LP} = 10$, which yields 100 ESVs. These variables are, however, subject to $(N_M + N_{LP})(N_M + N_{LP} + 1)/2$ orthonormalization conditions. These conditions are very difficult to use explicitly to reduce "the number of numbers" to be computed (it may be possible by introducing $(N_M + N_{LP})(N_M + N_{LP} - 1)/2$ Jacobi angles). Even thus reduced number of variables (=45) is superfluous. The reason is that the single determinant wave function is determined up to the subspace of the filled orbitals. Any unitary transformation applied separately to the filled and (of course) to the empty orbitals does not change the wave function. The numbers of variables necessary to describe these irrelevant transformations are equal, respectively, to $N_M(N_M - 1)/2$ and $N_{LP}(N_{LP} - 1)/2$. For this reason, the true number of independent variables necessary to describe the single determinant function with N_{LP} doubly filled and N_M empty orbitals is only $N_M \times N_{LP}(=24)$. The actual construction of this reduced set of variables is possible by eqs. (1.104)–(1.107). Taking the product of the lone pair geminals (the ionic asymptotic wave function) as a zero approximation for the Φ_{CLS} function, we set the operator P_0 projecting to the subspace lone pair HOs ($\dim \operatorname{Im} P_0 = N_{LP}$) as a starting point for constructing the parametrization of the subspace of the filled orbitals according to eq. (1.104). The projection operator P eq. (1.104) is in its turn the one-electron density entering the effective bond Hamiltonians eq. (3.2) for the bonding geminals and the semiempirical SLG energy expression eq. (3.69). Matrices V are obviously $N_{LP} \times N_M$ matrices organizing in a single entity the relevant ESVs describing the HFR Φ_{CLS} function and ensure the correct asymptotic behavior: if a ligand goes to infinity the corresponding row in the matrix V vanishes. One may check that in this case the corresponding off-diagonal matrix elements in the projection operator P remain zero and the diagonal ones remain unity as they were in the zero approximate projection operator P_0.

4.4.2. Effective Hamilton and Fock operators and DMM energy of the CLS

Now let us turn to the contribution of the closest surrounding of the metal ion in the complex which consists of the metal centered AOs and the donor atoms' lone pair HOs to the energy. For the fixed set of HOs the bond geminals (and other more general electron groups if any) included in Φ_{SLG} affect the effective Hamiltonian acting in the CLS carrier subspace only through the one-electron densities $P_m^{tt'}$ residing in the HOs ascribed to the bonds. The same applies to the effect of the Φ_{CLS} function upon the effective Hamiltonians for the bond geminals: only the one-electron densities in the lone pair HOs and in the metal AOs enter the expressions for the bond effective Hamiltonians. In that respect, the form of the effective Hamiltonians for the bond geminals eq. (4.1) does not change when the product of the LP functions in eq. (2.61) is replaced by Φ_{CLS}. The energy of the "organic" part of the complex is described by eq. (3.69) with the only variance that in the intraatomic intergeminal Coulomb terms (proportional to $g_{t_m t'_k}$) for the donor atoms whose LPs interact with the metal, the densities must be taken as diagonal matrix elements of the projection operator P instead of unity values characteristic for LPs in a pure "organic" environment (diagonal matrix elements of the P_0 projection operator). The same values must be used in eq. (2.79) for calculating the effective charges residing on the donor atoms.

We notice that the combination of methods which arises in the present context of metal complexes with organic ligands is very much the same as in the SLG/SCF setting for the π-electronic approximation as described in [33, 34]. Here as well, the "organic" part is described by numerous effective Hamiltonians for the isolated bonds and the π-system – by the effective Fock operator. Moreover, in the setting pertaining to the complexes, one may expect an even subtler situation when numerous HFR treated groups of electrons must be considered, e.g. when pyridine ligands with delocalized π-system lend their lone pairs used for the complex formation to the HFR treated CLS group. The one-electron part of the effective group Hamiltonian can be presented as

$$(4.21) \quad h^{\mathrm{eff}} = h_0^{\mathrm{eff}} + h'$$

where h' describes the one-electron transfers (resonance) between the LP HOs and the metal valence AOs. The effective operator h_0^{eff} can be defined as one commuting with the unperturbed projection operator P_0. This representation of the one-electron part of the Hamilton (Fock) operator stresses the possibility of the coordination bond formation. For the metal AOs the matrix elements of h_0^{eff} are particularly simple:

$$(4.22) \quad \left(h_0^{\mathrm{eff}}\right)_{\mu\mu} = U_{\mu\mu}^M + \sum_{B \neq M, F} \gamma_{MB} Q_B + \sum_F \gamma_{MF} \left(2 \sum_{r \in SLG \cap F} P_{rr} - Z_F\right)$$

$\mu = s, p$; the first summation extends to all atoms in the complex except the donor atoms involved in the resonance interaction with the metal atom (termed here as frontier atoms — F) and the metal atom itself. The second summation extends to those HOs on the frontier (donor) atoms which are responsible for the bond formation in the "organic" part of the complex. The diagonal matrix elements of h_0^{eff} for the m-th LP HOs residing on the F-th frontier atom have a form similar to that of the effective two-electron bond Hamiltonian eq. (3.2) with the obvious variance that the matrix elements of h_0^{eff} are of the one-electron operator:

$$
(4.23) \qquad
\begin{aligned}
\left(h_0^{\text{eff}}\right)_{l_m l_m}^F &= U_m^l + \sum_{B \neq F} \gamma_{MB} Q_B + 2 \sum_{t_{m_1} \in SLG \cap F} g_{l_m t_{m_1}}^F P_{m_1}^{tt} + \\
&\quad + \sum_{F'} \gamma_{FF'} \left(2 \sum_{r \in SLG \cap F'} P_{rr} - Z_{F'} \right) - \gamma_{MF} Z_M
\end{aligned}
$$

By this the one-electron part of the effective Fockian in the CLS carrier subspace is completely defined.

Now we address the average Coulomb interaction entering the Fock operator and the energy. It has the standard HFR form for the metal orbitals:

$$
(4.24) \qquad
\begin{aligned}
\Sigma_{ss}[P] &= g_{ss} P_{ss} + 2 g_{sp} \sum_p P_{pp} + 2 \sum_F \gamma_{MF} \sum_{r \in CLS \cap F} P_{rr}; \\
\Sigma_{pp}[P] &= g_{pp} P_{pp} + 2 g_{sp} P_{ss} + 2 g_{pp'} \sum_{p' \neq p} P_{p'p'} + 2 \sum_F \gamma_{MF} \sum_{r \in CLS \cap F} P_{rr}
\end{aligned}
$$

where the Coulomb parameters g_{ss}, g_{sp}, g_{pp} and $g_{pp'}$ relevant for the HFR description of the sp-shell of a metal atom have been introduced [35]. The Coulomb interactions between electrons in the LPs reduce to a standard contribution of the form:

$$
(4.25) \qquad
\begin{aligned}
\Sigma_{l_m l_m}[P] &= (l_m l_m | l_m l_m)^F P_m^{ll} + 2 \sum_{t_{m_1} \in CLS \cap F} g_{l_m t_{m_1}}^F P_{m_1}^{tt} + \\
&\quad + 2 \gamma_{MF} \left(P_{ss} + \sum_p P_{pp} \right) + 2 \sum_{F'} \gamma_{FF'} \sum_{r \in CLS \cap F'} P_{rr}
\end{aligned}
$$

The latter expression flows to the same value as in the SLG approximation for the free ligand, as for the LPs the following (HFR-type) relation naturally holds:

$$
(4.26) \qquad \Gamma_m^{ll} = (P_m^{ll})^2
$$

The HFR approximation implies also the off-diagonal matrix elements of the Coulomb mean field. They have the form:

$$
(4.27) \qquad \Sigma_{\mu l_m}[P] = -\gamma_{MF} P_{\mu l_m}
$$

By this the energy operator for the electronic group describing the properties of the metal ion and its closest surrounding is defined.

The energy corresponding to the single determinant wave function with the occupied subspace Im P is given by [36, 37]:

$$
(4.28) \qquad E_{\text{CLS}} = 2 \left(\text{Sp}\, h^{\text{eff}} P + \text{Sp}\, P \Sigma[P] \right)
$$

where h^{eff} is the one-electron part of the effective Hamiltonian which contains also the mean Coulomb field induced by electrons from other electron groups (in our case – with those in the bond geminals of the "organic" part of the complex) in the CLS and $\Sigma[P]$ is the average (mean-field) Coulomb electron-electron interaction of electrons described by the Φ_{CLS} function.

Finally we address possible approximations to the energy of the metal ion with its closest surrounding on the basis of eqs. (1.104) and (4.28). Inserting the series eq. (1.104) in eq. (4.28) and cutting the expansion at a particular overall order in V and V^+ results in approximation for the energy in the form of a polynomial with respect to the set of the ESVs characterizing the CLS electron group. The terms of odd overall order obviously correspond to the bond orders that appear between the LPs on the donor atoms and the metal ion. The necessity to take into account the variability of the metal-donor bond orders has been demonstrated recently in [31] in the context of an MM study. The terms of even overall order correspond to electron density transfers from LPs to metal ion. While considering the dative bond from the point of view of a single donor atom, we were restricting ourselves with the harmonic approximation relative to the corresponding parameter x. In the case of a metal complex, it may be insufficient since the first terms responsible for the interactions between electrons transferred from the donor atoms to the metal appear only in the fourth order with respect to V. On the other hand, the Coulomb interactions between the donor atoms themselves change in the second order in V. The amount of this variation may be about a couple of electron volts and can cause significant deviations from the "points-on-a-sphere" or any other model operating with interligand force fields which do not depend on the interaction with a "third" body – the central ion.

The general picture of the electronic structure of the CLS can be obtained with certain precision in terms of the 6×4 matrices V departing from the limit of separated metal and ligands. However, for our purpose of constructing a mechanistic model of metal complexes, it is more convenient to start from another unperturbed state: namely, that of a symmetric complex and to consider the effects of different perturbations. It is substantiated by the well-known archetypal symmetric polyhedra such as octahedra or tetrahedra in the stereochemistry of CCs. We describe a symmetric equilibrium configuration of an ML_6 complex and use it as the starting point for describing subtler effects of the substitution upon molecular geometry. To obtain this starting point, we have to solve the Hartree-Fock problem for the CLS of an octahedral CC. This approach has the advantage that for the octahedral symmetry, the MOs are completely described by two parameters equivalent to the coefficients x_{a_1} and $x_{t_{1u}}$ mentioned above in eq. (4.20). This problem can be equivalently reformulated directly in terms of the density matrix, which will be done for the sake of uniformity.

In the octahedral geometry, the orbitals of each entering symmetry appear no more than twice. For that reason, the problem of defining variables x_{a_1} and $x_{t_{1u}}$ (or their equivalents – see below) reduces to diagonalization of the 2×2 Fockian blocks corresponding to the respective irreducible representations Γ:

$$(4.29) \quad F^{\Gamma} = \begin{pmatrix} a_{\Gamma} & b_{\Gamma} \\ b_{\Gamma} & c_{\Gamma} \end{pmatrix}$$

The exact definition of the matrix elements of the Fockian for an HFR-treated group of electrons in the presence of other groups is given in [13] and [33] (and above).

The one-electron density matrix corresponding to the solution of the Hartree-Fock problem in the CLS is, like any Hartree-Fock density matrix, an operator (matrix) P

projecting to the occupied MOs:

$$P = x_{a_1}^2 \left|a_1^0\right\rangle\left\langle a_1^0\right| + y_{a_1}^2 \left|s\right\rangle\left\langle s\right| + x_{a_1}y_{a_1}\left(\left|s\right\rangle\left\langle a_1^0\right| + \left|a_1^0\right\rangle\left\langle s\right|\right) +$$

$$(4.30) \qquad + \sum_{\gamma=c,s} \left|e_g^0\gamma\right\rangle\left\langle e_g^0\gamma\right| + \sum_{\gamma=x,y,z}\left[x_{t_{1u}}^2\left|t_{1u}^0\gamma\right\rangle\left\langle t_{1u}^0\gamma\right| +\right.$$

$$\left. + y_{t_{1u}}^2\left|p_\gamma\right\rangle\left\langle p_\gamma\right| + x_{t_{1u}}y_{t_{1u}}\left(\left|p_\gamma\right\rangle\left\langle t_{1u}^0\gamma\right| + \left|t_{1u}^0\gamma\right\rangle\left\langle p_\gamma\right|\right)\right]$$

where the quantities x_Γ, y_Γ are defined after eq. (4.32) and the orbitals with the superscript "0" refer to the symmetry adapted combinations of the LP HOs χ_γ in the right hand side of eq. (4.20). The above expression can be further simplified by noticing that the normalization condition for the quantities x_Γ, y_Γ can be absorbed in a rational function of another (single) electronic structure variable for each Γ. Indeed, an operator projecting onto one-dimensional subspace of two-dimensional space has the form:

$$P_\Gamma = \begin{pmatrix} x_\Gamma^2 & x_\Gamma y_\Gamma \\ x_\Gamma y_\Gamma & y_\Gamma^2 \end{pmatrix} = \frac{1}{1+v_\Gamma^2}\begin{pmatrix} 1 & v_\Gamma \\ v_\Gamma & v_\Gamma^2 \end{pmatrix}$$

The projection operator eq. (4.30) is a direct sum of the 2×2 projectors with the appropriate values of v_Γ (in particular $v_{e_g} = 0$) taken in the required number of instances (one for each row γ of the irreducible representation Γ). The projection operator eq. (4.30) is one for the 12-electron complex. In a 14-electron complex, the P_{a_1} in the direct sum has to be replaced by the 2×2 identity matrix, thus reducing the number of ESVs to only one: $v_{t_{1u}}$.

Inserting the ground state projection operator in the Hartree-Fock expression for the energy of the CLS electron group we get:

$$(4.31) \quad E_{\mathrm{CLS}} = \left(2\,\mathrm{Sp}\,h^{\mathrm{eff}}P + \mathrm{Sp}\,P\Sigma[P]\right),\,\text{provided}$$

$$F_{\mathrm{CLS}} = h^{\mathrm{eff}} + \Sigma[P]$$

where h^{eff} is the one-electron part of the Fock operator and $\Sigma[P]$ is the self-energy part representing the electrostatic field induced by electrons in the CLS group upon each other. By this we arrive at an explicit expression for the energy in terms of the ESVs v_Γ. This is the closed expression for the energy required by the DMM methodology (the molecular geometry enters through the respective dependence of the Fockian matrix elements). Moreover, it is the rational function of the ESVs involved. This expression can be efficiently searched with respect to the relevant variables yielding the equilibrium geometry and corresponding electronic structure.

It is possible, however, to obtain analytical estimates for the equilibrium values of ESVs, which possess rather interesting properties. The simplest analytical expression representing the solution can be written for the product $x_\Gamma y_\Gamma$ which is expressed through the single parameter ζ_Γ:

$$\zeta_\Gamma = \frac{b_\Gamma}{c_\Gamma - a_\Gamma}$$

condensing all the necessary information:

$$(4.32) \qquad x_\Gamma^2 y_\Gamma^2 = \frac{1}{4}\left(1 - \frac{1}{1 + \zeta_\Gamma^2}\right)$$

If one is interested in the complex formation, then the limit $\zeta_\Gamma \ll 1$ has to be considered for long interatomic distances. In this case:

$$x_\Gamma^2 y_\Gamma^2 \approx \frac{1}{4}\zeta_\Gamma^2$$

The opposite limit $\zeta_\Gamma \gg 1$ describes the situation close to the equilibrium. In it the following estimate holds:

$$(4.33) \qquad x_\Gamma^2 y_\Gamma^2 \approx \frac{1}{4}\left(1 - \frac{1}{\zeta_\Gamma^2}\right)$$

These results, known for decades, have never been considered from the point of view of possible transferability of the off-diagonal density among different molecules as far as we know. This latter property is however a key to constructing any mechanistic model of PES as shown in [38] and in Chapter 3.

The situation described by eq. (4.33) differs in an important respect from the analogous results of [39] described in Section 3.2.1 for isolated two-center two-electron bonds characteristic of organic species. In the "organic" domain, the transferability of the off-diagonal element of the one-electron density matrix immediately brings up the transferability of the corresponding Coulson bond-order directly involved in the expression for the bond energy. The formula eq. (4.33), however, applies to the density matrix element in the basis of the symmetry adapted linear combinations of the LP HOs. They are not related to individual $M - L$ bonds, which are not even "observable" elements of the molecular electronic structure in the sense proposed by Ruedenberg [40] (in opposition with the two-center two-electron bonds in "organics"). By contrast, the stable (up to the second order in the presumably small parameters ζ_Γ^{-1}) values of the one-electron density matrix elements refer to a completely different element of the construction: to the three-dimensionally delocalized CLS group of electrons whose ESVs themselves possess necessary transferability properties which make them an "observable" component of the molecular electronic structure in the sense of [40]. The pragmatic outcome of this might be the replacing, in the vicinity of the equilibrium of the ESVs, either by the transferable value of $v_\Gamma = 1$ ($\Gamma = a_1, t_{1u}$) or by inserting the estimates eq. (4.33) and by this arriving at the PES as a function of the nuclear coordinates only. The described result applies however to the octahedral complexes only. The major task is to make this treatment useful for the analysis of the molecules of lower symmetry, which will be done in subsequent sections.

4.4.3. Numerical results on metal-ligand resonance interaction

The theory developed above serves largely to reformulate the results of the otherwise very old and traditional HFR-based treatment of the CLS known from many sources and textbooks to the form suitable for subsequent use in the context of DMM and hybrid QM/MM treatments. The problem setting is that the reduced DMM and/or QM/MM treatments are intended to reproduce the results of HFR-based semiempirical calculations of the CCs. In order to get a feeling of the real numerical results to be reproduced by the target reduced scheme, we consider octahedral complexes formed by metal ions with vacant valence s, p-AO's (like K^+, Ca^{2+}, Fe^{2+}) and six monodentate ligands each bearing one lone pair. According to semiempirical calculations, which are the reference points here, the effective charges of divalent cations do not exceed $+1.1$, which corresponds to the overall density transfer to the metal ion not exceeding $0.9\bar{e}$, which amounts to only $0.15\bar{e}$ of transfer per donor atom. This corresponds to the λ and/or ζ values of the orders of 10^{-2} which can be safely treated in a low order approximation. One can also notice that in the ab initio context the effective (Mulliken) charge of a divalent cation is even larger (reaching almost two with a precision of several per cent) so that the transfers from the donor atoms to the metal ions are even smaller than in the semiempirical treatment. This allows us to conclude that the simple perturbative estimates can be well acceptable. We stress once again that using a purely ionic model cannot provide any explanation for the mutual ligand influence. All its tiny features are the consequences of the electronic structure variations and ultimately of the wave-like behavior of electrons in molecules and for this reason the variables responsible for the latter, *e.g.* one-electron density matrices, must be retained in the theory. On the other hand, the smallness of the electron transfer allows us to keep only the lower powers in all expansions where the corresponding terms appear.

4.5. MECHANISTIC MODEL FOR STEREOCHEMISTRY OF COMPLEXES OF NONTRANSITION ELEMENTS

Now we can continue by constructing a mechanistic model of stereochemistry of CCs, which is able to reproduce the features of ligand influence. Before plunging into this, we explain what is expected to be constructed here from the theory. As we have mentioned many times, our main concerns are numerosity of the parameters necessary for the mechanistic description of (metal) CCs and, related to it, sophistication of the mutual ligand influence effects, which ultimately requires that amount of parameters to be introduced to get an acceptable description, which makes the entire enterprise eventually senseless. Our purpose is to theoretically derive DMM-like and MM-like descriptions covering metal complexes and by this to give an explanation and introduce systematization to this diversity. There is no chance to construct such a theory from scratch. Fortunately, Levin and Dyachkov (LD) have performed an exhaustive qualitative analysis of the interplay between substitutions and deformations, i.e. of the ligand influence in CCs of both transition and non-transition elements [41] with donor ligands for the most widespread coordination polyhedra: octahedron, tetrahedron, and planar square. Analysis performed in [41] on the basis of eq. (4.20) used as a zero approximation, reduces to qualitative reasoning on the properties of the solutions of the MO LCAO method in the restricted basis of functions.

Formally the model [41] is a specific case of applying the theory of perturbations of MOs to their special class of eq. (4.20).

The construction of the LD theory of the ligand influence evolves in terms of two key objects: the electron-vibration (vibronic) interaction operator and the substitution operator. The vibronic interaction in the present context is the formal expression for the effect of the system Hamiltonian (Fockian) dependence on the molecular geometry taken in the lower – linear approximation with respect to geometry variations. It describes coupling between the electronic wave function (or electron density) and molecular geometry.

The substitution operator H_S is defined by LD relative to a symmetric (octahedral) molecule ML_6. The substitution operator is somewhat more tricky. By definition it is the difference between the operator related to a substituted complex $ML_nXYZ...$ and its symmetric prototype:

$$(4.34) \quad H_S = H_{ML_nXYZ...} - H_{ML_6}$$

This definition, if taken literally, brings several questions. First, the electronic Hamiltonians entering eq. (4.34) are the functions of the nuclear coordinates of the respective complexes and thus their difference is a strongly singular operator. This concern is lifted by going to the representation eq. (4.19) for the wave function and by considering the above definition in the sense of the effective Hamiltonian for the CLS subsystem. In this case we deal with the matrix representation of the corresponding Hamiltonians written with respect to the same set of orbitals. By this the dependence of the Hamiltonians on the chemical composition of the species involved condenses in the matrices of the same dimensionality, which can be manipulated irrespective of their origin. The second question is how to sequentially define the orbitals χ_γ to be used to construct the MOs eq. (4.20) in a polyatomic system like a CC with organic ligands. Using our previous results we can conclude that for the ligands themselves ("organic" part) the SLG form of the trial wave function must be a relevant approximation. We performed a comparative study of electronic structures of simple amines and ethers on one hand and their polycyclic counterparts on the other, by the semiempirical SLG-MNDO method. The calculation results given in [13] show that the relevant parameters of electronic structure (the bond orders, electron densities on the bonding orbitals of the donor atoms, and the weights of the s-functions in the lone pairs), the low-molecular amines and ethers and their cyclic polyanalogs, are fairly close. So we can assume that the σ-orbitals χ_γ required for the Levin-Dyachkov construction can be extracted from an SLG based procedure for free ligands with a subsequent slight modification occurring throughout the complexation process. Our calculations on cyclic chelating ligands have been performed at more or less arbitrary conformation of the molecule at hand (NH_3, Me_3N, Et_3N, $MeEtNH$, $18ane(N)_6$). We found that the dispersion of the equilibrium values of all ESVs related to donor atoms entering the cyclic chelating ligands is always smaller than the dispersion of the analogous values in a series of ethers or amines ranging from water or ammonia to the corresponding alkyl di- or trisubstitutes, respectively. Thus the SLG form (together with its semiempirical implementation) seems to be a relevant approximation for treating free chelating agents like crown ethers or cyclic polyamines.

4.5.1. Perturbative analysis of the DMM model of CLS and its relation to LD theory of ligand influence

Now let us consider what is going to happen to the above DMM picture under the variation of composition (chemical substitution) and/or geometry, both reducing the symmetry of the CLS. An interplay between these two types of perturbation is the main concern in the LD theory of ligand influence.

4.5.1.1. DMM on nonsymmetrical coordination compounds

Any Fock operator can be represented as a sum of the symmetric one and of a perturbation which includes both the dependence of the matrix elements on nuclear shifts from the equilibrium positions and the transition to a less symmetric environment due to the substitution. To pursue this, we first introduce some notations. Let h' be the supervector of the first derivatives of the matrix of the Fock operator with respect to nuclear shifts δq counted from a symmetrical equilibrium configuration. By a supervector, we understand here a vector whose components numbered by the nuclear Cartesian shifts are themselves 10×10 matrices of the first derivatives of the Fock operator, with respect to the latter. Then the scalar product of the vector of all nuclear shifts $|\delta q)$ and of the supervector h' yields a 10×10 matrix of the corrections to the Fockian linear in the nuclear shifts:

$$(4.35) \quad (h' \mid \delta q) = \sum_i \frac{\partial h}{\partial q_i} \delta q_i$$

(Here we introduce the notation $(\cdots \mid \ldots)$ for the scalar product of vectors whose components are numbered by the Cartesian shifts of the nuclei). Next, let h'' be the supermatrix of the second derivatives of the matrix of the Fock operator with respect to the same shifts. As previously, we refer here to the supermatrix indexed by the pairs of nuclear shifts in order to stress that the elements of this matrix are themselves the 10×10 matrices of the corresponding second derivatives of the Fock operator with respect to the shifts. The contribution of the second order in the nuclear shifts can be given the form of the (super)matrix average over the vector of the nuclear shifts:

$$(4.36) \quad (\delta q \mid h'' \mid \delta q) = \sum_{ij} \frac{\partial^2 h}{\partial q_i \partial q_j} \delta q_i \delta q_j$$

Supplying this with the 10×10 matrix of the substitution operator

$$(4.37) \quad h^S = F_S^{\mathrm{CLS}} = F_{\mathrm{ML}_n \mathrm{XYZ}\ldots}^{\mathrm{CLS}} - F_{\mathrm{ML}_6}^{\mathrm{CLS}}$$

we get the "bare" perturbation of the effective Fock operator in the CLS carrier space as:

$$(4.38) \quad (h' \mid \delta q) + \frac{1}{2} (\delta q \mid h'' \mid \delta q) + h^S$$

This does not form the entire ("dressed") perturbation because, in case the electron density changes to the first order in the above perturbation, the Fock operator acquires additional perturbation through the variation of its self-energy part, which leads to the self-consistent perturbation. Thus the perturbed Fock operator can be written as:

$$(4.39) \quad F = F_0[P_0] + (h' \mid \delta q) + \frac{1}{2}(\delta q \mid h'' \mid \delta q) + h^S + \Sigma[\Delta P]$$

Here ΔP stands for the correction to the unperturbed projection operator P_0 to the occupied MOs, which in the case of the octahedral complexes is given by eq. (4.30). This serves as a prerequisite for performing the two remaining steps of the prescription of Section 3.1 of constructing a DMM description of CCs of arbitrary (low) symmetry and of the linear response theory based on it and leading to a strictly mechanistic description of this class of molecules.

To proceed further, we look at the perturbed density matrix. It was assumed to have the form

$$(4.40) \quad P = P_0 + \Delta P = P_0 + \sum_{n>0} P^{(n)}$$

where the correction ΔP can be expanded in terms of the matrices V satisfying the conditions:

$$P_0 V = 0; V P_0 = V; (1 - P_0)V P_0 = V;$$
$$P_0 V^+ = V^+; P_0 V^+(1 - P_0) = V^+; V^+ P_0 = 0$$

as follows [37]:

$$P^{(1)} = V + V^+$$
$$P^{(2)} = VV^+ - V^+V$$
$$P^{(3)} = -VV^+V - V^+VV^+$$
$$P^{(4)} = V^+VV^+V - VV^+VV^+$$

which can be continued. The matrices V are 4×6 matrices for 12-electron complexes and 3×7 matrices for 14-electron complexes, which organize into a single entity independent ESV's of the problem – the first order transition densities between the occupied and empty MOs of the unperturbed problem. One can check that only the even terms of the above expansion contribute to the effective charges residing on the atoms (orbital populations) of the CLS.

Inserting the expansion eq. (4.40) rewritten in terms of matrices V in the energy expression eq. (4.31) with the perturbed Fock operator eq. (4.39) yields a DMM model of the CC of an arbitrary symmetry since the transition densities V take account of all possible perturbations of the electronic structure, keeping the CLS a separate entity. The series eq. (4.40) in fact appears by expanding the closed expression for the projection operator:

$$P = (P_0 + V)(1 + V^+V)^{-1}(P_0 + V^+)$$

which involves the inversion of a 10×10 matrix and nowadays is not a great computational problem. On the other hand, it is possible to restrict oneself to a certain power in the expansion of eq. (4.40), getting to the polynomial model of the electronic structure of required accuracy.

It is easy to analyze the above model keeping the terms of the total order not higher than two in δq and V simultaneously and taking into account that under the spur sign the argument of the self-energy part Σ of the Fock operator can be interchanged with the matrix multiplier [37]. Using these moves we arrive at:

$$E_{\text{CLS}} = \underbrace{2\,\mathrm{Sp}[h_0 P_0] + \mathrm{Sp}[P_0 \Sigma(P_0)]}_{=E_0} + 2\,\mathrm{Sp}[F_0\,(V + V^+)] +$$

(4.41)
$$+ 2\,\mathrm{Sp}[(h' \mid \delta q)\,P_0] + 2\,\mathrm{Sp}[(h' \mid \delta q)\,(V + V^+)] +$$
$$+ \mathrm{Sp}[(V + V^+)\,\Sigma(V + V^+)] +$$
$$+ \mathrm{Sp}[(\delta q \mid h'' \mid \delta q)\,P_0] + 2\,\mathrm{Sp}[F_0\,(VV^+ - V^+V)]$$

At the equilibrium the terms linear in δq and $V + V^+$ vanish so that the electronic energy becomes:

(4.42)
$$E_{\text{CLS}} = E_0 + 2\,\mathrm{Sp}[(h' \mid \delta q)\,(V + V^+)] + \mathrm{Sp}[(V + V^+)\,\Sigma(V + V^+)] +$$
$$+ \mathrm{Sp}[(\delta q \mid h'' \mid \delta q)\,P_0] + 2\,\mathrm{Sp}[F_0\,(VV^+ - V^+V)]$$

which is a quadratic form with respect to the nuclear shifts and the ESVs V. The expectation value of the second derivatives of the one-electron part of the Fock operator with the operator P_0 projecting to the occupied MOs of the unperturbed system:

$$(\delta q \mid 2\,\mathrm{Sp}[h'' P_0] \mid \delta q) = (\delta q \mid D_0 \mid \delta q)$$

is nothing but the bare harmonic potential of the symmetric complex with the dynamic matrix D_0 acting on the nuclear shifts. Analogously the second order energy corrections with respect to V – the variation of ESVs describing one-electron density matrix:

(4.43)
$$2\,\mathrm{Sp}[F_0\,(VV^+ - V^+V)] + \mathrm{Sp}[(V + V^+)\,\Sigma(V + V^+)] = \frac{1}{2}\,\langle\langle V \mid \Lambda \mid V \rangle\rangle$$

turns out to be the quadratic form giving the electronic energy as a function of the variation of the one electron density matrix. The quantity Λ can be considered a superoperator (supermatrix) acting in the space of the 10×10 matrices taken as elements of a linear space (the Liouville space). The supermatrix Λ has four indices running through one-electron states in the carrier space of the CLS group. Then the formula

$$\langle\langle A \mid B \rangle\rangle = \mathrm{Sp}\,(A^+ B)$$

defines a scalar product in the Liouville space, which ultimately permits the notation used in eq. (4.43). The next move consists in forming a direct sum of the Liouville space of the matrices V which can be expanded over the basis formed by the matrix unities $|b\rangle\langle a|$ with a and b running over all basis states of the CLS carrier space and of the space spanned by the nuclear shifts. Extending the definition of the scalar product to this new space allows us to rewrite the spurs in eqs. (4.41) and (4.42) as scalar products in this new vector space. Then the two types of perturbations introduced above couple by the bilinear term:

$$(4.44) \quad 2\,\mathrm{Sp}[(h' \mid \delta q)\,(V + V^+)] = \langle\langle V \mid h' \mid \delta q) + (\delta q \mid h' \mid V \rangle\rangle.$$

This is nothing but the electron-vibration interaction in the chosen notation. The quantity h' is the three index supervector; acting on the vector of nuclear shifts they form the scalar product $(\ldots \mid \ldots)$ giving a 10×10 matrix, next forming a Liouville scalar product with matrix V. On the other hand, acting on the variations V of the density matrix by forming the Liouville scalar product h' produces a vector to be convoluted with that of nuclear shifts δq. With use of this set of variables the energy in the vicinity of the symmetric equilibrium point becomes:

$$(4.45) \quad E_{\mathrm{CLS}} = E_0 + \frac{1}{2}\,(\delta q\ \langle\langle V \ \left| \begin{matrix} D_0 & h' \\ h' & \Lambda \end{matrix} \right|\ \begin{matrix} \delta q) \\ V \rangle\rangle \end{matrix}$$

which is a quadratic form with respect to both the nuclear shifts and the ESVs. The substitution operator gives additional terms which also can be recast into the form of the scalar products in the Liouville space:

$$(4.46) \quad \begin{aligned} h^S &= w + w^+ \\ 2\,\mathrm{Sp}[h^S\,(V + V^+)] &= \langle\langle V \mid w \rangle\rangle + \langle\langle w \mid V \rangle\rangle \end{aligned}$$

With this notation the energy of the CLS becomes:

$$(4.47) \quad \begin{aligned} E_{\mathrm{CLS}} = E_0 &+ \langle\langle V \mid w \rangle\rangle + \langle\langle w \mid V \rangle\rangle + \\ &+ \frac{1}{2}\,(\delta q\ \langle\langle V\ \left| \begin{matrix} D_0 & h' \\ h' & \Lambda \end{matrix} \right|\ \begin{matrix} \delta q) \\ V \rangle\rangle \end{matrix} \end{aligned}$$

This can be treated as the minimal order of the DMM picture for the PES of the CCs of nontransition elements. It perfectly condenses all the necessary elements of the LD theory of the ligand influence and of the theory of vibronic interactions. The specificity of the "class" of compounds is fixed by the presence of the CLS group. The specificity of a "subclass" within this class is controlled by the number of electrons in the CLS which defines the specific form of the quantities P_0 and Λ. Both the geometry and the electronic structure of the substituted or/and deformed complex can be obtained (in the "harmonic" approximation) by taking derivatives of the above expression with respect to δq and V and setting these former equal to zero. Doing that, we see that the fixed deformation $|\delta q)$ and the substitution w result in the modification of the electronic structure as compared to the symmetric undeformed complex. The amount of the modification necessary to bring the system to the new equilibrium is given by the formula:

$$(4.48) \quad |V \rangle\rangle = \Lambda^{-1}[|\,h' \,|\delta q) + |w \rangle\rangle]$$

It is remarkable that the supermatrix Λ^{-1} is nothing [42] but the polarization propagator Π for the CLS subsystem calculated for the symmetric molecule. With this we get:

$$(4.49) \quad V = \Pi[|\,h' \,|\delta q) + |w \rangle\rangle]$$

This performs the announced program of obtaining a closed expression for the energy of the CC (or at least of its CLS) in terms of its geometry and electronic structure variables.

4.5.1.2. PES of coordination compound as derived from DMM

Now we can turn to deriving a true mechanistic (MM-like) model for CCs of non-transition element by excluding the ESVs V. Inserting eq. (4.49) in eq. (4.47) we get for the energy:

$$(4.50) \quad \frac{1}{2}\left[(\delta q\,|D|\,\delta q) + (\delta q|\,\langle\langle h'\,|\Pi|\,h'\rangle\rangle\,|\delta q) + \langle\langle w\,|\Pi|\,w\rangle\rangle + \right.$$
$$\left. (\delta q|\,\langle\langle h'\,|\,\Pi\,|\,w\rangle\rangle) + (\langle\langle w\,|\,\Pi\,|\,h'\rangle\rangle\,|\delta q)\right]$$

This expression contains in a condensed form all the results which are obtained in detail in [41], namely the theory of ligand influence which can be considered as one describing a response of molecular geometry to the chemical substitution. For example, optimizing the above expression with respect to $|\delta q)$ yields the response of the complex geometry to the substitution of the ligands. One easily gets the close expression for it:

$$(4.51) \quad |\delta q) = -D^{-1}\,|\langle\langle h'\,|\Pi|\,w\rangle\rangle)$$

Different ligands are characterized by their specific contributions to the Fock operator for the CLS group. In the simplest approximation adopted in [41] the ligand is characterized by its diagonal matrix element in the Fock operator, which is a true parameter of the model. The semiempirical SLG theory as applied to isolated ligands allows us to estimate these quantities related to the LPs and even provides formulae describing their dependence on the deformations of the "organic" bonds incident to the donor atom. However, it is important to mention that replacing one ligand by another in a CC (local perturbation) produces a nonlocal effect in the sense that it does not necessarily decrease with the distance from the perturbation location (as it will be described below).

The MM-like model of complexes of nontransition elements requires even less than is given by eq. (4.50): only the first and the second term in the first row. They represent the bare harmonic dependence of the energy on the nuclear shifts and the renormalizations of the respective harmonic constants due to adjustment of the electronic structure to these shifts:

$$(4.52) \quad D = D_0 + \langle\langle h'\,|\Pi|\,h'\rangle\rangle$$

As mentioned previously, the specifics of the central atoms in CCs are determined by the structure of the supermatrix Π, which is in its turn predefined by the structure of the carrier space of the CLS group and by the number of electrons in it. Indeed, the supermatrix Π of the polarization propagator is particularly simple in the basis of the eigenstates of the Fock operator F_0. Its matrix elements then are:

$$(4.53) \quad \Pi_{ii'jj'} = \frac{\delta_{ii'}\delta_{jj'}}{\epsilon_i - \epsilon_j}$$

where the subscripts ii' run over all occupied MOs and the subscripts jj' run over the vacant ones. In this basis the elements V_{ji} of the matrix V and of its conjugate by definition represent the transition densities between the i-th occupied and the j-th empty MO. They are numerical coefficients at the matrix unities $|j\rangle \langle i|$ being the basis vectors of the Liouville space. In terms of the Liouville space the superoperator Π can be written:

$$(4.54) \quad \Pi = \sum_{\substack{i \in \mathrm{occ} \\ j \in \mathrm{vac}}} \frac{|i \to j\rangle\rangle \langle\langle i \to j|}{\epsilon_i - \epsilon_j}$$

($|i \to j\rangle\rangle$ is the Liouville space notation for the matrix unity $|j\rangle \langle i|$) which allows the straightforward use of the scalar product formulae with the notion that:

$$(4.55) \quad \langle\langle i \to j | i' \to j' \rangle\rangle = \langle i | i' \rangle \langle j | j' \rangle = \delta_{ii'} \delta_{jj'}$$

The simplest approximate description of Π corresponds to what is known as the frontier orbitals approximation, where only the highest occupied and lowest unoccupied MOs (HOMO and LUMO, respectively) are involved. Within it one gets:

$$(4.56) \quad \Pi_{hh'll'} = -\delta_{hh'} \delta_{ll'} \left(\varepsilon_\mathrm{H} - \varepsilon_\mathrm{L} \right)^{-1}$$

where subscripts hh' run over the orbitals in the HOMO manifold (they may be degenerate in the highly symmetric case) and ll' do the same in the possibly degenerate LUMO manifold.

The given formulae contain all the necessary results, but cannot be easily qualitatively interpreted. The necessary interpretation has been done by Levin and Dyachkov and is based on clarifying the interplay of the effects produced by substitution and vibronic operators upon the solution of the Hückel-like problem in the 10-dimensional orbital carrier space using symmetry considerations. This will be done in the next section.

4.5.1.3. Symmetry adapted formulation

Using the variables introduced in the previous sections, the symmetry analysis of [41] can be reformulated as follows. The deformation of the molecule of a CC $|\delta q\rangle$ is a vector with the components referring to the individual nuclear shifts:

$$(4.57) \quad |\delta q\rangle = \sum_i \delta q_i \, |i)$$

For a symmetric (say, octahedral) molecule, it may be rewritten using the symmetry adapted nuclear shifts:

$$(4.58) \quad |\delta q\rangle = \bigoplus_{\Gamma\gamma} \delta q^{\Gamma\gamma} \, |\Gamma\gamma)$$

where Γ and γ refer respectively to the irreducible representation of the symmetry group and its row (in the case of a degenerate irreducible representation). In an octahedral complex, if only the shifts leading to the $M - L$ ($M - X$) bond lengths variation are concerned, the symmetry classification suffices to label all possible collective shifts which can be either of a_{1g}, e_g, or t_{1u} symmetry. They can be explicitly written through the nuclear shifts of the individual ligands according to:

$$(4.59) \quad |a_{1g}) = \frac{1}{\sqrt{6}}\left[|x_{L_x}) - |x_{L_{-x}}) + |y_{L_y}) - |y_{L_{-y}}) + |z_{L_z}) - |z_{L_{-z}})\right]$$

$$|e_g s) = \frac{1}{2}\left[|x_{L_x}) - |x_{L_{-x}}) - |y_{L_y}) + |y_{L_{-y}})\right]$$

$$|e_g c) = \frac{1}{2\sqrt{3}}\left[2|z_{L_z}) - 2|z_{L_{-z}}) - |x_{L_x}) + |x_{L_{-x}}) - |y_{L_y}) + |y_{L_{-y}})\right]$$

$$|t_{1u}\gamma) = \frac{1}{\sqrt{2}}\left[|\gamma_{L_\gamma}) + |\gamma_{L_{-\gamma}})\right]$$

The meaning of the notation for the individual nuclear shifts is that $|\gamma_{L_{\pm\gamma}})$ represents a unit shift in the positive direction along the γ axis of the ligand located at the $\pm\gamma$ semiaxis of the coordinate frame.

A remarkable feature is that the derivative of the one-electron part of the Fock operator with respect to the symmetry adapted nuclear shift $\delta q^{\Gamma\gamma}$ (an operator acting on the one-electron states in the CLS carrier space) itself transforms according to the irreducible representation Γ and its row γ. That means that applying the deformation $|\Gamma\gamma)$ to a complex results in a perturbation of the Fock operator having the same symmetry $\Gamma\gamma$. This allows us to write the vibronic operator in a symmetry-adapted form:

$$(4.60) \quad (h' \mid \delta q) = \sum_{\Gamma\gamma} \delta q^{\Gamma\gamma} \left(h'_{\Gamma\gamma} \mid \Gamma\gamma\right)$$

Finally, the substitution operator can be expanded as a sum of symmetry-adapted components. For example, in the octahedral complex, single substitution $ML_6 \rightarrow ML_5X$ results in the substitution operator:

$$(4.61) \quad h^S = \frac{1}{\sqrt{6}}h^S_{a_{1g}} + \frac{1}{\sqrt{3}}h^S_{e_g c} + \frac{1}{\sqrt{2}}h^S_{t_{1u}z}$$

As we see, for the symmetric system all the elements of the present picture are classified according to irreducible representations of the relevant symmetry group – O_h. For example, the energies defining the polarization propagator depend on Γ_H and Γ_L, but not on the rows γ_H and γ_L of the involved irreducible representations. Using the symmetry notation for the polarization propagator allows us to realize its rôle as a selection mechanism for interference of different perturbations. As mentioned, in the frontier orbitals approximation, the only energy parameter is the energy gap $\varepsilon_H - \varepsilon_L$. The polarization propagator thus acquires the form

$$(4.62) \quad \Pi = -(\varepsilon_H - \varepsilon_L)^{-1} \sum_{\gamma_H, \gamma_L} |\gamma_H \rightarrow \gamma_L\rangle\rangle \langle\langle\gamma_H \rightarrow \gamma_L|$$

It is obvious that the superoperator Π acts as a projection operator in the Liouville space, cutting out those components of the 10×10 transition density matrices which mix γ_H state with the γ_L state, which is only possible if the symmetries of the perturbations of both the symmetry of deformation Γ_{def} and the symmetry of substitution Γ_S satisfy the selection rule:

$$(4.63) \quad \Gamma_{\text{def}}, \Gamma_{\text{S}} \subset \Gamma_{\text{H}} \otimes \Gamma_{\text{L}}$$

i.e. both enter the expansion of the tensor product of the irreducible representations of the frontier orbitals.

4.5.2. Applications

4.5.2.1. Off-diagonal elastic constants for stretching of bonds incident to the central atom

Up to this point, our main concern was to reformulate the results of the LD ligand influence theory in the DMM form. Its main content was the symmetry-based analysis of the possible interplay between two types of perturbation: substitution and deformation, controlled by the selection rules incorporated in the polarization propagator of the CLS. The mechanism of this interplay can be simply formulated as follows: substitution produces perturbations of different symmetries which are supposed to induce transition densities of the same symmetries. In the frontier orbital approximation, only those densities among all possible ones can actually appear, which have the symmetry which enters into decomposition of the tensor product $\Gamma_{\text{H}} \otimes \Gamma_{\text{L}}$ to the irreducible representations. These survived transition densities then induce the geometry deformations of the same symmetry.

The deformation (nuclear shifts) may play the same rôle as substitution. Inducing a deformation of some symmetry leads to the appearance of the transition densities of the corresponding symmetry. The same selection rule as that for the substitution makes only the symmetry component entering into decomposition of the tensor product $\Gamma_{\text{H}} \otimes \Gamma_{\text{L}}$ to survive and to induce the deformation of the same symmetry. For example: the z-shift of the apical ligand expands as:

$$(4.64) \quad |z_{L_z}) = \left[\frac{1}{\sqrt{6}} |a_{1g}) + \frac{1}{\sqrt{3}} |e_g c) + \frac{1}{\sqrt{2}} |t_{1u} z) \right]$$

Thus it may produce the transitional densities of the a_{1g}, $e_g c$, and $t_{1u} z$ symmetries. At this point selection rules pertinent to the frontier orbitals approximation enter: for the 12-electron complexes the symmetries of the frontier orbitals are $\Gamma_{\text{H}} = e_g$ and $\Gamma_{\text{L}} = a_{1g}$, the tensor product $\Gamma_{\text{H}} \otimes \Gamma_{\text{L}} = e_g \otimes a_{1g} = e_g$ contains only the irreducible representation e_g so that the selection rules allow only the density component of the $e_g c$ symmetry to appear. In its turn this density induces additional deformation of the same symmetry. That means that in the frontier orbitals approximation, only the elastic constant for the vibration modes of the symmetry e_g is renormalized. This result is to be understood in terms of individual nuclear shifts of the ligands in the trans- and cis-positions relative to the apical one. They, respectively, are:

$$(4.65) \quad |z_{L_{-z}}) = -\left[\frac{1}{\sqrt{6}} |a_{1g}) + \frac{1}{\sqrt{3}} |e_g c) - \frac{1}{\sqrt{2}} |t_{1u} z) \right]$$

$$|x_{L_x}) = \left[\frac{1}{\sqrt{6}} |a_{1g}) - \frac{1}{2\sqrt{3}} |e_g c) + \frac{1}{2} |e_g s) - \frac{1}{\sqrt{2}} |t_{1u} x) \right]$$

Combining all this we obtain for the off-diagonal constant, coupling the individual shifts of the ligands in the *trans*-positions to each other, as:

$$(4.66) \quad \frac{1}{3}\left(e_g c \left| \left\langle \left\langle h'_{e_g c} \left| \Pi \right| h'_{e_g c} \right\rangle \right\rangle \right| e_g c \right)$$

and for the off-diagonal constant coupling the individual shifts of the ligands in the cis-positions to each other we get

$$(4.67) \quad -\frac{1}{6}\left(e_g c \left| \left\langle \left\langle h'_{e_g c} \left| \Pi \right| h'_{e_g c} \right\rangle \right\rangle \right| e_g c \right)$$

By contrast, for the 14-electron complexes (nontransition nonmetals) the symmetries of the frontier orbitals are: $\Gamma_{\mathrm{H}} = a_{1g}$ and $\Gamma_{\mathrm{L}} = t_{1u}$ and the tensor product $\Gamma_{\mathrm{H}} \otimes \Gamma_{\mathrm{L}} = a_{1g} \otimes t_{1u} = t_{1u}$ so that only the transition density corresponding to the representation t_{1u} survive. Analogous moves allow us to conclude that the off-diagonal elastic constant for stretching the *trans*-bonds has the form:

$$(4.68) \quad -\frac{1}{2}\left(t_{1u}z \left| \left\langle \left\langle h'_{t_{1u}z} \left| \Pi \right| h'_{t_{1u}z} \right\rangle \right\rangle \right| t_{1u}z \right)$$

whereas that for the *cis*-bonds vanishes.

This allows us to make some predictions concerning the off-diagonal elastic constants, depending on the electron count in their CLS. Due to the different symmetry properties of the polarization propagator in these two cases (and according to the LD picture which ultimately explains the qualitative difference in the stereochemistry of the 12- and 14-electron complexes) the off-diagonal constant coupling the shifts of the ligands in the *trans*- and *cis*-positions to each other in the 12-electron case is expected to have a different sign. The sign of the off-diagonal coupling of the *trans*-positioned ligands in the 14-electron case is expected to be the same as that for the *cis*-positioned ligands in the 12-electron case, whereas the coupling of the shifts of the *cis*-positioned ligands in the 14-electron case is expected to be small.

4.5.2.2. Medium range off-diagonal elastic constants

In the previous section we obtained some estimates for the off-diagonal harmonic terms coupling the stretchings of different $\mathrm{M} - \mathrm{L}$ bonds incident to the central atom. The employed treatment can be extended to other types of off-diagonal terms. They originate as well from the $h'\Pi h'$ term in the general energy expression. The traditional MM picture tends to avoid the appearance of such off-diagonal terms and tries to represent the energy as a sum of force fields attributed to local elements of the molecular structure such as bonds, etc. This implies the strictly local character of the underlying electronic structure. It is also easy to understand from a pragmatic point of view, as including long-range type-specific terms in addition to those already introduced makes the entire parametrization too complicated. On the other hand, if the electronic structure is physically formed not by local elements such as two-center bonds, this must be reflected in the corresponding force fields. Incidentally, the CCs possess delocalized structure elements – the CLS – where one-electron states are extended over all atoms forming it. In such a situation, one has to expect some medium range off-diagonal harmonic couplings i.e. specific effective coupling between the deformations occurring at the separations usually not included in the MM-like consideration. Using this technique, it is possible to get estimates of such "off-diagonal" elements of the harmonic molecular potential, mediated by the metal atom, the very existence of which in the PES expansion is difficult to imagine, if only not to follow an non-informative idea that "all must be included". As an illustrative example, we consider a model two-coordinated linear compound, for which chemical examples are provided by those of Cu^+, Ag^+, or Hg^{2+}. In the context of the standard MM analysis, it is assumed that the interactions between the atoms separated by more than three bonds are not specific and must be taken into account as nonbonded "fields" using the Lennard-Jones potentials. Meanwhile, using the technique presented above, it can be easily shown that in the case of the above metal complexes, there are specific interactions of noticeable magnitude which, according to the standard scheme, must be classified as the 1–5 interactions (those between the atoms separated by four bonds).

Let us consider a (metal) ion bearing as previously four vacant (one s and three p) orbitals. As previously, we assume that ligand molecules are represented by one LP each. In the case of linear coordination (z-axis is the molecular axis) and assuming that in the equilibrium state the LPs are directed along the bonds between the donor atoms and the metal atom, the symmetry adapted combinations of the LPs have the form:

$$(4.69) \quad \left| a_{\pm}^{(0)} \right\rangle = \frac{1}{\sqrt{2}} \left(|u\rangle \pm |l\rangle \right)$$

According to [43] (see also Section 2.4.1) the LP HOs $|u\rangle$ and $|l\rangle$ (upper and lower positions on the z-axis relative to the central atom) are composed of s- and p-orbitals of the donor atom which are directed along the unit vectors $\vec{e}_u$ and $\vec{e}_l$:

$$(4.70) \quad \begin{aligned} |u\rangle &= s|s_u\rangle + \sqrt{1-s^2}\, |p_{\vec{e}_u}\rangle \\ |l\rangle &= s|s_l\rangle + \sqrt{1-s^2}\, \left|p_{\vec{e}_l}\right\rangle \end{aligned}$$

(with the obvious sense of s as of a coefficient of the s-orbital in the expansion of the corresponding HO). Using these definitions and the symmetry considerations, it is easy to identify nonvanishing matrix elements of the Fock operator acting in the CLS:

$$(4.71) \quad \begin{aligned} \left\langle \sigma \left| h \right| a_+^{(0)} \right\rangle &= \sqrt{2} \left(\beta_{\sigma\sigma}^{DM} s + \beta_{\zeta\sigma}^{DM} \sqrt{1-s^2} \right) \neq 0 \\ \left\langle \zeta \left| h \right| a_-^{(0)} \right\rangle &= \sqrt{2} \left(\beta_{\sigma\zeta}^{DM} s + \beta_{\zeta\zeta}^{DM} \sqrt{1-s^2} \right) \neq 0 \end{aligned}$$

where $\beta_{\sigma\sigma}^{DM}$, $\beta_{\zeta\sigma}^{DM}$, $\beta_{\sigma\zeta}^{DM}$, and $\beta_{\zeta\zeta}^{DM}$ are the resonance (one-electron hopping) integrals in the diatomic coordinate frame for the pair metal-donor atoms and where we denote by σ and ζ respectively the s- and p-states of the metal and donor atoms, having the σ symmetry with respect to the molecular axis (linear coordination).

The nontrivial one-electron eigenstates of the effective Fock operator for this CLS have the form:

$$\text{occupied}: |a_+\rangle = y_+ |\sigma\rangle + x_+ \left|a_+^{(0)}\right\rangle, |a_-\rangle = y_- |\zeta\rangle + x_- \left|a_-^{(0)}\right\rangle$$

$$\text{(4.72)}$$

$$\text{empty}: |a_+^*\rangle = -x_+ |\sigma\rangle + y_+ \left|a_+^{(0)}\right\rangle, |a_-^*\rangle = -x_- |\zeta\rangle + y_- \left|a_-^{(0)}\right\rangle$$

Two more states of the π-symmetry ($|\xi\rangle$ and $|\upsilon\rangle$) on the metal ion remain unchanged as in the free metal ion and both are empty. The frontier orbitals here are the $|a_-\rangle$ (HOMO) and those in the π-manifold ($|\xi\rangle$ and $|\upsilon\rangle$ – LUMO).

Now let us assume that the LPs belong to polyatomic ligands. Then a valence angle MDX with a vertex at a donor atom D is one of the geometry variables of the molecules in the standard MM setting. We shall estimate the magnitude of the indirect (CLS mediated) interactions between variations of these valence angles. Further consideration evolves as follows. We assume that the LPs are rigidly attached to the ligands. Then changing the valence angle MDX by $\delta\chi_u$ ($\delta\chi_l$) yields the corresponding nonvanishing angle between the vector $\vec{e}_u$ ($\vec{e}_l$) and the molecular axis. It turns on the resonance interaction between this LP and the $|\xi\rangle$ state of the metal atom (we assume that either the ligand LPs or the metal atom itself stays in the $(\xi\zeta)$ plane). The corresponding matrix elements are:

$$\text{(4.73)} \quad \langle \xi |h| u\rangle = \beta_{\pi\pi}^{DM} \sqrt{1-s^2} \sin \delta\chi_u$$

$$\langle \xi |h| l\rangle = \beta_{\pi\pi}^{DM} \sqrt{1-s^2} \sin \delta\chi_l$$

where $\beta_{\pi\pi}^{DM}$ is the resonance (one-electron hopping) parameter for the pair of states of the metal and donor atoms, which have π-symmetry with respect to the molecular axis. The derivatives of these matrix elements (and of the Fock operator itself) with respect to $\delta\chi_u$ and $\delta\chi_l$ are:

$$\text{(4.74)} \quad \left\langle \xi \left| \frac{\partial h}{\partial \chi_u} \right| u \right\rangle\Bigg|_{\delta\chi_r=0} = \left\langle \xi \left| \frac{\partial h}{\partial \chi_l} \right| l \right\rangle\Bigg|_{\delta\chi_l=0} = \beta_{\pi\pi}^{DM} \sqrt{1-s^2}$$

$$\left\langle \xi \left| \frac{\partial h}{\partial \chi_l} \right| u \right\rangle = \left\langle \xi \left| \frac{\partial h}{\partial \chi_u} \right| l \right\rangle = 0$$

The deformation coordinates $|\delta\chi_u\rangle$ and $|\delta\chi_l\rangle$ apparently transform according to the ξ-th row of the representation π and can be further combined into the symmetric and antisymmetric adapted coordinates with respect to the symmetry plane perpendicular to the molecular axis:

$$\text{(4.75)} \quad |\delta\chi_+) = \frac{1}{2}\left(|\delta\chi_u) + |\delta\chi_l)\right)$$

$$|\delta\chi_-) = \frac{1}{2}\left(|\delta\chi_u) - |\delta\chi_l)\right)$$

The individual deformation coordinates recover from the relations:

$$\text{(4.76)} \quad |\delta\chi_u) = (|\delta\chi_+) + |\delta\chi_-))$$
$$|\delta\chi_l) = (|\delta\chi_+) - |\delta\chi_-))$$

Assembling the relevant terms (those producing the antisymmetric ξ-transition densities in the CLS) we get for the off-diagonal interaction of two valence angles the following expression:

$$K\delta\chi_u\delta\chi_l$$

$$(4.77) \qquad K \propto \frac{\beta_{\pi\pi}^2 x_-^2 \left(1 - s^2\right)}{4\left(\epsilon_p - \epsilon_L\right)}$$

whose numerical value can be estimated as follows: for the sp^3 of the donor atoms $s^2 = \frac{1}{4}$, the weight x_-^2 of the antisymmetric combination of the ligand LP states in the corresponding HOMO can be safely estimated as $\frac{2}{3}$ so that with the energy gap $(\epsilon_p - \epsilon_L)$ of about 5 eV and the same value of $\beta_{\pi\pi}$ we arrive at the estimate for K of 0.7 eV/rad^2 which can be treated as, if not a large, at least a noticeable specific contribution of the 1–5 type.

4.5.3. Discussion

It is a widespread point of view in the MM community that the MM represents a "practical" alternative to standard quantum chemical treatments of molecular structure. On this basis, the quantum mechanical models are taken as excessively complex and superfluous, compared to the problems to be solved. The problem, however, is that in the absence of such models, it is difficult to estimate to what extent each specific problem possibly fits into some MM scheme or by contrast requires some essentially quantum mechanical approach to be solved. On the other hand, practical needs stipulate the interest in developing some MM-like models for wider classes of molecules as compared to "organic" ones, for which the standard MM treatment is by many examples proved to be valid. The key point is that behind any "classical" MM picture, there is always a fairly quantum view of molecular electronic structure. As it has been shown in [13] and Chapter 3, it is possible to imagine and successfully construct more general mechanistic models of molecular potentials (PES) than the usually accepted "balls-and-springs" models of the standard MM. The derivation in [13] and Chapter 3 is based on the concept of electron group dating back to McWeeny [37] and on the "observability" of these groups introduced by Ruedenberg [40]. In these terms one can state that classical MM of organic molecules implies that two-electron groups describing bonds are "observable" i.e. well defined stable groups spanning the molecular electronic structure. Then the moves described in Section 3.1 result in a fairly mechanistic picture of interacting atomic tetrahedra representing the sets of orthogonal HOs which can be further reduced to the standard MM with the externally i.e. independently defined force field parameters. The problems faced when extending any MM-like description to another class of molecules is the lack of understanding of the pertinent electronic group structure of the wave functions characteristic of the new classes of compounds to be included in the MM domain. In this section we employ the representation of the electronic structure of CCs in the form of the GF product and develop a mechanistic picture of their PES involving some necessary elements of the electronic structure description through the ESVs v_Γ and V. This approach can be qualified as deductive molecular mechanics (DMM) of the CLS group of electrons specific for the octahedral environment.

For other types of coordination, an analogous picture can be developed, which may be useful provided the electronic structure of the molecule at hand can be described using the corresponding CLS group. Then, using the perturbation theory, the ESVs have been excluded from consideration, thus yielding the estimates for the parameters of the force fields of a more traditional form.

The models thus built remain mechanistic ones, but they naturally take into account those important features of the electronic structure, which in a standard formulation, would require innumerable parameterizations for more and more tricky force fields, whose form remains without any fundamental basis. For example, off-diagonal elastic constants obtained thus do not assume the angular dependent form like

$$(4.78) \quad K \sim \sin 2\theta$$

proposed in [8] (θ stands for the valence angle between the bonds incident to the central ion), but suggest the existence of some more or less stable ratio between the constants describing the coupling of the *cis-* and *trans-*positioned ligands. Also, the estimates obtained allow us to relate the sign and other characteristics of these off-diagonal constants with the chemical nature of the central atom, which is a complex problem even for classical MM.

This analysis shows the weakness of all tentative attempts to include metals in "classical" MM. Within the classically looking picture, possible influence is attributed to charge redistributions among other possibilities. In fact, the charge variations are the quantities of the second order in the ESVs V, whereas the energy in the DMM picture depends on the first power of V. This affects the entire structure of the theory, where the polarization propagator supermatrix becomes the key player defining the generalized elastic properties of molecular electronic structure, expressed in terms of the ESVs V in the harmonic approximation. Of course this treatment is parallel to the random phase approximation (see e.g. [37]). It is also fair to say that polarization propagators were in use when analyzing the substitution effects in the CCs at an early stage of these studies [44–46]. However, in these papers the polarization propagator was used within the reactivity indices paradigm: i.e. to estimate some elements of the density matrices considered as "indices of influence" rather than the molecular energy/geometry itself. The general vibronic approach of [47] adopted in [41] stresses the possibility of explicit expression for the PES of substituted compounds, but does not underline the importance of the polarization propagator.

4.6. INCORPORATING D-METALS INTO MOLECULAR MECHANICS. MODELS OF SPIN-ACTIVE COMPOUNDS

Finally we arrive at the promised hybrid QM/MM construct intended to extend an MM-like treatment to transition metal complexes (TMCs). Despite the fact that in the literature [8, 48–53] various MM constructions are considered as effective methods for modeling PES of TMCs, it is noteworthy that in the case of TMCs the very basic characteristics of electronic structure comprising the basis of MM may be questioned. An important feature specific for the TMCs is the presence of the partially filled

d-shell on the metal ion, which produces a variety of electronic states on the complex of different total spin and spatial symmetry in a relatively narrow energy range close to the ground state energy. Geometry dependence of these energies may be rather confusing, which results in the existence of the areas in the nuclear coordinate space where the PESs belonging to different electronic terms closely approach each other and even intersect, leading to experimentally observed spin transitions [54–57] or Jahn-Teller distortions [47]. Thus, the very problem of including the transition metals in the MM context implies a certain contradiction: if several close in energy (or even crossing) electronic terms are present, there is no object for the MM modeling in a strict sense, since there is no unique PES of such a molecule. Indeed, as it is mentioned in [58, 64] and in Chapter 3, the physical pre-condition for successful construction and use of MM theories for common organic molecules is that their electronic excited states are well separated from the respective ground states on the energy scale. Only one quantum state of their electronic system is experimentally observed in 'organics' at ambient conditions and the MM (in fact a classical) description becomes valid. The behavior of the metal valence d-shell is sufficiently quantum: several electronic states may appear in a narrow energy range close to its ground state and this quantum feature requires special care, not reducible to a simple adjustment of the form and parameters of force fields, no matter how sophisticated they are. These features of the electronic structure of TMCs can be clearly observed in many cases. The results on blue copper proteins with approximately trigonal-bipyramidal coordination of the copper ion as reviewed in [52] may serve as one of the most recent examples. The Cu^{2+} cation is a Jahn-Teller ion due to the spatial degeneracy of its respective ^2E_g and $^2T_{2g}$ ground state terms in the octahedral and tetrahedral environments. The latter Jahn-Teller instability is inherited also by the trigonal bipyramidal environment, where the ground state is ^2E due to the electron count in the d-shell of the Cu^{2+} cation. Clearly the spatial degeneracy of the ground state is the limiting case of the closeness of electronic terms on the energy scale. This degeneracy is lifted when the molecular geometry deviates from the symmetrical arrangement and this is the content of the Jahn-Teller theorem (see for details [47]; an original and concise proof is given in [59]). Technically the Jahn-Teller instability manifests itself in the presence of multiple minima on the PES, having a close total energy. It must be understood, however, that these minima arise as a result of the sufficiently quantum behavior of the d-shell of the Cu^{2+} cation which, as it has been noticed previously, in a certain sense prevents the use of the classical MM picture. Otherwise one should try to develop an artificial force field with multiple minima. This is, however, wrong: the true picture is a result of superposition of multiple PESs for different electronic states, which may be simple by itself, whereas the complexity comes from their superposition.

A plausible way out of this situation has been proposed by R. Deeth ([60] and references therein). In order to handle quantum behavior of the d-shell he suggested adding the ligand filed stabilization energy (LFSE) term to the MM energy eq. (2.124). The LFSE in [60] is taken as a sum of the orbital energies of the d-orbitals, calculated in the angular overlap approximation (AOM – see Section 2.4.2.1) whose parameters are assumed to be linearly dependent on the internuclear separation between the metal and donor atoms. Applying such a model apparently eases many complications inherent to the MM of TMCs since the LFSE is a purely quantum contribution to the energy. The Jahn-Teller effect in Cu^{2+} compounds must be perfectly covered within such a setting. On the other hand, the LFSE is by construction a sum of one-electron energy contributions, whereas the energy of the d-shell is greatly dependent on the two-electron Coulomb interactions, particularly for relative energies of the states of different total spins and spatial symmetries. Bringing the latter into the MM context requires a much more developed and refined theory than that of [60], which will be explained below.

Turning in this context to the main topic of our interest, namely, modeling of the spin active TMCs, we notice that the above considerations apply to them to a large extent. The change of the spin state of a complex takes place when at least two electronic states (differing by the value of the total spin) have their respective minima at quite similar geometries of the complex at hand so that their respective total energies become equal at some intermediate geometry. As in the case of the Jahn-Teller Cu^{2+} or Co^{2+} cations in that of the spin-active ions (e.g. d^6 Fe^{2+}) the unique PES of the complex does not exist and at least two of them (the low-spin – LS – for $S = 0$ and the high-spin – HS – for $S = 2$) must be considered. Previously the MM force fields using different parameter sets for different spin states of the central atom were in use [61], but due to the absence of a predictive force they are considered to be obsolete now. However, the basic principles of their construction do not differ from those force fields which explicitly use different parameter sets for axial and equatorial ligands in the Cu^{2+} complexes [8] as the latter are well designed to imitate, by means of a classical potential, sufficiently quantum features of the TMC's electronic structure. Thus one can expect different sets of parameters for four-coordinate complexes of the Ni^{2+} ion, which must be tetrahedral in the triplet states and square planar in the singlet states. In this respect example [62] is very remarkable. The authors try to construct the MM potential capable of describing transformation between the square pyramidal and two trigonal bipyramidal forms of the pentacoordinate [Ni(acac)$_2$py] complex (acac stands for acetylacetone, py – for pyridine ligands). To do so, a specially designed force field is employed, which depends on the L–Ni–L$'$ angle and possesses two minima at 90° and 120° separated by a barrier higher than 5 eV (500 kJ/mole). This clearly indicates some problems which can be revealed by a simple analysis: the trigonal bipyramidal forms of the complex are obviously [63] triplet (two d-levels degenerate in the trigonal field filled by two electrons) whereas the square pyramidal form may well be singlet. This spin switch has to take place somewhere along the rearrangement reaction coordinate but it can by no means be described by the pure MM picture. Clearly, any approach employing the LFSE is not capable of describing such a low-symmetric and potentially correlation-dependent situation. However, the idea of employing LFSE is correct: the energy of the d-shell must be taken into account. The EHCF method described in Section 2.4.2 is capable of reproducing even tiny affects of its geometry dependence and distinguishes states of different spin and symmetry, which is not possible in the ultimately one-electron LFSE model.

4.6.1. EHCF *vs.* LFT and AOM

The success of the EHCF method (Section 2.4.2) in reproducing the crystal field from geometry data and ligand electronic structure as described by the semiempirical QC procedure, poses a question about the possible relation between the EHCF method and the successful parametrization scheme for the LFT, the already mentioned AOM. As it has been noticed, the ionic model of CFT yields the estimates of the crystal field parameters of very low quality. In order to overcome this shortcoming and also to take into account the diversity of the ligands in chemistry, the AOM is used for parametrizing the effective crystal field. Its main drawback (as in the case of the ionic model) is that it is not possible to obtain independently estimates of its parameters theoretically. As it will be shown, a local version of the EHCF method EHCF(L) derived and tested in [58] and [64] represents an effective tool allowing us to independently estimate the AOM parameters with precision. The derivation reduces to two unitary transformations applied to the orbitals involved in the EHCF construct. The first one is from the basis of canonical MOs (CMOs) of the l-system used in eq. (2.122) to the basis of localized one-electron states representing characteristic features of the ligand electronic structure — such as the presence of lone pairs on the donor atoms.

The EHCF theory [65] relates the dominating covalent contribution to the effective crystal field to the properties of the delocalized canonical MO of the l-system. Following the calculations presented in [65–69] the covalent contribution yields about 80–90% to the splitting of the d-electrons eq. (2.122). The remaining 10–20% is provided by the Coulomb interactions with the effective charges on the ligand atoms.

We concentrate on the expression for the covalent part of the crystal field $W_{\mu\nu}^{cov}$ and transform it to the form coinciding with the AOM and relate the parameters of the latter with the electronic structure of the ligands. To do so, we perform a unitary transformation of the canonical MOs of the l-system $|l\rangle$ – the eigenstates of the Fock operator – to the localized MO $|L\rangle$ separately for the occupied and vacant canonical MO:

$$(4.79) \quad |L\rangle = \sum_{l} |l\rangle \langle l \mid L\rangle$$

Summation over l in eq. (4.79) is extended to either only occupied or only vacant MOs. Here $\langle l \mid L\rangle$ are the coefficients of the l-th canonical MO in the expansion of the L-th local MO. The expansion coefficients of the localized MOs over the canonical MOs are the invariants of the molecular electronic structure as they do not change under the rotation of the coordinate frame. Expansion of the LMOs over AOs has the form:

$$(4.80) \quad \langle \alpha \mid L\rangle = \sum_{l} \langle \alpha \mid l\rangle \langle l \mid L\rangle$$

where $\langle \alpha \mid L \rangle$ is the coefficient of the α-th AO in the expansion of the L-th LMO. In the literature, a variety of methods of localization is present. They are all in principle suitable for finding the coefficients $\langle l \mid L \rangle$. In [64] we used the method based on the max Ψ^4 procedure [70], which is technically most feasible.

The max Ψ^4 procedure yields states which are strongly localized. Those that are localized largely on the donor atoms can be identified with the LPs. In the frame of the accepted method of localization the main contribution to the LPs is provided by the orbitals of the donor atom ($\sim$93% come from the s- and p-AO of the donor atom), $4s$- and $4p$-AOs of the metal give $\sim$5%. The contribution of other AOs does not exceed 2%. Also, the LPs are well localized in energy. In the case of the $[\mathrm{Fe(py)}_6]^{2+}$ all the contributions to the LPs of pyridine fall into four energy intervals not wider than 0.1 a.u. each giving respectively 14%, 19%, 30%, and 29% (totally 92 %) of the LP weight. In the case of the $[\mathrm{Co(NH_3)}_6]^{2+}$ and $[\mathrm{Fe(H_2O)}_6]^{2+}$ complexes 94% of the LP weight falls into two such intervals (16% and 78%, respectively).

In the basis of the LMOs the resonance integrals take the form:

$$(4.81) \quad \beta_{\mu L} = \sum_l \langle l \mid L \rangle \beta_{\mu l}$$

Inserting $\beta_{\mu l}$ into the expression for $W_{\mu\nu}^{cov}$ eq. (2.122) (the resonance integrals $\beta_{\mu l}$ and $\beta_{\mu L}$ refer correspondingly to the basis sets formed by canonical and localized MOs) we get:

$$(4.82) \quad W_{\mu\nu}^{cov} = \sum_{LL'} \beta_{\mu L} \beta_{\nu L'} \{ G_{LL'}^{ret}(I_d) + G_{LL'}^{adv}(A_d) \}$$

The unitary transformation from the basis of the CMOs to the basis of the LMOs of the l-system does not change the covalent contribution to the effective crystal field. According to numerical estimates the resonance integrals $\beta_{\mu L}$ between d-AO and LPs of the donor atoms by $10 \div 100$ times overcomes the resonance integrals between d-AO and any other LMOs and thus dominates the resonance interaction of the d- and l-systems. So, as it has been shown in [71], restricting the summation in eq. (4.82) by the sum of diagonal elements ($L' = L$) over only the LPs results in error in the estimated splitting of the d-levels of 0.1 eV. This precision is comparable to that of the EHCF method itself. This estimate is described by the formula:

$$(4.83) \quad W_{\mu\nu}^{cov} = \sum_{\Lambda} \sum_{L \in \Lambda} \beta_{\mu L} \beta_{\nu L} G_{LL}^{adv}(A_d)$$

where Λ enumerates the ligands and subscript L, the LPs located on the donor atoms of the Λ-th ligand and it is taken into account that due to the fact that only the occupied LMOs contribute to the LPs, only the advanced Green's function enters into the answer. By this we arrive at the local formulation of the EHCF theory – EHCF(L).

Now we are in a position to establish the relation between the AOM and EHCF. According to the AOM ([72] and Section 2.4.2.1) the crystal field is the sum of the static and dynamic contributions:

$$(4.84) \quad V_{\mu\nu} = \langle d_\mu | V | d_\nu \rangle = (V_{\mu\nu})_{stat} + (V_{\mu\nu})_{dyn}$$

On the other hand, the matrix elements of the crystal field in the AOM are expressed in terms of the cellular expansion of eq. (2.97). The relation between the EHCF and the AOM is that the matrix elements defined by the eqs. (2.97), (2.121), (4.83), and (4.84) must be equal. We identify the Coulomb contribution of the EHCF eq. (2.121) with the static contribution of the AOM and the covalent contribution $W_{\mu\nu}^{cov}$ of the EHCF method eq. (4.83) with the dynamic contribution to the AOM. Next we identify the subscript l (for the cells) with the subscript Λ (for the ligands). Accordingly we can rewrite eq. (4.83) as

$$(4.85) \quad W_{\mu\nu}^{cov} = \sum_{\Lambda} (v_{\mu\nu}^{\Lambda})_{dyn}$$

where we set

$$(4.86) \quad (v_{\mu\nu}^{\Lambda})_{dyn} = \sum_{L \in \Lambda} \beta_{\mu L} \beta_{\nu L} G_{LL}^{adv}(A_d)$$

and analogously for the static contribution:

$$(4.87) \quad (v_{\mu\nu}^{\Lambda})_{stat} = \sum_{L \in \Lambda} Q_L V_{\mu\nu}^{L}$$

Now we find the covalent (dynamic) contribution to the AOM parameters $e_{\lambda\lambda'}^{l}$. Inverting the relation between the matrix of the contribution $\mathbf{v}^{\Lambda}$ to the crystal field and the matrix of the AOM parameters $\mathbf{e}^{\Lambda}$ – this is the second unitary transformation of the two mentioned in the beginning of this section – we get:

$$(4.88) \quad \mathbf{R}^{\Lambda} \mathbf{v}^{\Lambda} \mathbf{R}^{\Lambda+} = \mathbf{e}^{\Lambda}$$

Separating the dynamic and static contributions we get:

$$(4.89) \quad \begin{aligned} \mathbf{R}^{\Lambda}(\mathbf{v}^{\Lambda})_{dyn}\mathbf{R}^{\Lambda+} &= (\mathbf{e}^{\Lambda})_{dyn} \\ \mathbf{R}^{\Lambda}(\mathbf{v}^{\Lambda})_{stat}\mathbf{R}^{\Lambda+} &= (\mathbf{e}^{\Lambda})_{stat} \end{aligned}$$

where from:

$$(4.90) \quad (e_{\lambda\lambda'}^{\Lambda})_{dyn} = \sum_{\mu\nu} R_{\lambda\mu}^{\Lambda} [\sum_{L \in \Lambda} \beta_{\mu L}^{\Lambda} G_{LL}^{adv}(A_d)^{\Lambda} \beta_{\nu L}^{\Lambda+}] R_{\nu\lambda'}^{\Lambda+}$$

where $\beta_{\mu L}^{\Lambda}$ is the resonance integral between the L-th LMO and μ-th d-AO of the metal in the laboratory coordinate frame. We express them through the components of the vector of the resonance integrals between the d-AOs of the metal and the L-th LMO $\mathbf{t}^{L}$ in the diatomic coordinate frame (DCF):

$$(4.91) \quad t_{\lambda}^{L} = \sum_{\mu} R_{\mu\lambda} \beta_{L\mu}$$

so that eq. (4.90) takes the invariant form:

$$(4.92) \quad e_{\lambda\lambda'}^{\Lambda} = \sum_{L \in \Lambda} t_{\lambda}^{L} G_{LL}^{adv}(A_d) t_{\lambda'}^{L+}$$

The advanced Green's function $G_{LL}^{adv}(\epsilon)$ for the local state L in eq. (4.83) is given by

$$(4.93) \quad G_{LL}^{adv}(\epsilon) = -\sum_l \frac{n_l \, |\langle l \mid L \rangle|^2}{\epsilon - (g_{dl} - \epsilon_l)}$$

where g_{dl} is the interaction energy between d-electron and the electron on the l-th MO, and ϵ_l is the energy of the l-th CMO of the l-system in the TMC.

Formula eq. (4.92) defines the AOM parameters in terms of the quantities which can be calculated in the frame of the local version of the EHCF – EHCF(L) – method. The matrix of the AOM parameters $\mathbf{e}^\Lambda$ is determined by the form of the vector of the resonance integrals $\mathbf{t}^L$, $L \in \Lambda$ in the DCF. Using the quaternion notation $(s_L, \vec{v}_L)$ in eq. (3.58) for the expansion of the LP over AOs of the donor atom (here we accept that each donor atom contributes one s- and three p-AOs to the basis of the AOs of the complex) we easily express the vector t_λ^L through the resonance integrals between d-AOs of the metal and the AOs of the donor atom $\beta_{\mu\alpha}$ in the same DCF ($\mu = z^2, x^2 - y^2, xz, xy, yz$ $\alpha = \sigma, \xi, v, \zeta$):

$$t_\sigma^L = s_L \beta_{z^2\sigma} + (\vec{e}, \vec{v}_L)\beta_{z^2\zeta}$$

$$(4.94) \quad t_{\pi x}^L = (\vec{e}_\xi, \vec{v}_L)\beta_{xz,\xi}$$

$$t_{\pi y}^L = (\vec{e}_v, \vec{v}_L)\beta_{yz,v}$$

where $\vec{e}$ is the unit vector directed from the donor atom to the metal atom (the z ort of the DCF), and vectors $\vec{e}_\xi$ and $\vec{e}_v$ (the x- and y-orts of the DCF). The integrals of the δ-symmetry do not appear if only the d-orbitals are not included into the basis of the ligand AOs. In this notation, it is obvious that the coefficient of the s-function in the LP expansion is invariant under the spatial rotations. The coefficients LMO-AO for the p-AO in the DCF $\vec{v}_L$ are not invariants of the molecular electronic structure. However, they can be expressed in terms of the invariant 3-vector of the coefficients of the p-AO contributions to the LP $\vec{v}_L^{(0)}$ in some ligand-fixed coordinate frame (LFCF). It does not coincide either with the laboratory frame or with the DCF, but is rigidly attached to the ligand, like a system of principal axes of its inertia tensor. However, any other frame fixed at the ligand is acceptable. Then the relation between $\vec{v}_L$ and $\vec{v}_L^{(0)}$ takes the form:

$$(4.95) \quad \vec{v}_L = \mathcal{R}_\Lambda \vec{v}_L^{(0)}$$

where $\mathcal{R}_\Lambda$ is the rotation matrix superimposing the DCF with the LFCF fixed at the ligand Λ. Equation (4.95) thus transforms invariant expansion coefficients of the LPs in the LFCF to the analogous coefficients in the DCF defined by the complex geometry. With these precautions the entire picture remains invariant with respect to rotations of the molecule as a whole.

Now we restrict our consideration to those complexes where each donor atom bears a single LP. We consider the geometries where the z axis of the LFCF coincides with the z axis of the diatomic coordinate frame and the x- and y-axes in the two systems are parallel. Under these assumptions the components t_λ^L of the vector $\mathbf{t}^L$ differ from zero only for $\lambda = \sigma$. The component t_σ^L of the vector t^L for each LP has the form of eq. (4.94). If the ligand rotates around the donor atom so that θ is the polar angle between the z axes of the DCF and the LFCF, and ϕ is the azimuthal angle between the projection of the z axis of the LFCF on the xy-plane of the DCF and the x axis of the LFCF, the following holds:

$$(4.96) \quad \begin{aligned} t_\sigma^L &= s_L \beta_{z^2\sigma} + \vec{v}_{Lz}^{(0)} \beta_{z^2\zeta} \cos\theta + \vec{v}_{Lx}^{(0)} \beta_{xz\xi} \sin\theta \cos\phi + \\ &\quad + \vec{v}_{Ly}^{(0)} \beta_{yz\upsilon} \sin\theta \sin\phi \end{aligned}$$

Taking into account all the above, we can write down the expression for the AOM parameter e_σ of the ligand Λ as a function of the polar and azimuthal angles θ and ϕ of the ligand Λ relative to the axis z directed from the donor atom to the metal atom:

$$(4.97) \quad e_\sigma^\Lambda = (t_\sigma^L)^2 G_{LL}^{adv}(A_d)$$

where L denotes the LP of the ligand Λ. In eq. (4.97) the resonance integrals $\beta_{z^2\sigma}$ and $\beta_{z^2\zeta}$ in the DCF depend only on the separation between the metal and donor atoms. The contributions of the different MOs of the l-system to the LP are fixed by the electronic structure of the l-system and reflected by the Green's function in the right-hand side.

In [71] the parameters $e_\sigma = e_{\sigma\sigma}$ and $e_\pi = e_{\pi\pi}$ have been found following eq. (4.97). The results indicate acceptable agreement between the calculated and experimental values of the $10Dq$ parameters for octahedral complexes $[Ni(NH_3)_6]^{2+}$ and $[Co(NH_3)_6]^{2+}$. The agreement in terms of $10Dq$ both with the experiment and with the EHCF calculation taken to be precise for the purpose of testing the local version of the EHCF is in the range of 1000 cm^{-1}, which is of the order of the error of the EHCF method itself.

An important feature of the EHCF(L) theory is that it allows us to estimate the crystal field in terms of the local ESVs of the ligands. This can be done for arbitrary geometry of the complex, which is a prerequisite for developing a hybrid QM/MM method.

4.6.2. Hybrid EHCF/MM method

In this section we finally arrive at the construction mentioned in the introduction to this chapter: the hybrid method allowing us to incorporate the transition metal ions in the otherwise classical MM context. This is done using the local version of the EHCF method. The EHCF methodology allows us to perform systematic calculations of the crystal field for various ligand environments. The results of these calculations are in fair agreement with the experimental data, particularly with respect to the spin multiplicity of the ground states of the complexes. In their respective simple versions, the EHCF/X methods treat the electronic structure of the ligands within a semiempirical approximation X. These methods are not, however, well suited to conduct the systematic studies on PES of TMCs. Further application of the EHCF methodology would be to develop a method for the calculation of PESs of TMCs. To do so we notice that the CNDO or INDO parameterizations for the ligands are probably accurate enough for the charge distribution in the ligands and the orbital energies at fixed experimental geometries, although they do not suit for geometry optimizations (or more generally for searching PES) of TMC. Nevertheless, the EHCF method can be adapted for the PES search in a more general framework of the hybrid QM/MM methodology. This finally allows us to "incorporate" a quantum description of TMC into the "classical" methodology of MM and provide the necessary flexibility for quantum/classical junction (see below).

The EHCF formalism allows us to separate electronic variables in the d-shell, which require correlated and quantum description from the electrons in the rest of the complex. Then the EHCF assumes the HFR approximation for the electronic wave function of the l-system in order to establish the necessary parameters of the electronic structure. According to [37] (see also Section 1.7.2) the total electronic energy of the n-th state of a system with the wave function of eq. (2.103) is

$$(4.98) \quad E_n = E_d^{\text{eff}}(n) + E_l$$

where $E_d^{\text{eff}}(n)$ is the energy of the n-th state of the effective Hamiltonian for the d-shell in the crystal field. For estimating the total energy E_n for the complex in n-th state in [64] we proposed the replacement of the energy of ligands E_l by its E_{MM} estimate calculated in some MM approximation. Then the expression for the PES of the state n becomes:

$$(4.99) \quad E_n = E_d^{\text{eff}}(n) + E_{MM}$$

This coincides with the prescription of Section 1.7.2 and eq. (1.256) provided one can notice that in row (1) the core-core energy of the latter equation totally vanishes for the d-shell and in row (6) the renormalization of the Coulomb interaction in the d-shell due to its interaction with the polarization in the environment is omitted at the EHCF stage. In addition, to be in close relation to the general QM/MM scheme as derived in Chapter 1, eq. (4.99) is also a very natural and intuitively transparent way of combining MM and EHCF [64], allowing us to calculate energies of low-energy electronic states of the d-shell $E_d^{\text{eff}}(n)$ and the ligand energy E_{MM} for different nuclear configurations of TMC. This allows us to obtain approximate PES for various electronic states of the d-shell of TMC in a single setting.

The proposed approach is a family member of the general QM/MM techniques (see Section 2.6), which were invented with the general purpose of treating different parts of the polyatomic systems at different levels of theory. The general setting of this theory is discussed in detail in the previous chapters of this book. The main difference between the standard QM/MM technique as employed for treating TMCs and the present one is that the majority of them require as a desirable feature, the possibility to extend the subsystem to be treated on a quantum level as much as possible. This is considered a medication against the uncontrollable errors introduced by incautious cutting of the entire electronic system, in parts treated by QM and MM techniques respectively. The hybrid EHCF/MM technique uses an opposite approach: it tries not to extend, but to reduce the QM subsystem as much as possible, just to the size which is responsible for the truly quantum behavior of the systems under study. The intersubsystem frontier is then treated in such a way that the interactions between the quantally and classically treated parts are sequentially taken into account. As the true quantum effects – the low-energy excited states in TMCs, – are localized physically in the d-shell, we restrict the true quantum description to these latter. This is related to the very understanding of the concept of "quantum" relevant to the present problem, which we have already mentioned at the beginning: in organic chemistry, one normally deals with the ground state PES only, which on the energy scale is well separated from the lowest excited state. This is the deepest physical reason why the classical (MM) description is possible for organics. The TMCs differ from that picture due to the existence of the low-energy excitations in the d-shell accessible in experiment, and this is the reason why it must be treated on a quantum level.

The technical problem is of course to develop an adequate form of the intersubsystem junction for the case when the quantum system is represented by the d-shell. This is done using the EHCF(L) technique described above. In the EHCF(L), the effective crystal field in agreement with the general theory of Section 1.7.2 is given in terms of the l-system Green's function. The natural way to go further with this technique is to apply the perturbation theory to obtain estimates of the l-system Green's function entering eqs. (4.83) and/or (4.92). That is what we shall do now.

The bare Green's function for the l-system in the state when the metal is taken out has the block-diagonal form:

$$(4.100) \quad \mathbf{G}_{00}^{l} = \bigoplus_{\Lambda} \mathbf{G}_{0}^{\Lambda}$$

Nonvanishing blocks $\mathbf{G}_{0}^{\Lambda}$ correspond to separate ligands Λ containing the unperturbed diagonal Green's function matrix elements $(G_{0}^{\Lambda}(\epsilon))_{LL}^{adv}$ corresponding to the LP L located on the ligand Λ:

$$(4.101) \quad (G_{0}^{\Lambda}(\epsilon))_{LL}^{adv} = \lim_{\delta \to 0^{+}} \sum_{l \in \Lambda} \frac{\left|\langle l \mid L \rangle^{\Lambda}\right|^{2} n_{l}}{\epsilon - \epsilon_{\Lambda l}^{(0)} + i\delta}$$

where $\langle l \mid L \rangle^{\Lambda}$ are the same expansion coefficients as in eq. (2.122) but for the LP of the separate ligand Λ, and $\epsilon_{\Lambda l}^{(0)}$ is the l-th MO energy of that same free ligand. Then eq. (4.83) contains the Green's function $(G_{0}^{\Lambda}(\epsilon))_{LL}^{adv}$ of the free ligand and the summations in eq. (4.83) are performed over the separate ligands Λ and their LPs indexed by subscript L.

The Coulomb interaction between the ligands themselves and between each of them and the metal ion when turned on, does not break the block diagonal structure of the bare Green's function $\mathbf{G}_{00}^{l}$. The approximate Green's function for the l-system conserves the form of eq. (4.100), but the poles now must be equal to the orbital energies of the ligand molecules in the Coulomb field induced by the central ion and by other ligands ($\Lambda' \neq \Lambda$) rather than to those of the free ligands.

The simplest description of the effect of the central ion on the surrounding ligands reduces to that of the Coulomb field affecting the positions of the poles of the Green's function (orbital energies) of the free ligand. According to [73], the effect of the Coulomb field upon the orbital energies is represented by:

$$(4.102) \quad (G^\Lambda)^{-1} = (G_0^\Lambda)^{-1} - \Sigma^{(f)}$$

where G_0^Λ is the Green's function for the free ligand and the self-energy term $\Sigma^{(f)}$ is due to the external Coulomb field. The perturbed Green's function G^Λ within the first order has the same form as G_0^Λ but its poles are shifted by diagonal elements of the self-energy matrix $\Sigma_{ll}^{(f)}$:

$$(4.103) \quad \begin{aligned} \epsilon_l &= \epsilon_l^{(0)} + \Sigma_{ll}^{(f)} \\ \Sigma_{ll}^{(f)} &\approx \sum_{A \in \Lambda} \rho_{lA} \delta h_A \end{aligned}$$

where ρ_{lA} is the partial electron density of the l-th CMO of the ligand Λ on the A-th atom of the ligand:

$$(4.104) \quad \rho_{lA} = \sum_{\alpha \in A} |\langle l \mid \alpha \rangle|^2$$

where $\langle l \mid \alpha \rangle$ are the l-th MO LCAO coefficients of the free ligand, and the core Hamiltonian perturbation δh_A is:

$$(4.105) \quad \delta h_A = -e^2 \left(\frac{(Z_M - n_d)}{R_A} + \sum_{\substack{\Lambda' \neq \Lambda \\ A' \in \Lambda'}} \frac{Q_{A'}}{R_{AA'}} \right)$$

The atomic quantities δh_A are equal to the perturbations $\delta h_{\alpha\alpha}$ of the corresponding core Hamiltonian matrix elements in the ligand AO basis. This is so because within the CNDO approximation [74] accepted in [58], for the description of the l-system, the quantities $\delta h_{\alpha\alpha}$ are the same for all $\alpha \in A$.

According to Section 1.7.2 the polarization effects in the ligand sphere must be taken into account as well. For this, the metal ion is considered as a point charge equal to its oxidation degree. This setting coincides with the sparkle model [75]. Within models of that type, semi-empirical HFR calculation is performed for the ligands of the complex placed in the electrostatic field induced by the central ion with its formal charge ('sparkle'). Within models of this family, the electron distribution changes when the ligand molecules are put into the field. We model this by using the molecular polarizability rather than by equilibration of effective atomic electronegativities (see Section 2.5). The preference of the latter models is their greater theoretical soundness as the wave-like behavior of electrons is retained in the structure of the atomic mutual polarizability matrices (see below) although it is lost in the equilibration schemes.

According to the polarizability definition, the difference between the effective charge on the atom A in the complex and that in the free ligand is:

$$(4.106) \quad \begin{aligned} \delta Q_A = Q_A - Q_A^0 &= \sum_B \Pi_{AB} \delta h_B = \\ &= \sum_B \Pi_{AB} \left(\delta h_B^0 + \sum_{C \neq B} \Gamma_{AC} \delta Q_C \right) \end{aligned}$$

where Π_{AB} is the atomic mutual polarizability and δh_A^0 is that defined by eq. (4.105). The above expression takes into account the renormalization of the field felt by a given atom, which appears due to the variation in all the charges located on other atoms. In the vector notation, this acquires the form:

$$(4.107) \quad \delta\mathbf{Q} = \mathbf{Q} - \mathbf{Q}^0 = \Pi(\delta\mathbf{h}^0 + \Gamma\delta\mathbf{Q})$$

where the components of vectors and matrices are indexed by atoms. Resolving the last equation for $\delta\mathbf{Q}$ yields:

$$(4.108) \quad \begin{aligned} \delta\mathbf{Q} &= (1 - \Pi\Gamma)^{-1}\Pi\delta\mathbf{h}^0 \\ \delta\mathbf{Q} &= \Pi\delta\mathbf{h}^0 + \sum_{n=1}^{\infty}(\Pi\Gamma)^n\Pi\delta\mathbf{h}^0 = \sum_{n=1}^{\infty}\delta\mathbf{Q}^{(n)} \end{aligned}$$

Though procedures for inverting matrices of the dimensionality equal to the number of atoms are admitted in modern MM schemes (see again Section 2.5) using the electronegativity equilibration and directed to the systems with significant charge redistribution as described in [76] we consider such a procedure to be too resource consuming and restrict ourselves by several lower orders with respect to Π in the expansion. Then the term $\Pi\delta\mathbf{h}^0$ corresponds to the first order perturbation by the Coulomb field induced by the metal ion and bare (non-polarized) ligand charges. The second order term corresponds to the perturbation due to the Coulomb field induced by the mutually polarized (up to the first order) charges:

$$(4.109) \quad \begin{aligned} \delta\mathbf{Q}^{(1)} &= \Pi\delta\mathbf{h}^0, \\ \delta\mathbf{Q}^{(2)} &= \Pi\Gamma\Pi\delta\mathbf{h}^0 \end{aligned}$$

The charges thus obtained are used for calculating the $\Sigma_{ll}^{(f)}$ term and for renormalizing the orbital energies by eq. (4.103).

To complete the picture, calculating mutual polarizabilities relevant to the EHCF/MM context can be found in [77]. The atom-atom mutual polarizability matrix Π has a block-diagonal form:

$$(4.110) \quad \Pi = \bigoplus_{\Lambda}\Pi^{\Lambda}$$

where Λ enumerates the ligands.

To evaluate Π^{Λ} we consider first the mutual atomic orbital polarizabilities $\Pi_{\alpha\beta}^{\Lambda(0)}$:

$$(4.111) \quad \Pi_{\alpha\beta}^{\Lambda(0)} = \frac{\delta P_{\alpha\alpha}}{\delta h_{\beta\beta}}$$

where α, β enumerate AOs. Corresponding mutual atomic polarizabilities $\Pi_{AB}^{\Lambda(0)}$ are:

$$(4.112) \quad \Pi_{AB}^{\Lambda(0)} = \frac{\delta Q_A}{\delta h_B}$$

where A, B enumerate atoms. Turning to the difference $\delta P_{\alpha\alpha}$ of the electron density on the αth AO of atom A and renormalizing $\delta h_{\beta\beta}$ accordingly to eq. (4.106) results in:

$$(4.113) \quad \delta P_{\alpha\alpha} = \sum_{B \neq A} \sum_{\beta \in B} \Pi_{\alpha\beta}^{\Lambda(0)} \delta h_{\beta\beta} = \sum_{B \neq A} \sum_{\beta \in B} \Pi_{\alpha\beta}^{\Lambda(0)} \left(\delta h_{\beta\beta}^0 + \sum_{\mu} \tilde{\gamma}_{\alpha\mu} \delta P_{\mu\mu} \right)$$

$$\delta \mathbf{Q} = \mathbf{\Pi}^{\Lambda(0)} (\delta \mathbf{h}^0 + \tilde{\gamma} \delta \mathbf{Q})$$

where $\tilde{\gamma}_{\alpha\mu}$ is intraligand (Λ) two-electron Coulomb integral that in the CNDO approximation has the form:

$$(4.114) \quad \tilde{\gamma}_{\alpha\beta} = (1 - \delta_{\alpha\beta})\gamma_{\alpha\beta} + \delta_{\alpha\beta}\gamma_{\alpha\alpha}/2$$

The coefficient one-half at the diagonal interaction element in the above expression reflects the fact that in the HFR approximation for the closed electron shell system, only that half of the electron density residing at the α-th AO contributes to the energy shift at the same AO, which corresponds to the opposite electron spin projection. Then the expression for the renormalized mutual atomic polarizability matrix $\mathbf{\Pi}^{\Lambda}$ can be obtained:

$$(4.115) \quad \mathbf{\Pi}^{\Lambda} = \left(1 - \tilde{\gamma}\mathbf{\Pi}^{\Lambda(0)}\right)^{-1} \mathbf{\Pi}^{\Lambda(0)}$$

The calculation according to eq. (4.115) involves inversion of the matrix but only of the dimensionality equal to the number of atoms in the ligand Λ.

Finally, according to the general formulae given, say, in [27], the matrix element of the bare orbital mutual polarizability entering eq. (4.111) is given by:

$$(4.116) \quad \Pi_{\alpha\beta}^{\Lambda(0)} = 4 \sum_{k \in occ} \sum_{l \in vac} \frac{c_{l\alpha} c_{l\beta} c_{k\alpha} c_{k\beta}}{\epsilon_k - \epsilon_l}$$

$$\Pi_{AB}^{\Lambda(0)} = \sum_{a \in A} \sum_{\beta \in B} \Pi_{\alpha\beta}^{\Lambda(0)}$$

where α, β are the AO's indices, k, ϵ_k and l, ϵ_l are, respectively, the occupied and vacant MO's indices and orbital energies, and $c_{l\alpha}$ are the MO LCAO coefficients for the free molecule of the ligand Λ.

This is the method for constructing the renormalized polarizability matrix $\mathbf{\Pi}^{\Lambda}$ for the ligand Λ. The form of the total matrix $\mathbf{\Pi}$ for the whole TMC is given by eq. (4.110). Using this matrix, we can obtain renormalized atomic charges by eq. (4.108). This model can be called the perturbative sparkle (PS) model. Specifically, PSn approximation level of the PS model stands for using the n-th order charge corrections by the series eq. (4.108), while PS itself stands for the exact expression with the inverse matrix in the second row of the same equation. Then, eqs. (4.106)–(4.109) comprise the perturbative form of the sparkle model of the l-system's electronic structure. The proposed procedure improves the junction between the EHCF(L) method, playing the role of the QM procedure and the MM part, as shown below, where details of the calculations performed within this approximation are given. This illustrates the implementation of the general prescription for taking into account charge redistribution in the M-system of the complex system given in Section 1.7.2

Appropriate test objects for this approach are the spin-active TMCs. Spin isomerism i.e. existence of the same complex in different spin states under different conditions (temperature and/or pressure) is observed in d^4–d^8 compounds of the first transition row metals. They are most pronounced for the iron(II) complexes with nitrogen donor atoms. In their case the lengths of the Fe-N bonds change by more than 10% of the bond length in the low-spin (LS) complex. The local version of the EHCF method, combined with various MM techniques, was implemented and used for the analysis of the molecular geometries of complexes of iron (II) [58, 77, 78]. The satisfactory agreement in the description of complexes geometry with different spin is achieved only when the effect of the electrostatic field of the metal ion on the ligands has been taken into account through the electrostatic polarization of the ligands within perturbative sparkle model. Two aspects are important here: first, the ECF/MM approach based on eq. (4.99) allows the use of a single MM potential for the ligands for all electronic states of the metal center. The "different ionic radii" for the ions of the different spin states have clearly become obsolete. Second, the MM potentials for transition metal atoms extracted from structural data on TMCs cannot be used directly in eq. (4.99) as they include implicitly, effects of the d-shells. This means that the problem is somehow reloaded on fitting the parameters of a single set relative to metal ions of each chemical sort. The harmonic approximation does not suffice for the MM part of the energy. Thus for the Fe-N bond stretching force field, some potential with a finite dissociation limit must be used. Estimates of parameters of the crystal field for a series of complexes of iron (II) and cobalt (II) (both LS and HS ground states) were obtained. Totally, 35 six-coordinated iron complexes with mono- and polydentate ligands, containing both aliphatic and aromatic donor nitrogen atoms (mixed complexes with different types of donor nitrogen atoms and different spin isomers of one complex are included in this number) and ten cobalt complexes also with different types of donor nitrogen atoms and with coordination numbers ranging from four to six have been considered. Deviations of calculated bond lengths Fe-N and Co-N from the experimental values are quite well described by the normal distribution. The parameters of that distribution were the following: the mean value (average deviation over the data set): $\mu = 0.037$ Å and the mean square deviation $\sigma = 0.054$ Å in the case of Fe(II) complexes, and the mean value, $\mu = 0.017$ Å and $\sigma = 0.044$ Å in the case of Co(II) complexes. The above values seem to be quite acceptable for the entire set of data. They however somehow mask an inherent bias of the proposed approach. In the iron(II) complexes the distances for the HS complexes are systematically underestimated, whereas those in the LS and the Fe-N bond lengths come out slightly overestimated. The parameters of the fit of the empirical distribution function of deviations restricted to the LS complexes are $\mu = 0.011$ Å and $\sigma = 0.034$ Å and those restricted to the HS complexes are $\mu = -0.023$ Å and $\sigma = 0.054$ Å. The reason seems to be in the inherent "stiffness" of the Morse potential. In order to avoid this, another MM bond stretching potential for metal-ligand bonds in Fe(II) complexes had been tested:

$$(4.117) \quad E_{NR}(r) = \frac{a}{r} + \frac{b}{r^5} + \frac{c}{r^9}$$

This form was originally proposed by Níketić and Rasmussen (NR) in their version of the CFF force field [79]. The NR potential can be characterized as a softer potential than the Morse one in the following sense. Two potentials are both three-parametric so that a one-to-one correspondence can be established between them by proclaiming the potentials of the two forms to be equivalent if the well depth, minimum position, and elasticity constants K_{NR} and K_M expressed through the a, b, and c parameters of the NR potential eq. (4.117) or the D_0 and α parameters of the Morse potential, respectively, coincide:

$$(4.118) \qquad r_0^4 = \frac{-5b - \sqrt{25b^2 - 36ac}}{2a}$$

$$(4.119) \qquad D_0 = \frac{a}{r_0} + \frac{b}{r_0^5} + \frac{c}{r_0^9}$$

$$K_{NR} = \frac{a}{r_0^3} + \frac{15b}{r_0^7} + \frac{45c}{r_0^{11}}$$

$$K_M = D_0 \alpha^2$$

One can see that $E_{NR}(r) < E_{Morse}(r)$ for all values; $r > r_0$ so that the NR potential approaches the asymptotic slower than the Morse potential. One may identify the a parameter value with the Coulomb interaction of some effective charges. These effective values in Fe(II) complexes with nitrogen-containing ligands are $Q_{\text{Fe}} = 1.757\,\bar{e}$ and $Q_N = -0.293\,\bar{e}$; the latter is close to the real CNDO charges on the donor atoms obtained in the EHCF calculations [65].

Using the NR potential equivalent to the Morse potential fitted in [77] allows us to level out the quality of description of the LS and HS complexes of Fe^{2+}.

4.6.3. Discussion

In this section we try to demonstrate that the problems faced by most empirical and MM techniques, when applied to modeling TMCs, have deep roots in the specific features of the electronic structure of the latter and in approximations which tacitly drop the necessary elements of the theory required to reproduce these features of the former. Of course, the EHCF approach, the success story of which is described in detail in Section 2.4.2 is not completely isolated from other methods. In general, various CAS techniques must be mentioned in relation to it. The characteristic feature uniting these two otherwise very different approaches is the selection of a small subset of one-electron states, followed by performing adequately complete correlation calculation restricted to this smaller subset. The general problem with such approaches is that usually it is taken for granted that the HFR MO LCAO is a good source for obtaining the states to be used in the correlated calculation. Two pitfalls can be expected and actually occur on this route. The first is that in the TMCs the HFR MO LCAOs can be difficult to obtain or those obtained are of a poor quality. The second is that even if the MO LCAOs are obtained correctly, they provide too delocalized a picture of electron distribution. In terms first proposed by J.-P. Malrieu [80] and then extensively used by P. Fulde [81], it is equivalent to saying that in the HFR solution for the TMC, the number of electrons in the d-shell that fluctuate too much may be the correct average (integer) value. In both cases, the limited CI (CAS) techniques are applied to improve a very poor zero approximation. With the input supplied by the HFR approximation, taking only five MOs of appropriate symmetry to model the d-shell may be too naïve since the number of states to be included in the CI formation to reduce the excessive fluctuations can be much larger. Going to the one-electron states obtained from the canonical MO LCAOs by some localization technique may be useful, but numerically expensive. The EHCF here advantageously uses the fact that the exact wave function of the TMCs most probably corresponds to very high localization of electrons in the d-shell, which enables taking their delocalization into account as a perturbation. Among other approaches based on a similar vision of the situation in TMCs, [82] and [83] must be mentioned.

Another group of approaches can be described as an attempt at using the DFT to evaluate the parameters of the CFT/LFT theory. In this respect, [84] and [85] must be mentioned. The latter in a sense follows the same line as the old semiempirical implementation [86] where the MOs for the TMC molecule are first obtained by an approximate HFR-like procedure and then a CI is done in some restricted subspace of the latter. In some sense, this approach is also similar to the EHCF model, with the general difference that the one-electron states used to construct the complete CFT/LFT manifold are taken "as is" from the KS calculation. In this case, one can expect some difficulties while selecting the MOs to be introduced into the set of those to be used in constructing the CI (it is not obvious whether simple energy/symmetry criteria allow one to select the necessary manifold of the KS orbitals to reproduce the states in the d-shell; and what can be done when the symmetry is low?) Also, the degree of delocalization of the KS orbitals may interfere with the evaluation of the CFT/LFT parameters from the results of the DFT calculation. It seems that this is precisely what happened in [85] where the values of the Racah parameters turned out to be strongly underestimated as compared to the values known to fit the experiment within the CFT/LFT model, indicating by this the excess delocalization of the KS orbitals as compared to that necessary to reproduce the experimental data.

Generally one can notice that almost every review on computational chemistry of TMCs starts from a sort of "triple denial" of the old CFT/LFT approaches as being pertinent to something which was happening "once upon a time". Our point of view on the CFT/LFT picture is absolutely different. It more or less corresponds to that given in the brilliant introduction to [84]. The clear-cut conclusion to be derived from there and also from our experience and derivation is that the CFT/LFT picture keeps track of the very physical picture of the low-energy spectrum of the TMCs. Whatever discrepancy there may be between the results obtained by QC methods, however refined they may be, and those appearing from the CFT/LFT must be considered failures of the QC rather than "age effects" of the CFT/LFT. It is the purpose of a QC study to reproduce results obtained within the CFT/LFT paradigm and it is not easily reachable and in many cases has not been reached yet. This idea was the leading one in our studies on TMCs from the very beginning and its adequate formal representation in terms of the group functions and the Löwdin partition technique provided a crucial step forward, which allowed the numerical implementation of the EHCF method [65]. It immediately solved the problem of constructing a semi-empirical description of the TMCs, which has remained unsolved for 30 years. The cost of this was rejecting the HFR form of the wave function of the TMC, which, in the present context, cannot be considered as a big loss. The further development of this approach and realizing its deeper relation to the general QM/MM setting helped in evolving the corresponding EHCF/MM hybrid scheme. The latter is in relation to those proposed by Deeth [60] and Berne [87]. Both involve the d-shell energy as an additional contribution to that of the MM scheme and use the AOM model with inter-polated parameters to estimate the latter. In the case of the approach [60] there are two main problems. The first is that the AOM parameters involved are assumed to depend only on the interatomic separation between the metal and donor atoms. This is obviously an oversimplification as from eq. (4.92) it is clear that the lone pair ori-entation is of crucial importance. This is taken into account in the EHCF/MM method and in the SIBFA-LF method [88] as well. The second major flaw is the absence of any correlation in describing the d-shell in [60]. This precludes a correct description of the switch between different spin states of the open d-shell, although in some sit-uations, different spin states can be described uniformly. This is also corrected in the EHCF/MM method.

References

1. I.I. Tchernyaev. Reports of the Institute of Platinum of AN SSSR, **4**, 243, 1926 [in Russian].
2. A.L. Tchougréeff and A.M. Tokmachev. *J. Comp. Meth. Sci. Eng.* **2**, 309, 2002.
3. A.L. Tchougréeff. *J. Mol. Struct. THEOCHEM*, **630**, 243, 2003.
4. A.M. Tokmachev and A.L. Tchougréeff. *Int. J. Quant. Chem.* **96**, 175, 2004.
5. U. Burkert and N.L. Allinger. *Molecular Mechanics*, ACS, Washington, 1982.
6. T.A. Halgren. *J. Comp. Chem.* **17**, 490, 1996; *ibid.*, **17**, 520, 1996; *ibid.*, **17**, 553, 1996; T.A. Halgren and R.B. Nachbar. *ibid.*, **17**, 587, 1996; *ibid.*, **17**, 516, 1996.
7. N.L. Allinger, K. Chen and J.-H. Lii. *J. Comp. Chem.* **17**, 642, 1996; N. Nevins, K. Chen and N.L. Allinger. *ibid.*, **17**, 669, 1996; N. Nevins, J.-H. Lii and N.L. Allinger. *ibid.*, **17**, 695, 1996; N. Nevins and N.L. Allinger. *ibid.*, **17**, 730, 1996; N.L. Allinger, K. Chen, J.A. Katzenellenbogen, S.R. Wilson and G.M. Anstead. *ibid.*, **17**, 747, 1996.
8. P. Comba and T. Hambley. *Molecular Modeling of Inorganic Compounds*. VCH, NY, 1995.
9. V. Burton, R. Deeth, C. Kemp and P. Gilbert. *J. Am. Chem. Soc.* **117**, 8407, 1995; R. Deeth and V. Paget. *J. Chem. Soc. Dalton Trans.*, 537, 1997.
10. I.V. Pletnev and V.L. Melnikov. *Koord. Khim.*, **23**, 205, 1997 [in Russian]; *Russ. Chem. Bull.* **7**, 1278, 1997.
11. I.B. Bersuker. *Electronic Structure and Properties of Coordination Compounds* [in Russian] Khimiya, Moscow, 1976.
12. S. Shaik and A. Shurki. *Angew. Chem. Int. Ed.*, **38**, 586, 1999.
13. A.L. Tchougréeff. Deductive molecular mechanics as applied to develop QM/MM picture of dative and coordination bonds. *J. Mol. Struct. (THEOCHEM)*, **632**, 91, 2003.
14. G. Tiraboschi, M.-C. Fournié-Zaluski, B.-P. Roques and N. Gresh. *J. Comp. Chem.*, **22**(10), 1038, 2001; F. Rogalewicz, G. Ohanessian and N. Gresh, *ibid.*, **21**(11), 963, 2000.
15. A.M. Tokmachev and A.L. Tchougréeff. *Int. J. Quant. Chem.*, **85**, 109, 2001.
16. B.P. Hay, L. Yang, J.-H. Lii and N.L. Allinger. *J. Mol. Struct. (THEOCHEM)*, **428**, 203, 1998.
17. M.T.M. Koper. *Modern Aspects of Electrochemistry No 36* (ed. C.G. Vayenas, B.E. Conway and R.E. White) Kluwer, Academic, Plenum, New York, 2003.

18. A.M. Tokmachev and A.L. Tchougréeff. Unpublished.
19. M. Möllhoff and U. Sternberg. *J. Mol. Model.*, **7**, 90, 2001.
20. L. Pauling. *General Chemistry*, W.H. Freeman, San Francisco, 1958.
21. J. Kozelka. In *Metal Ions in Biological Systems*, vol. 33 (eds. A. Sigel and H. Sigel) Marcel Dekker New York, 1996.
22. J. Golebiowski, V. Lamare, M.T.C. Martins-Costa, C. Millot and M.F. Ruiz-López. *Chem. Phys.* **272**, 47, 2001.
23. A.V. Soudackov, A.L. Tchougréeff and I.A. Misurkin. *Russ. J. Phys. Chem.*, **68**, (1994) 1135; ibid **68** (1994) 1142, A.V. Sinitsky, M.B. Darkhovskii, A.L. Tchougréeff and I.A. Misurkin. *Int. J. Quant. Chem.*, **88**, 370, 2002.
24. M. Fuxreiter, Ö. Farkas and G. Náray-Szabó. Protein Eng. **8**, 925, 1995.
25. R. Cini, D.G. Musaev, L.G. Marzilli and K. Morokuma. *J. Mol. Struct. (THEOCHEM)*, **392**, 55, 1997.
26. O.L. Malta, H.J. Batista and L.D. Carlos. *Chem. Phys.*, **282**, 21, 2002.
27. T.E. Peacock. *Electronic Properties of Aromatic and Heterocyclic Molecules*, Academic, London, 1965.
28. B. Hay. *Coord. Chem. Rev.*, **126**, 177, 1993.
29. J.H. Van Vleck. *J. Chem. Phys.*, **3**(12), 803, 1935.
30. J.H. Van Vleck. *J. Chem. Phys.*, **3**(12), 807, 1935.
31. U. Sternberg, F.-T. Koch, M. Bräuer, M. Kunert and E. Anders. *J. Mol. Model.*, **7**, 54, 2001.
32. J. Owen. *Proc. Roy. Soc. A.*, **227**(1169), 183, 1955.
33. A.L. Tchougréeff and A.M. Tokmachev. *Int. J. Quant. Chem.*, **106**, 571, 2006.
34. M.B. Darkhovskii, A.M. Tokmachev and A.L. Tchougréeff. *Int. J. Quant. Chem.*, **106**, 2268, 2006.
35. L. Di Sipio, E. Tondello, G. De Michelis and L. Oleari. *Chem. Phys. Lett.*, **11**, 287, 1971.
36. R. McWeeny and B.T. Sutcliffe. *Methods of Molecular Quantum Mechanics.*, Academic, London, 1969.
37. R. McWeeny. *Methods of Molecular Quantum Mechanics.* (2 ed.) Academic, London, 1992.
38. A.L. Tchougréeff. *DSc Thesis* [in Russian] Karpov Institute Moscow, 2004.
39. A.M. Tokmachev and A.L. Tchougréeff. Transferability of parameters of strictly local geminals' wave function and possibility of sequential derivation of molecular mechanics, *J. Comp. Chem.*, **26**, 491, 2005.
40. K. Ruedenberg. *Rev. Mod. Phys.*, **34**, 326, 1962.
41. A.A. Levin and P.N. Dyachkov. *Electronic Structure and Transformations of Heteroligand Molecules* [in Russian] Nauka, Moscow, 1990; A.A. Levin and P.N. Dyachkov, *Heteroligand Molecular Systems: Bonding, Shapes and Isomer Stabilities*, Taylor and Francis, New York, London. 2002.
42. V.I. Pupyshev. Private communication.
43. A.L. Tchougréeff and A.M. Tokmachev. *Int. J. Quant. Chem.*, **100**, 667, 2004.
44. V.I. Baranovskii and O.V. Sizova. *Teor. i Eksp. Khim.*, **10**, 678, 1974 [in Russian].
45. N.A. Popov. *Koord. Khim.*, **1**, 732, 1975 [in Russian].
46. N.A. Popov. *Koord. Khim.*, **2**, 1155, 1976 [in Russian].
47. I.B. Bersuker. *Jahn-Teller Effect and Vibronic Interactions in Modern Chemistry*, Nauka, Moscow, 1987 [in Russian].
48. A.K. Rappé, C.J. Casewit, K.S. Colwell, W.A. Goddard III and W.M. Skiff. *J. Am. Chem. Soc.*, **114**, 10024, 1992.
49. A.K. Rappé, K.S. Colwell and C.J. Casewit. *Inorg.Chem.*, **32**, 3438, 1993.
50. B.P. Hay. *Coord. Chem. Rev.*, **126**, 177, 1993.
51. B.P. Hay. L. Yang, J.-H. Lii, N.L. Allinger. *J. Mol. Struct. (THEOCHEM)*, **428**, 203, 1998.
52. P. Comba and R. Remenyi. *Coord. Chem. Rev.*, **238–239**, 9, 2003.
53. H. Erras-Hanauer, T. Clark and R. Van Eldik. *Coord. Chem. Rev.*, **238–239**, 233, 2003.
54. P. Gütlich. *Struct. Bond.*, **44**, 83, 1981.
55. E. König, G. Ritter and S.K. Kulshreshtha. *Chem. Rev.*, **85**, 219, 1985.

56. E. König. *Struct. Bond.*, **76**, 53, 1991.
57. P. Gütlich and H.A. Goodwin eds. *Spin Crossover in Transition Metal Compounds*, vols. **233–235** of *Topics Curr. Chem.*, Springer, NY, 2004.
58. M.B. Darkhovskii and M.G. Razumov, I.V. Pletnev and A.L. Tchougréeff. *Int. J. Quant. Chem.*, **88**, 588, 2002.
59. V.I. Pupyshev. *Int. J. Quant. Chem.*, **104**, 157, 2005.
60. R.J. Deeth. *Coord. Chem. Rev.*, **212**(1), 11, 2001.
61. H.M. Marques, O.Q. Munro, N.E. Grimmer, D.C. Levendis, F. Marsicano, G. Pattrick and T. Markoulides. *J. Chem. Soc. Faraday Trans.*, **91**(2), 1741, 1995.
62. C. Daul, S. Niketić, C. Rauzy and C.-W. Schlöpfer. *Chem. A. Eur. J.*, **10**, 721, 2004.
63. M.B. Darkhovskii and A.L. Tchougréeff. Unpublished.
64. M.B. Darkhovskii and A.L. Tchougréeff. *Khim. Fiz.*, **18**(1), 73, 1999.
65. A.V. Soudackov, A.L. Tchougréeff and I.A. Misurkin. *Theor. Chim. Acta*, **83**, 389, 1992.
66. A.V. Soudackov, A.L. Tchougréeff and I.A. Misurkin. *Zh. Fiz. Khim.*, **68**(7), 1256, 1994.
67. A.V. Soudackov, A.L. Tchougréeff and I.A. Misurkin. *Zh. Fiz. Khim.*, **68**(7), 1264, 1994.
68. A.V. Soudackov, A.L. Tchougréeff and I.A. Misurkin. *Int. J. Quant. Chem.*, **57**, 663, 1996.
69. A.V. Soudackov, A.L. Tchougréeff and I.A. Misurkin. *Int. J. Quant. Chem.*, **58**, 161, 1996.
70. P.G. Perkins and J.J.P. Stewart. *J. Chem. Soc. Faraday Trans.(II)*, **78**, 285, 1982.
71. M.B. Darkhovskii and A.L. Tchougréeff. *Zh. Fiz. Khim.* **74**, 360, 2000; *Russ. J. Phys. Chem.*, **74**, 296, 2000.
72. M. Gerloch and R.G. Wooley. *Prog. Inorg. Chem.*, **31**, 371, 1983.
73. A.A. Abrikosov, L.P. Gor'kov and I.Y. Dzyaloshinskii. *Quantum Field Theoretical Methods in Statistical Physics*, Pergamon, Oxford, 1965.
74. J.A. Pople and D.L. Beveridge. *Approximate Molecular Orbital Theory*, McGraw-Hill, New York, 1970.
75. A.V.M. de Andrade, N.B. da Costa Jr., A.M. Simas and G.F. de Sá. *Chem. Phys. Lett.*, **227**, 349, 1994.
76. P. Ren and J.W. Ponder. *J. Comput. Chem.*, **23**, 1497, 2002.
77. M.B. Darkhovskii and A.L. Tchougréeff. *J. Phys. Chem. A.*, **108**, 6351, 2004.
78. M.B. Darkhovskii, I.V. Pletnev and A.L. Tchougréeff. *J. Comp. Chem.*, **24**, 1703, 2003.
79. S. Niketic and K. Rasmussen. *The Consistent Force Field: A Documentation*, vol. 3 of *Lecture Notes in Chemistry*, Springer, NY, 1977.
80. J.-P. Malrieu. In *The Concept of the Chemical Bond. Theoretical Models of Chemical Bonding, Part 2*, ed. by Z.B. Maksić, Springer, Heidelberg, 1989.
81. P. Fulde. *Electron Correlations in Molecules and Solids*, (3 ed.) Springer, Berlin, 1995.
82. G.J.M. Janssen and W.C. Nieuwpoort. *Int. J. Quant. Chem.*, **22**, 679, 1988.
83. L. Seijo and Z. Barandiaran. The ab initio model potential method: A common strategy for effective core potential and embedded cluster calculations. In J. Leszczynski, (ed), *Computational chemistry: Reviews of Current Trends*, **4**, pp. 55–152, World Scientific, Singapore, 1999.
84. C.E. Schäffer and C. Anthon. *Coord. Chem. Rev.*, **226**, 17, 2002.
85. M. Atanasov, C.A. Daul and C. Rauzy. *Chem. Phys. Lett.*, **367**, 737, 2003.
86. M.C. Zerner. Intermediate neglect of differential overlap calculations on the electronic spectra of transition metal complexes. In N. Russo, D.R. Salahub. (ed), *Metal-Ligand Interactions: Structure and Reactivity*, Kluwer, Dordrecht, 1996, pp. 493–531, NATO ASI Workshop.
87. C.J. Margulis, V. Guallar, E. Sim, R.A. Friesner and B.J. Berne. *J. Phys. Chem. B.*, **106**, 8038, 2002.
88. J.-P. Piquemal, B. Williams-Hubbard, N. Fey, R.J. Deeth, N. Gresh and C. Giessner-Prettre. *J. Mol. Struct. (THEOCHEM)*, **632**, 91, 2003.

CONCLUSION. REMAINING PROBLEMS

In this book we have developed a general theory of hybrid methods of molecular modeling of complex molecular systems, including the QM/MM ones. Nowadays the term theory is somewhat diluted: it frequently applies to assemblies of formulae, used together for performing calculations. This method of doing things caused many doubts many years ago, not only in chemistry. Yet at the dawn of the computer era (1956!) one of the leading experts in operation research qualified it in the following words: 'Mechanitis is the occupational disease of one who is so impressed with modern computing machinery that he believes that a mathematical problem, which he can neither solve nor even formulate, can readily be answered, once he has access to a sufficiently expensive machine' [1]. We prefer, therefore, the more old-fashioned understanding of 'theory' – as of the sequential derivation starting from some general principles, applying various approximations (and trying to estimate their precision) and arriving by this to the formulae which may be used for calculation, but first of all for qualitative analysis. This probably can serve as a remedy against mechanitis.

Two general elements of the theory presented in this book were the *derivation* of the GF form of the approximate trial wave function for electrons in the modeled molecular system and the derivation of the mechanistic models of PES based on the GF representation. Although the specific forms of the wave functions of groups and thus the electronic structure variables either classically approximated or quantally treated may be very different from those considered in this book, the general methodology given here is completely universal. It can be applied to whatever class of molecular systems for developing a theory on the basis of the modeling methods described in this book. It has to be understood that diversity of chemistry is first of all the diversity of types of the specific forms of group multipliers in the GF functions representing both the diversity of the observed types of chemical behavior and also a diversity of the sets of ESVs adequately describing this behavior.

A couple of immediate targets for applying the general methodology proposed in this book can be indicated, based on an analysis of the literature. The first candidate might be the valence bond approach to the analysis of stereochemistry of compounds of heavy transition metals (e.g. tungsten) proposed in a series of works [2–4]. It is based on the concepts of bonding and of hybridization of s- and d-AOs of the metal atom in different ligand environments. These works explore a physical situation different from that in "first-row" TMC, but similar to the "organic" model. Bonding in these compounds may be qualified as covalent and the entire setting as "organometallic" due to direct involvement of the d-orbitals in valence bond formation, which establishes corresponding conditions for possible hybridization schemes. Although in works [2–4] the correct stereochemistry is loaded upon the overlap between the potentially nonorthogonal HOs, it must be equally possible to reproduce it with some one-center energy of an sd-hybridized transition metal atom and orthogonal hybrids. A further object of interest may be the (deductive) molecular mechanics of metallocenes, which can be developed analogously to that of octahedral coordination compounds, but, of course, the symmetry based considerations must be modified accordingly.

The general perspective may be formulated as that of addressing the diversity of the conceivable types of wave functions of various electron groups, which is by no means restricted by those already analyzed or only mentioned in the present book. Some collection of those not considered here can be found e.g. in [5], but even it should not be thought to exhaust the possible diversity. An analysis of the possible forms of electronic group functions is very promising. One can foresee that extending it to groups representing the chemically active parts of molecular systems – reaction centers as proposed in [6] characteristic of that or an other chemical transformation, can provide an inexhaustible source of chemical imagination and pave the way to a true qualitative understanding of chemistry.

References

1. B. Kupman. *Oper. Research.*, **4**, 442, 1956.
2. D.M. Root, C.R. Landis and T. Cleveland. *J. Am. Chem. Soc.*, **115**, 4201, 1993.
3. C.R. Landis, T. Cleveland and T.K. Firman. *J. Am. Chem. Soc.*, **117**, 1859, 1995.
4. T.K. Firman and C.R. Landis. *J. Am. Chem. Soc.*, **123**, 11728, 2001.
5. J. Michl and V. Bonacic-Koutecký. *Electronic Aspects of Organic Photochemistry*, Wiley, New York, 1990.
6. N. Epiotis. *Structural Theory of Organic Chemistry*, Springer, NY, 1977.

INDEX

infinite bond-length asymptotic, 281
normalized to unity, 14–16
nuclear, 13
of TMCs, 329, 330
trial, 17, 18, 278, 292, 333

YETI force field, 170
Young tableaux, 126

ZDO approximation, 107
ZINDO (INDO/S) scheme, 121

Printed in the United Kingdom
by Lightning Source UK Ltd.
131902UK00001B/64-69/P